MAGNETOSPHERIC PHENOMENA IN ASTROPHYSICS

AIP CONFERENCE PROCEEDINGS 144

RITA G. LERNER
SERIES EDITOR

MAGNETOSPHERIC PHENOMENA IN ASTROPHYSICS

LOS ALAMOS 1984

EDITORS:
RICHARD I. EPSTEIN &
WILLIAM C. FELDMAN
LOS ALAMOS NATIONAL
LABORATORY

AMERICAN INSTITUTE OF PHYSICS NEW YORK 1986

L.C. Catalog Card No. 86-71149
ISBN 0-88318-343-9
DOE CONF-8408121

Printed in the United States of America

Pstack
SCIMON

Table of Contents

IV. Physical Processes

Foreword

The Earth and Space Science Division of Los Alamos National Laboratory sponsored a five-day workshop on Magnetospheric Phenomena in Astrophysics during August 1984 in Taos, New Mexico. The participants at this workshop examined areas of space physics and astrophysics which share some common physics or phenomenology. The organizers hoped that by providing a forum for examining the regions of overlap between these two subjects, the workers in each specialty could gain better perspectives on their research and might be able to start up synergistic collaborations with workers in the other specialty. Since an important goal of the workshop was cross-disciplinary communication, the speakers were encouraged to be pedogogical and focus on the fundamentals of their subject areas so that the non-specialist could readily follow their presentations. In both their talks and their final manuscripts the workshop participants attempted to adhere to these guidelines. These proceedings thus contain lucid descriptions of plasma physics and magnetohydrodynamic phenomena in the solar system and in astronomical systems. This volume can be used as an introduction to these fields by scientists and graduate students in space science, astrophysics and plasma physics.

The first section of the volume contains the overview lectures by Roger Blandford and Jean Jacques Aly. Blanford's first chapter discusses particle acceleration processes in the solar system, in the Galaxy and in radio galaxies. Fermi acceleration at strong shock waves is described in some detail since this appears to be a particularly promising process for accelerating electrons in radio galaxies and accelerating the galactic cosmic rays. In his second chapter, Blandford examines hydrodynamic aspects of active galactic nuclei and addresses the questions of how radio jets are confined and how energy can be extracted from a central black hole in a galactic nucleus. At the request of the organizers, Aly undertook the task of summarizing the topics in magnetohydrodynamic (MHD) theory that are often used in the study of magnetospheric phenomena. This extensive chapter, which was prepared with Steve McMillan's help, is a rich guide to both the standard MHD theory and to the newer developments.

The second section deals with solar system phenomena. Jack Gosling describes the observations and theory of solar wind streams. The understanding developed in modeling these streams should be applicable in the study of winds from other rotating objects. Magnetic reconnection has been proposed as a mechanism for releasing magnetic free energy in a host of astronomical systems ranging from solar flares to gamma-ray bursts. Feldman examines the evidence that magnetic reconnection occurs in the terrestrial magnetosphere, where theory and *in situ* measurements appear to be in good agreement. Collisionless plasma shock waves are commonly assumed to be ubiquitous features of the astronomical landscape, even through the direct evidence for them is scanty. Thomsen presents an overview of the present understanding of collisionless shocks and the particle acceleration they produce. Baker discusses substorms in the Earth's magnetosphere and emphasizes their gross features such as the energy budgets, time scales, and conversion efficiencies for particle acceleration. He discusses possible analogues of substorm-like behavior in astrophysical systems. Jovian magnetospheric processes and planetary radio emission is described in two chapters by Goertz. He explains Jupiter's rotation power processes such as auroral uv emission, radio waves and charged particle energization. He also explains how Jupiter's moon Io generates the Alfven waves which produce that planet's decametric radio waves. In his second chapter Goertz discusses the observed coherent radio emission from the Earth and Saturn as well as from Jupiter.

The third section of the book deals with astronomical phenomena. Imamura discusses oscillatory instabilities in shock waves. These instabilities could reconcile puzzling observations of supernova shock waves during their radiative stage. Córdova, Ladd and Mason describe how ultraviolet spectrophotometry of two dwarf novae uncover the presence of high velocity winds. Hawley and Smarr present their result on high resolution numerical simulations of matter swirling on black holes. These studies reveal several new and unexpected features such as the existence of the centrifugal barrier which plays an important role in shock heating the inflowing matter. Lovelace, Mehanian, Mobarry and Sulkanen derived the general theory for axisymmetric, ideal-MHD flows in the limit of weak gravitational fields. They use this theory to calculate self-consistently the equilibria of a thin, magnetized disk orbiting around an aligned, rotating, magnetized star. Eilek describes the particle acceleration by Alfven wave turbulence in radio galaxies. She shows that under certain conditions the competition between acceleration and synchrotron losses leads to asymptotic particle distributions which are in agreement with those inferred from observations.

The fourth section contains the articles which discuss physical processes. Field discusses the concept of magnetic helicity and explains how the relative helicity is a measure of the amount of linkage of two fields which emerge from a photosphere. He shows how helicity may dissipate much more slowly than magnetic energy. Quest describes his numerical results on the existence and properties of high-Mach-number shocks. He finds that these shocks become turbulent and may result in strong electron heating.

The scientific organizing committee consisted of R. I. Epstein, W. D. Evans, and W. C. Feldman from Los Alamos and F. K. Lamb from the University of Illinois. The workshop secretary was S. A. Fradkin. The Earth and Space Sciences Division of the Los Alamos National Laboratory provided funding for this workshop.

Richard I. Epstein
William C. Feldman

Los Alamos National Laboratory
April 1986

ASTROPHYSICAL PARTICLE ACCELERATION

R. D. Blandford
Theoretical Astrophysics 130-33, California Institute of Technology, Pasadena, CA 91125

ABSTRACT

The observed properties of high energy particles in different astrophysical environments are briefly summarized. It appears that cosmic rays are freely produced with high (up to 10 per cent) efficiency by supersonic flows. At relativistic energies, the integral distribution function of the accelerated particles (i.e., the number of particles with energy in excess of some value E) generally has a power law form with slope lying in the range ~1.2 to 1.8. Various particle acceleration schemes are examined. Fermi acceleration at strong shock fronts seems a particularly promising process. A simple test particle analysis suggests that this mechanism can fulfill the necessary energetic and spectral requirements of galactic cosmic rays. More detailed studies of shock acceleration including the mediating influence of the accelerated particles and the generation of scattering Alfvén modes together with models of the gas subshock give inconclusive results. Future progress depends upon combining the results of *in situ* observations of interplanetary shock waves with numerical simulations of high Mach number, quasi-parallel subshocks to obtain more complete models of astrophysical shocks.

1. INTRODUCTION

(i) Scope of the problem

Astrophysical particle acceleration proceeds in a wide variety of environments under a diverse range of conditions presumably by an equally diverse set of mechanisms. Within the heliosphere, we find double layers and wave energization (e.g., in the terrestrial aurorae), explosive tearing mode reconnection (e.g., in the magnetotail), and Fermi acceleration (operating at the planetary bow shocks, the traveling interplanetary shocks associated with solar flares and the co-rotating interaction regions) amongst many other mechanisms, both real and imagined.

As we venture (at least figuratively) beyond the solar system (passing through the solar wind termination shock where the "anomalous" component of cosmic rays is believed to be accelerated) we encounter galactic cosmic rays being accelerated within the interstellar medium. Additional components of the cosmic ray spectrum may be accelerated by pulsars and at a hypothetical galactic wind termination shock. In addition, synchrotron-emitting relativistic electrons are accelerated within expanding supernova remnants. Venturing even further away, we observe that relativistic electrons are created copiously by active galactic nuclei and accelerated locally within extended extragalactic radio sources and the intergalactic media of some rich clusters of galaxies.

(ii) General Characteristics

(a) Efficiency

It is surely unlikely that one mechanism be responsible for particle acceleration within all of these inaccessible regions. Nevertheless, some features of astrophysical acceleration are so common as to be prerequisites of candidate general processes. Firstly, particle acceleration is usually extremely efficient in the sense that the fraction of the apparently available free energy that is channeled into high energy particles can approach (and may exceed) 10 per cent. To give some examples, the best estimate of the cosmic ray power of our galaxy (outlined below) is $\sim 3\times 10^{40}\,\mathrm{erg\,s^{-1}}$. This is to be compared with the average supernova power for which the standard estimate is 10^{51} erg every 30 years or $10^{42}\,\mathrm{erg\,s^{-1}}$.[1] The most powerful solar flares create $\sim 10^{32}$ ergs of energetic ions. This energy has to be stored in magnetic form in a volume $\lesssim 10^{31}\,\mathrm{cm^3}$. A coronal field strength of ~ 50 G gives a total magnetic energy of $\sim 10^{33}$ erg and indicates that the apparent efficiency is ~ 10 per cent. A third example is furnished by the extended extragalactic radio sources. The theory of synchrotron radiation (which is almost certainly the relevant radio emission mechanism) allows us to convert the observed volume emissivity into a lower bound on the combined pressure of the relativistic electrons and magnetic field.[2] This lower bound is obtained when the particles and field are in rough energy equipartition. These sources must be confined by the pressure exerted by thermal plasma and we can use X-ray observations to limit its value. The end result is that in many cases, the minimum internal pressure approaches or even exceeds the maximum external pressure. As the radio sources are expanding (and consequently cooling) the relativistic electrons must be accelerated locally and the efficiency of this acceleration must therefore often approach unity. Arguments like this convince us that bulk kinetic or magnetic energy is readily degraded into high energy particles.

(b) Power-law distribution functions

We can often measure directly or use observations to infer indirectly the form of the particle distribution function of the accelerated particles. Frequently this has a power-law form. The interstellar spectrum of galactic cosmic rays has a differential energy spectrum $N_E \propto E^{-2.65}$ (where $N_E dE$ is the number of particles per unit volume with total energy E in the internal dE) between a few GeV and the "knee" in the spectrum at $\sim 10^5$ GeV where the spectrum steepens to a slope ~ 3.5. (Above the "ankle" at $\sim 10^{10}$ GeV, the spectrum flattens again to a slope ~ 2.2 to the highest energies observed ($\sim 10^{11}$ GeV)[1,3] (see Fig. 1).) In fact, we can use the observed energy-dependence of the ratio of primary to secondary cosmic rays (i.e., those accelerated in the sources to those created by spallation reactions en route to us) to argue that the residence time of ~ 3–100 GeV cosmic rays within the galaxy decreases with increasing energy $\propto E^{-0.4}$ implying that the injection spectrum of the lower energy cosmic rays is $S_E \propto E^{-2.2}$ similar to that of the highest energy particles. This has been widely interpreted as implying that similar mechanisms create the low energy cosmic rays within our galaxy and the high energy ($\gtrsim 10^9$ GeV) particles within intergalactic space (from which there is no escape!). (A third, intermediate energy, 10^5–10^9 GeV component may also be necessary.) Cosmic ray electrons in the energy range 10–100 GeV are found to have a flux ~ 3 per cent that of the cosmic ray protons and are observed throughout the galaxy through their synchrotron emission. Much larger local concentrations of relativistic electrons are

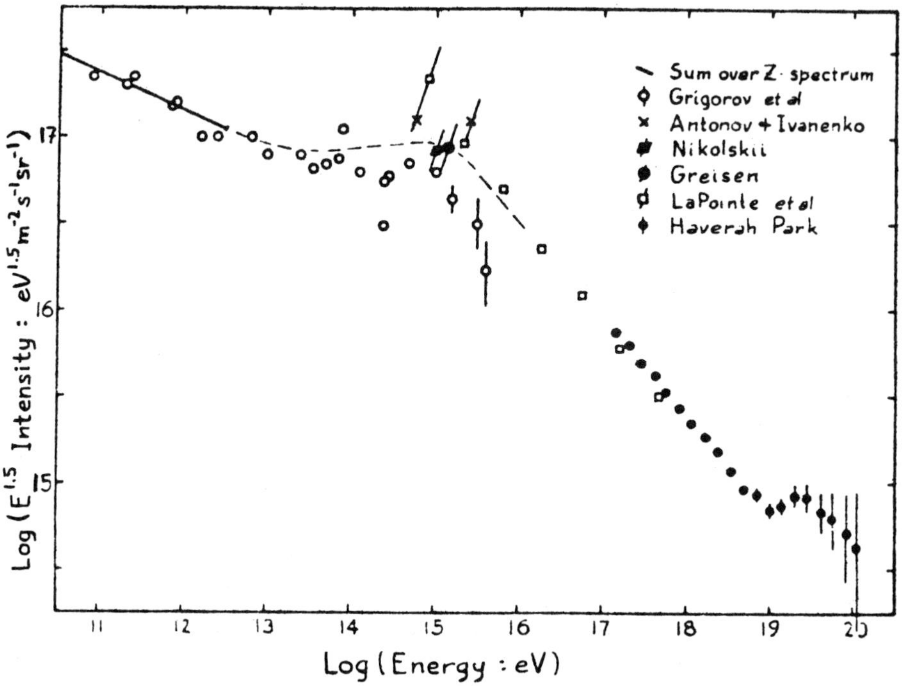

Fig. 1. The energy spectrum of cosmic rays at energies above 100 GeV taken from Meyer.[4] The spectrum is modulated by the effects of the solar wind below ~3 GeV. The spectrum has a logarithmic slope ~2.65 from ~3 GeV to ~10^5 GeV. Note the "knee" and "ankle" features at ~10^5 Gev and 10^{10} GeV.

observed within expanding supernova remnants again through their synchrotron emission (Fig. 2). (If the spectral index, $-d\ln I_\nu / d\ln\nu$, of the radiation intensity, I_ν, is denoted α and the source is optically thin and uniformly magnetized, then the underlying electron distribution function is $N_E \propto E^{-(2\alpha+1)}$.[2] We observe that $\alpha \sim 0.5-0.75$ so $N_E \cong E^{-2.0-2.5}$.)

This last method can also be applied to extended extragalactic radio sources (see Blandford, these proceedings). Excluding those sources that are either self-absorbed or radiatively cooled, it appears that the measured spectral indices (defined to be $\alpha = -d\log S_\nu / d\log\nu$) are clustered around the value $\alpha \sim 0.7$. The particle distribution functions are thus similar in shape to those inferred within our interstellar medium (as well as the interstellar media of external galaxies).

We have much less confidence in the details of the emission mechanisms within the nuclei of galaxies, but here again, for example, we observe power-law spectra ($\alpha \sim 1$ in the infrared regions of many quasars and $\alpha \sim 0.7$ in the X-ray range for Seyfert galaxies (Fig. 3). The infra-red and X-ray continua are probably separate components. They may be due to either synchrotron or inverse Compton emission in which case efficient acceleration of power law distribution functions with $N_E \propto E^{-(2.4-3)}$ is again required.

(c) Selective acceleration

Further clues about astrophysical particle acceleration come from abundance measurements.[1] If we compare the elemental abundances of galactic cosmic rays as inferred at their sources with solar system abundances, then (relative to carbon) hydrogen and helium are quite deficient and iron is overabundant. There is probably also a correlation of the abundance enhancement over the solar system abundance with first ionization potential. Isotopic anomalies are apparent in a few elements, most notably Ne, Mg, and Si. Similar trends have been observed in solar cosmic ray abundances. A complete theory of cosmic ray acceleration must explain how these differences come about.

The electron-proton ratio is also of interest. The value of 0.03 found in galactic cosmic rays may not be universal. Lower bounds on the electron-proton ratio $\gtrsim 0.3$ can be estimated in the Crab Nebula (see Fig. 4) and some of the extragalactic radio sources. These limits are generally derived by comparing the lower bound or the scan of the electron and the magnetic pressure from the theory of synchrotron radiation with an upper bound or the total pressure. In the case of the Crab Nebula this arises from the requirement that the acceleration of the wisps not exceed the observed value.[8] In the case of extragalactic radio sources, this comes from a limit on the pressure of the surrounding gas from X-ray observations.[9]

(iii) References

These two lectures are intended to be of a "pedagogic" nature. I shall attempt to explain some basic issues in astrophysical particle acceleration theory and refer the interested reader to current research monographs and texts for original references and discussion of more controversial topics. The most useful general reference discussing particle acceleration is Melrose.[10] A good introductory text that places the following discussion into its astronomical context is Longair.[2] Recent conference reports containing relevant papers have

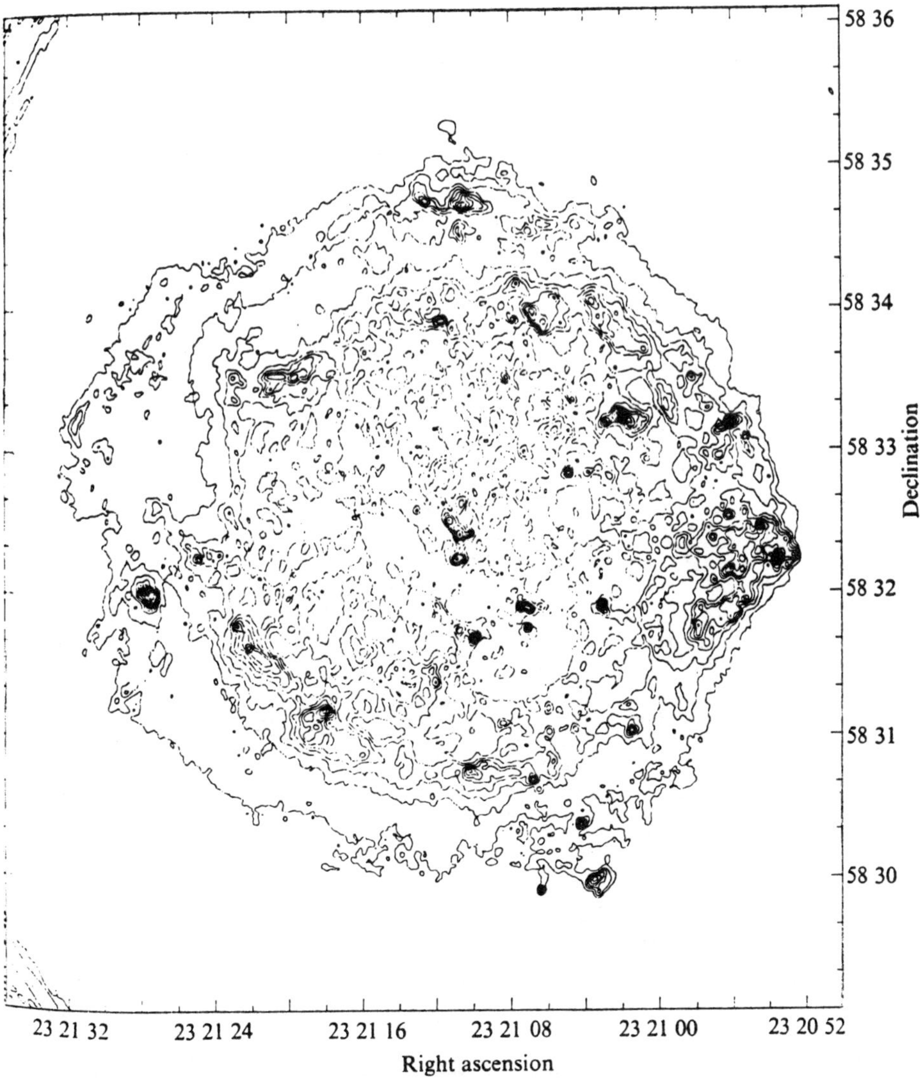

Fig. 2. 5 GHz radio map of the young supernova remnant Cassiopeia A from Bell, Gull and Kenderdine.[5] The blast waves which surround this remnant may be responsible for accelerating most galactic cosmic rays. Note the "knots" of high surface brightness in which there must be efficient particle acceleration.

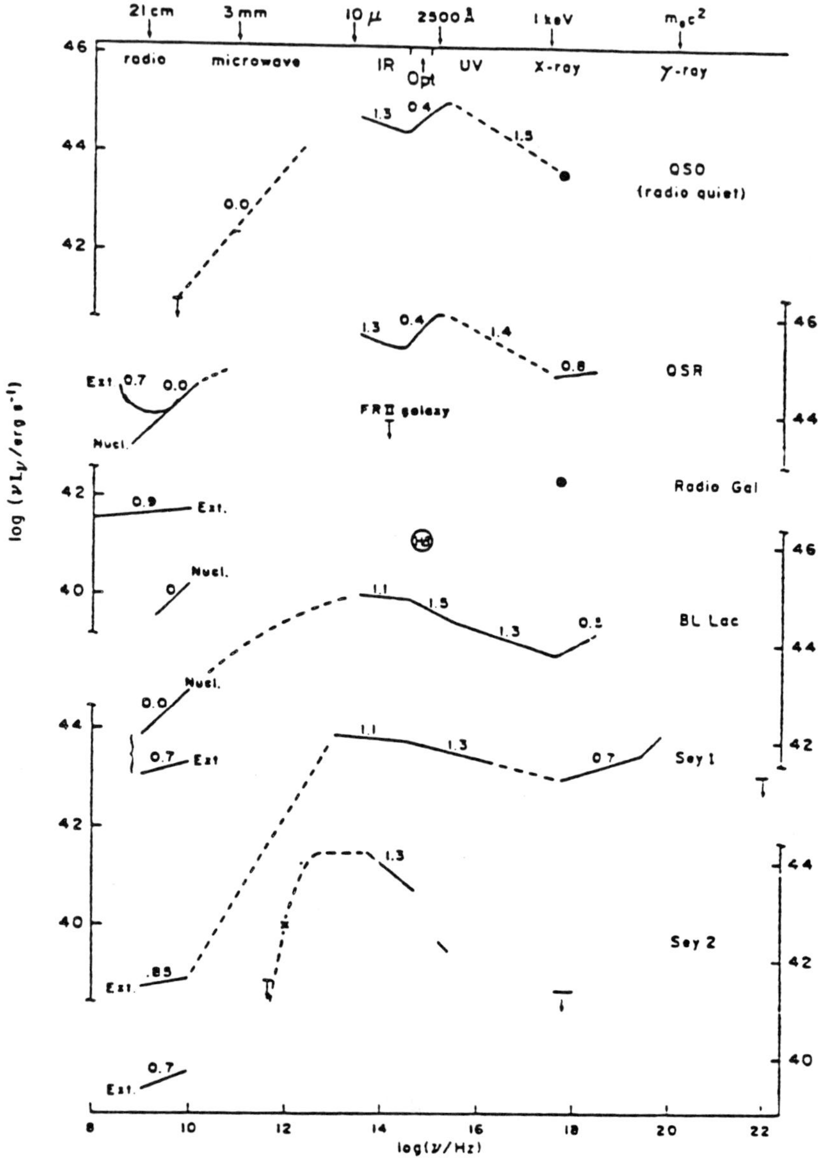

Fig. 3. Schematic spectra (plotted as power per log frequency bandwidth) for members of different classes of active galactic nucleus. The diagram is adopted from a similar one in Phinney.[6] The various portions of the spectrum are labeled by their spectral indices, α. The variations in these spectral indices are substantial.

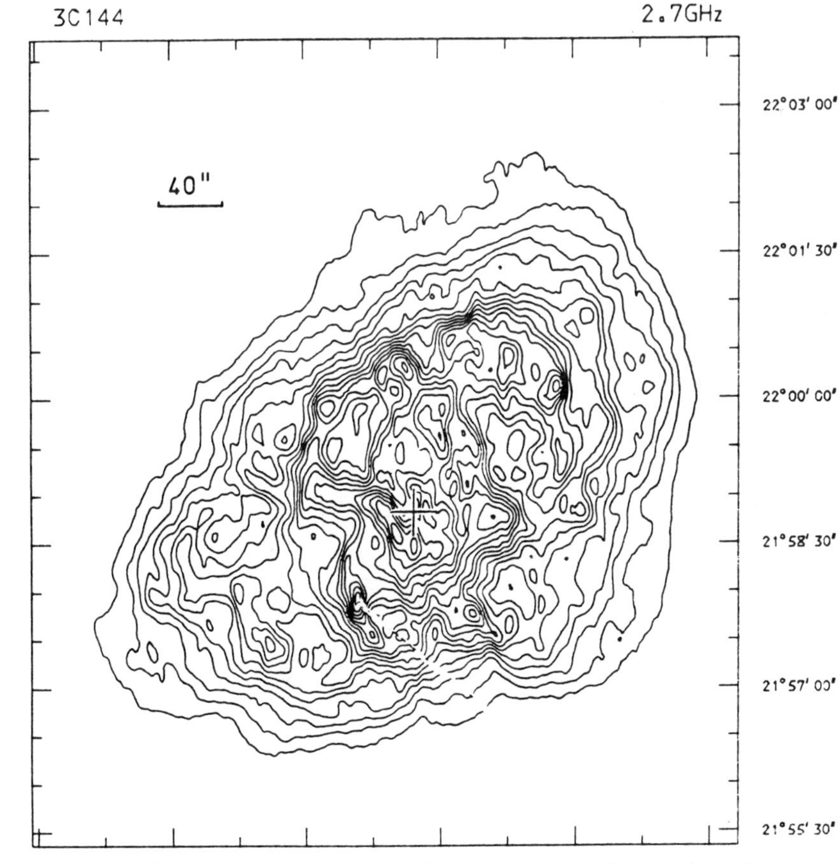

Fig. 4. 2.7 GHz radio map of the Crab Nebula adapted from Swinbank and Pooley.[7] The Crab Nebula is the prototypical filled or "plerionic" supernova remnant. The synchrotron-emitting electrons and the magnetic field are believed to derive from the central pulsar marked by a +, through a relativistic hydromagnetic electron-positron wind and may be accelerated by a strong shock wave that forms when the momentum flux in the wind balances the ambient nebular pressure.

been edited by Arons, Max, and McKee[11] and Guyenne and Levy.[12]

2. SOME PARTICLE ACCELERATION MECHANISMS

(i) Diffusion in momentum space

In many particle acceleration schemes, including the famous Fermi mechanism, high energy particles gain (and lose) energy in small steps. The simplest formal description is in terms of the momentum space distribution function, $f(\mathbf{p},\mathbf{x},t)$ where $f d^3p d^3x$ is the number of particles with momentum $\mathbf{p}$ in the element d^3p located within the spatial volume element d^3x. If the momentum space distribution function is isotropic then the energy distribution function is $N_E = 4\pi p E f(p,\mathbf{x},t)/c^2$ where E is the relativistic energy.

In the Fokker-Planck formalism,[13] the probability that an electron changes its momentum by $\Delta\mathbf{p}$ in time interval Δt is presumed to depend on the current values of $\mathbf{p}$ and $\mathbf{x}$ and not on the complete past history of the particle. In other words, we are dealing with a Markov process. The distribution function at time $t + \Delta t$ is then given by

$$f(\mathbf{p},\mathbf{x}+\mathbf{v}\Delta t, t+\Delta t) = \int d^3\Delta p\, \psi(\mathbf{p}-\Delta\mathbf{p},\Delta\mathbf{p}) f(\mathbf{p}-\Delta\mathbf{p},\mathbf{x},t) \tag{2.1}$$

where $\psi(\mathbf{p}'-\Delta p,\Delta\mathbf{p})d^3\Delta p$ is the element of probability for changing the momentum $\mathbf{p}$ to $\mathbf{p}+\Delta\mathbf{p}$ in time Δt. We expand both the integrand and the left hand side in a Taylor series and use the fact that $\int d^3\Delta p\, \psi(\mathbf{p},\Delta\mathbf{p}) = 1$ to obtain the Fokker-Planck equation

$$\frac{\partial f}{\partial t} + \mathbf{v}\cdot\frac{\partial f}{\partial \mathbf{x}} = \frac{\partial}{\partial \mathbf{p}}\cdot\left\{-\left\langle\frac{\Delta\mathbf{p}}{\Delta t}\right\rangle f + \tfrac{1}{2}\frac{\partial}{\partial \mathbf{p}}\cdot\left\langle\frac{\Delta\mathbf{p}\Delta\mathbf{p}}{\Delta t}\right\rangle f\right\} \tag{2.2}$$

where the Fokker-Planck coefficients are

$$\left\langle\frac{\Delta\mathbf{p}}{\Delta t}\right\rangle = \frac{1}{\Delta t}\int d^3\Delta p\, \psi(\mathbf{p},\Delta\mathbf{p})\Delta\mathbf{p} \tag{2.3}$$

and $$\left\langle\frac{\Delta\mathbf{p}\Delta\mathbf{p}}{\Delta t}\right\rangle = \frac{1}{\Delta t}\int d^3\Delta p\, \psi(\mathbf{p},\Delta\mathbf{p})\Delta\mathbf{p}\Delta\mathbf{p}\,. \tag{2.4}$$

This equation describes a biased random walk of the particles in momentum space.

In Fermi processes, particles gain and lose energy by scattering. There is a simplification when the recoil of the scatterer can be ignored. (This is usually true.) In this case, the principle of detailed balance assures us that

$$\psi(\mathbf{p},-\Delta\mathbf{p}) = \psi(\mathbf{p}-\Delta\mathbf{p},\Delta\mathbf{p})\,. \tag{2.5}$$

We again make a Taylor expansion and integrate over $d^3\Delta p$ to obtain

$$\frac{\partial}{\partial \mathbf{p}} \cdot \left\{ \left\langle \frac{\Delta \mathbf{p}}{\Delta t} \right\rangle - ½ \frac{\partial}{\partial \mathbf{p}} \cdot \left\langle \frac{\Delta \mathbf{p} \Delta \mathbf{p}}{\Delta t} \right\rangle \right\} = 0 .$$

Or, if the Fokker-Planck coefficients vanish for small values of $\mathbf{p}$ as is generally true,

$$\left\langle \frac{\Delta \mathbf{p}}{\Delta t} \right\rangle = \frac{1}{2} \frac{\partial}{\partial \mathbf{p}} \cdot \left\langle 2 \frac{\Delta \mathbf{p} \Delta \mathbf{p}}{\Delta t} \right\rangle . \tag{2.6}$$

The Fokker-Planck equation then simplifies to

$$\frac{\partial f}{\partial t} + (\mathbf{v} \cdot \nabla) f = \frac{\partial}{\partial \mathbf{p}} \cdot \mathbf{D}_{pp} \cdot \frac{\partial f}{\partial \mathbf{p}} \tag{2.7}$$

where the momentum space diffusion coefficient is

$$\mathbf{D}_{pp} = ½ \left\langle \frac{\Delta \mathbf{p} \Delta \mathbf{p}}{\Delta t} \right\rangle . \tag{2.8}$$

As we might expect, Fick's law of diffusion applies in momentum space if the scattering process is time-reversible. Terms describing loss and escape can be added to the right hand side of equation (2.7). Further discussion of equation (2.7) and its properties can be found in Melrose.[10]

As a first, simple example of a Fermi process, suppose that the scatterers are hard spheres, moving with speed V. The particle distribution function will be isotropized after a few scatterings and so we are only interested in the $p_{\|}p_{\|}$ component of the tensor

$$D_{p_{\|}p_{\|}} = \frac{pE \langle V^2 \rangle}{3Lc^2} \tag{2.9}$$

where L is the collision mean free path. (The hard spheres may be replaced by moving magnetic clouds but neither abstraction seems to be a good model of the behavior of an astrophysical plasma.)

A second, more physically relevant example is given by the Kompaneets equation in which the *photon* distribution function $n(\nu, \mathbf{x}, t)$ evolves under isotropic and spatially homogeneous conditions by Compton scattering off hot electrons of temperature T_e.[14]

$$\frac{\partial n}{\partial t} = n_e \sigma_T c \cdot \frac{1}{\nu^2} \frac{\partial}{\partial \nu} \nu^4 \left\{ \frac{kT_e}{m_e c^2} \frac{\partial n}{\partial \nu} + \frac{h}{m_e c^2} (n + n^2) \right\} . \tag{2.10}$$

The first term on the right hand side, which dominates when $\langle \nu \rangle \ll kT_e / h$, has the form of the right hand side of equation (2.7) and describes Fermi acceleration of the soft photons by the hot electrons. (The second term takes account of the electron Compton recoil and the third term describes induced scattering.) We give a few more examples below.

We can verify that equation (2.7) conserves particles by integrating over momentum space. Similarly, we can take the energy moment and integrate by parts twice to make the identification

$$\left\langle \frac{\Delta E}{\Delta t} \right\rangle = \frac{\partial}{\partial \mathbf{p}} \cdot \left\{ \frac{\mathbf{v}}{2} \cdot \left\langle \frac{\Delta \mathbf{p} \Delta \mathbf{p}}{\Delta t} \right\rangle \right\}.$$

$$= \mathbf{v} \cdot \left\langle \frac{\Delta \mathbf{p}}{\Delta t} \right\rangle + \frac{(E^2 c^2 \delta_{ij} - p_i p_j c^4)}{E^3} \cdot \left\langle \frac{\Delta \mathbf{p} \Delta \mathbf{p}}{\Delta t} \right\rangle_{ij} . \qquad (2.11)$$

Note that this does not equal $\mathbf{v} \cdot \langle \Delta \mathbf{p} / \Delta t \rangle$ as might at first be guessed because an unbiased random walk in momentum space will also produce a net increase in energy. The diffusion coefficient in energy space is related to that in momentum space through $D_{EE} = v^2 D_{p_\parallel p_\parallel}$.

In the case of hard sphere scattering, $\langle \Delta E / \Delta t \rangle = (4p \langle v^2 \rangle / 3L)$. At relativistic energy $dE/dt \propto E$ as in the traditional Fermi process. However, $dE/dt \propto E^{1/2}$ at non-relativistic energy.[15]

(ii) Fermi acceleration

In the traditional Fermi process introduced above, the acceleration is said to be second order because the mean energy gain, given by equations (2.9), (2.11) is proportional to the mean square scatterer speed.[2] To first order in the ratio of the speed of the scatter to that of the particles, the particles are just as likely to lose as to gain energy. It is only to second order in the ratio that the particles gain energy. (We discuss below a first-order acceleration scheme.)

The method that Fermi proposed for creating a power law was to note that (for relativistic particles at least) the rate of gain of energy was proportional to energy. The energy of an individual particle will then *increase* exponentially with time. However, if the acceleration region (e.g. our galaxy) is finite then the probability of a particle being accelerated for some long time will *decrease* exponentially with time. The net result is a power law distribution function.

We can formalize this using the Fokker-Planck equation by assuming that the major loss process is particle escape and adding a term $-f/\tau_e$ to the right hand side of equation (2.7). We seek a stationary, spatially homogeneous solution to this equation when there is a source of low energy particles. If $D_{p_\parallel p_\parallel} = p^2/\tau_a$ and τ_e, τ_a are energy independent, then the required solution is a power law $f(p) \propto p^{-q}$, where

$$q = \frac{3}{2}\left[1 + \left(1 + \frac{4\tau_a}{9\tau_e}\right)^{1/2}\right]. \qquad (2.12)$$

The production of a power law is a consequence of the competition between acceleration and escape. This is a feature of many theories of particle acceleration.

Unfortunately, there are several difficulties with the mechanism in its present form. (Most of these were acknowledged by Fermi.) The most obvious is that in order to accelerate a distribution with a preferred value of q, say 4.5, then the ratio τ_a/τ_e must also have a preferred value, in this case 6.75. It is not clear how this can always be arranged in a variety of different environments. In fact the assumption that τ_e is energy independent is suspect. We know that it is untrue in the interplanetary and interstellar media. Furthermore, as we shall see when considering acceleration by wave modes it is not in general true that the rate of gain of energy is proportional to energy. Under these circumstances even more contrivance is necessary to get a particular power-law distribution. Finally, if the energy gain were proportional to the energy, we would expect that most of the energy be absorbed by those particles that account for most of the energy density, i.e., the thermal particles. In other words, we must suppress the acceleration of the lowest energy particles.

Analogous comments can be made about Comptonization models that account for power law photon spectra directly through hot electron scattering.

(iii) Flow processes

Fermi acceleration becomes more plausible as a general mechanism if some way can be found to couple the acceleration to the escape. One way to achieve this is through a continuous flow process. Suppose, for example,[16,17] that there is a flow of plasma through an accelerating region where a delta function distribution of monoenergetic relativistic electrons is created impulsively (e.g. at a shock front). Suppose further that some wave turbulence is simultaneously created and that all of this energy is ultimately absorbed by the particles as they flow out of the region. If the waves accelerate the particles by a Fermi process they will give then a power law distribution with slope q.[16] (This imposes certain restrictions on the nature of the acceleration process.) The fractional mean energy gain per particle in absorbing the wave energy is then $(q-4)^{-1}$ assuming that $q > 4$. So if, for example, it turned out that the initial wave turbulence always contained twice as much energy density as the particles, then ultimately the power law slope would be fixed at $q = 4.5$.

(iv) Wave acceleration

A more physically complete description of stochastic particle acceleration is made possible by analyzing the magnetic disturbances into a Fourier spectrum of wave modes. The basic procedure is to consider a single mode with wave-vector **k** and treat its electromagnetic fields as causing small perturbations on the particle motions. The associated linear perturbation in the particle distribution function is then calculated from the collisionless Vlasov equation. This perturbation can then be used to compute an expression for the current density as an integral (or sum of integrals) over particle momentum space. This integral has a resonant denominator $(\omega - k_{\|}v_{\|} - n\Omega)^{-1}$ where $\|$ refers to the direction of the magnetostatic field, n is an integer, and Ω is the (relativistic) Larmor frequency, which vanishes when the unperturbed particle motion resonates with the wave, i.e., when $\omega - k_{\|}v_{\|} = n\Omega$ for some n. We can use the Plemelj formula to replace the resonant-denominator by a term $\propto i\pi\delta(\omega - k_{\|}v_{\|} - n\Omega/\gamma)$ and thereby obtain the current density $\mathbf{j}$ in phase with the wave electric field $\mathbf{E}$ as a sum over possible resonances. Finally, we make the random phase approximation to sum over all waves and then obtain the rate of energy transfer $\mathbf{E} \cdot \mathbf{j}$ from the waves to the particles (or vice versa if the particle distribution function is unstable).

This (quasi-linear) procedure is clearly only appropriate if the waves have small amplitude and are incoherent. The particle acceleration rate can generally be converted using equation (2.11) to a Fokker-Planck equation of the form (2.7) where D_{pp} is proportional to the resonant wave energy density.[10,18,19]

(a) Alfvén waves

A simple example is furnished by Alfvén waves propagating along the field. (Longer expressions have been given for obliquely propagating waves.) Here the only resonance has $n = +1$ or -1, depending on the charge of the particle because the perturbing is a pure sinusoid without harmonics. As we are concerned with waves propagating below the ion Larmor frequency, we generally approximate the resonance condition by $k_{\|} \simeq \Omega / v_{\|} = eB/p_{\|}c$. That is to say the waves resonate with particles whose Larmor radii are comparable with their wavelengths unless their pitch angles are close to 0 or $\pi/2$. If we think about the scattering from a quantum mechanical point of view, then there is a stimulated emission and absorption of Alfvén wave photons of momentum $\hbar k$. Now the ratio of energy to momentum in the waves is $a = B(4\pi\rho)^{-1/2}$, the Alfvén speed, whereas that of the resonant particles is $\sim v$ which is usually much larger. This implies that Alfvén waves are far more effective at scattering particles in pitch angle than they are at accelerating them. (The interaction of resonant particles with Alfvén waves is actually more subtle than this argument might suggest.[20])

To be quantitative we can compute Fokker-Planck coefficients for pitch angle (θ) scattering and diffusion in total momentum p. In the first case

$$\nu \equiv \left\langle \frac{\Delta\theta^2}{\Delta t} \right\rangle \simeq \left(\frac{\pi}{4}\right)\left(\frac{8\pi k\,\varepsilon_k}{B^2}\right)\Omega \qquad (2.13)$$

where $\varepsilon_k(k)$ is the energy density of waves with resonant wave vector $k = eB/pc$. This is only an approximate expression because we have averaged over the pitch angle θ instead of recognizing that as the pitch angle increases to 90°, the frequency of the resonant waves must also increase. (In fact as Alfvén waves do not propagate above the ion Larmor frequency, there is a formal difficulty in scattering through $\theta = 90°$, particularly in a high beta plasma. However, it now appears that non-linear effects will allow easy passage through $\theta = 90°$.[21] Equation (2.13) simply describes a random walk in pitch angle with step $\delta\theta \sim (\delta B/B)_{res}$ every Larmor period.

The associated momentum space diffusion coefficient is similarly shown to be[22]

$$D_{pp} = \frac{4}{3}p^2\left(\frac{a}{v}\right)^2\nu . \qquad (2.14)$$

The time it takes to change the momentum by a factor ~2 is therefore longer than the time to scatter in pitch angle by $O(v/a)^2$ as explained above.

Most schemes for implementing Alfvén wave acceleration have relied on creating an inertial range cascade (e.g. with the "Kolmogorov" spectrum, $\varepsilon_k \propto k^{-5/3}$) and incorporating this within some time-dependent scheme (cf. Eilek this volume).

(b) Magnetosonic waves

Fast magnetosonic modes may also propagate in a magnetized plasma. These waves have a compressive part and may therefore be Landau damped at the $n=0$ resonance ($\omega = k_{\|}v_{\|}$). Only the parallel component of momentum changes in this interaction. However, it is generally assumed that a low level resonant Alfvén wave turbulence will maintain isotropy in the particle distribution function. In this case the particles random walk in momentum space with step $\delta p \sim p(\delta B/B)_{res}$ every wave period. More formally

$$D_{pp} \simeq \frac{\pi}{48}p^2\left(\frac{a}{v}\right)\int \frac{dk\,\omega\varepsilon_k}{\varepsilon_m} \tag{2.15}$$

where we have assumed that the isotropy of the waves is maintained and $\varepsilon_m = B^2/8\pi$ is the magnetic energy density.[17] Note that all relativistic particles are essentially resonant with the same waves propagating perpendicular (to within an angle $\sim a/v_{\|}$) to the ambient field direction. An $n = 0$ resonance depends upon the velocity of the particle in contrast to an $n = 1$ resonance which depends upon the rigidity or momentum per unit charge. As the acceleration rate is proportional to the energy, the mechanism resembles a traditional Fermi process. Unfortunately, thermal electrons can also Landau damp the waves unless $\beta = 8\pi p/B^2$ is much smaller than unity.

(c) Sound waves

A different type of acceleration can operate in a high β plasma. It is simplest to regard the fast mode as a longitudinal sound wave that is being damped by thermal conduction in the background medium.

Suppose that there is some higher frequency turbulence that scatters the particles at rate $\nu \gg \omega$. This will confer fluid-like properties on the plasma and also fix the rate at which the waves are conductively damped. Now if the scattering rate decreases sufficiently rapidly with increasing energy, then heat may be carried from a high pressure region to a low pressure region most efficiently by the high energy particles. The particle acceleration rate associated with the wave damping can be calculated using quasi-linear theory to be

$$D_{pp} = \frac{p^2v^2}{60\nu(p)\varepsilon_0}\cdot\int dk k^2\varepsilon_k \tag{2.16}$$

where ε_0 is the thermal energy density of the background medium[19,23]

(d) Electromagnetic waves

Under relativistic conditions (e.g. perhaps in a galactic nucleus or a supernova remnant with a central pulsar like the Crab Nebula), the dominant waves may be essentially low frequency electromagnetic modes. If the wave phase velocity exceeds c and the gyrofrequency of the static field is small enough that linear resonances are unimportant, then we may still get particle acceleration through non-linear resonance. (Of course, acceleration by linear resonances is also quite possible under these conditions. However, as our aim is to illustrate different possibilities, we shall just consider the non-linear resonances.)

In classical language, wave 1 with amplitude E_1 induces an oscillatory component in an electron's motion with speed $\delta v \sim (eE_1/m\omega_1\gamma)$. This perturbation couples with the magnetic field $B_2 \sim E_2$ of a second wave satisfying the resonance condition ($\omega_1 - \mathbf{k}_1 \cdot \mathbf{v} = \omega_2 - \mathbf{k}_2 \cdot \mathbf{v}$) to create a secular kick in the momentum $\delta p_2 \sim (e^2 E_1 E_2/\gamma mc\,\omega_1\omega_2)$. These kicks add stochastically. Summing over all pairs of waves satisfying the resonance condition gives a momentum space diffusion coefficient (specialized for simplicity to ultra-relativistic particles)[24]

$$D_{p_\parallel p_\parallel} = \frac{4\pi^3 re^2 c^2}{\gamma^2} \int_0^\infty dk_1 \int_{k_1}^\infty \frac{dk_2 \varepsilon_k \varepsilon_k (\omega_1 - \omega_2)^2}{\omega_1^2 \omega_2^3} \qquad (2.17)$$

We can just as easily think of this process quantum mechanically. In this case, the waves are undergoing induced Compton scattering from state 1 to state 2 and back again. The Compton recoil is responsible for the net (classical electromagnetic) acceleration rate.

(v) Reconnection

A quite different acceleration mechanism is associated with magnetic reconnection (see articles by Lamb and by Feldman in these proceedings, for a discussion of some of the hydromagnetic aspects of reconnection).[25] One important and well-studied instance of magnetic reconnection is triggered by explosive tearing mode instabilities in the earth's magnetotail and produces substorms (see Baker, these proceedings).[26,27] have emerged from these studies; that the efficiency of production of supra-thermal particles can approach ten per cent and that the maximum energy per charge accelerated can exceed the total potential difference across the reconnecting region. The importance of this mechanism for astrophysics is that it might operate wherever there is a shear, in particular in jets and accretion disks. However, we do not yet know if there are ever circumstances when most of the energy dissipated in this type of flow is channeled into the ultrarelativistic electrons as observations of jets at least seem to mandate.

(vi) Electrostatic acceleration

Of course, all acceleration mechanisms require an electric field to do work on the particles. However, this is usually a frame-changing field, $\mathbf{E} = -(\mathbf{v}\times\mathbf{B})/c$ as required by MHD. Under some circumstances, MHD may break down and substantial potential difference may be maintained along the magnetic field (in particular as a result of unipolar induction). This cannot be removed by Lorentz transformation as $\mathbf{E}\cdot\mathbf{B}$ is invariant. An axisymmetric spinning magnetized object generates a potential difference between its poles and its equator of order

$$V \sim \Omega R^2 B / c \sim 3\times 10^{20} (\Omega R / c)(R / 10\,\mathrm{km})(B / 10^{12} G)\,\mathrm{volts} \tag{2.18}$$

where Ω is the angular velocity, R is the characteristic size, and B is the surface field strength on the object. When the inertial effects of the surrounding plasma are unimportant, the associated power dissipation is $P \sim V^2 / Z_0$ where $Z_0 \equiv 377\,\mathrm{ohm}$ is the impedance of free space. Some examples of unipolar induction are given in Table 1.

TABLE 1. The unipolar induction mechanism in a selection of astronomical objects.

Object	B Gauss	R cm	Ω rad s^{-1}	ε V	I amp	P erg s^{-1}
Jupiter/Io	0.1	10^{10}	10^{-4}	10^{6}	10^{6}	10^{19}
Saturn	0.1	10^{10}	10^{-4}	10^{5}	100	10^{14}
Radio pulsar	10^{12}	10^{4}	10	10^{13}	10^{11}	10^{33}
Crab pulsar/nebular	10^{12}	10^{5}	200	3×10^{16}	3×10^{14}	10^{38}
Massive black hole	10^{4}	10^{14}	10^{-4}	10^{20}	3×10^{18}	10^{45}

It is possible that many cosmic rays with energy $\gtrsim 10^5$ GeV are accelerated in this way within a pulsar magnetosphere. It is also possible that much of the particle acceleration within active galactic nuclei be essentially electrostatic.[28]

3. ACCELERATION BY ASTROPHYSICAL SHOCK WAVES

(i) Background

The idea that shock waves might be responsible for the acceleration of cosmic rays is not particularly new. (An early bibliography and a clear discussion of some of the issues involved is contained in Parker;[29] see also Ref. 1.) However, shock wave acceleration has become much more prominent as a mechanism over the past seven years and there is much recent theoretical work. Amongst several recent reviews are articles by Axford,[1] Drury,[30], and Blandford and Eichler.[31]

The reasons for a resurgence of interest in shock wave acceleration are two-fold. Firstly it was realized that, in the test particle limit at least, shock waves could transmit a power law distribution function of high energy particles the slope of which was constrained by kinematical considerations to have roughly the commonly observed range of values. Secondly, space physicists have observed this type of acceleration clearly operating at traveling interplanetary shock fronts. For the quasi-parallel shocks at least, the rate is in rough accord with theoretical predictions.[32]

Before we describe the mechanism of Fermi acceleration at a shock front, we must first discuss three topics; the evidence for spatial diffusion, the kinetic equation obeyed by the cosmic rays, and the gas dynamical jump conditions at a shock front.

(ii) Spatial diffusion

Cosmic rays appear to be diffusing through the galaxy. They are observed to be isotropic to roughly one part in 10^4 which is quite inconsistent with free streaming from the sources. Now we know from the ratio of secondary to primary cosmic rays that the mean grammage traversed by particles with a few GeV energy before they escape from the galaxy is $\lambda_{CR} \sim 5\,\mathrm{g\,cm^{-2}}$. The galactic disk at the solar neighborhood has a column density $\lambda_d \sim 10^{-3}\,\mathrm{g\,cm^{-2}}$ which is consistent with the measured isotropy. (The ratio of these two grammages can also be used to estimate the luminosity of the galaxy in cosmic rays, $L_{CR} \sim e_{CR}c(\lambda_d/\lambda_{CR})A_d \sim 3\times10^{40}\,\mathrm{erg\,s^{-1}}$ where e_{CR} is the cosmic ray energy density and A_d is the area of the disk.)

It has generally been argued that the cosmic rays are scattered by resonant Alfvén waves as described above. In this case an estimate of the spatial diffusion coefficient is

$$D_{\parallel\parallel} = \frac{v^2}{3\nu}. \tag{3.1}$$

The subscripts $\parallel$ denote that this is the component of the diffusion tensor parallel to the field. (Diffusion perpendicular to the magnetic field is generally unimportant except perhaps in the case of the earth's bow shock.[33])

(iii) Convection-diffusion equation

When cosmic rays are strongly scattered, their distribution function $f(p,\mathbf{x},t)$ will be isotropic. However, there will be spatial gradients in $f(p,\mathbf{x},t)$ when the velocity of the scatterers $\mathbf{u}(\mathbf{x},t)$ (assumed for the moment to be identical to that of the background fluid) varies. The evolution of the distribution function is governed by the convection-diffusion equation

$$\frac{\partial f}{\partial t} + \mathbf{u}\cdot\frac{\partial}{\partial \mathbf{x}}f - \frac{\partial}{\partial \mathbf{x}}\cdot\mathbf{D}\cdot\frac{\partial f}{\partial \mathbf{x}} = \frac{1}{3}\frac{\partial \mathbf{u}}{\partial \mathbf{x}}\cdot\frac{\partial f}{\partial \ln p}. \tag{3.2}$$

A rigorous derivation of this equation[22,31] is lengthy and we will just use heuristic arguments to justify it. The first and third terms on the left hand side of equation (3.2) can be recognized as the usual (spatial) diffusion equation. When $\mathbf{u} = 0$, the momenta of the particles will be unchanged when they are scattered by the Alfvén waves.

When the scatterers move we must make two changes. Firstly, we must replace the partial derivative $\partial f/\partial t$ by the convective derivative $\partial f/\partial t + \mathbf{u}\cdot\partial f/\partial\mathbf{x}$. Secondly, we must add a term which allows for the fact that the particle momenta change as the fluid expands or contracts. If we ignore the diffusion for the moment, then the momentum will decrease adiabatically in inverse proportion with the mean distance between scatterers in accordance with elementary statistical mechanics, $p \propto \rho^{1/3}$. Hence,

$$\frac{d\ln p}{dt} = \frac{1}{3}\frac{d\ln\rho}{dt} = -\frac{1}{3}\nabla\cdot\mathbf{u} \tag{3.3}$$

where we have used the equation of continuity for the background fluid. As we are using a phase space distribution function we must allow for the convection in momentum space as well as configuration space. So still ignoring diffusion,

$$\frac{Df}{Dt} = \frac{\partial f}{\partial t} + \mathbf{u} \cdot \frac{\partial f}{\partial \mathbf{x}} + \frac{dp}{dt}\frac{\partial f}{\partial p} = 0 . \tag{3.4}$$

After substituting equation (3.3) and replacing the diffusion term we recover the full equation (3.2).

(iv) Collisionless shocks

It has generally been assumed that shock fronts in collisionless astrophysical plasmas have a thickness of typically a few ion Larmor radii, far smaller than any relevant macroscopic length scales. As such, the background flow should satisfy the Rankine-Hugoniot relations. These relations express the conservation of mass, momentum, and energy.[34] If we denote conditions just ahead of and behind a normal shock by the subscripts −, + respectively, then

$$\rho_- u_- = \rho_+ u_+$$

$$p_- + \rho_- u_-^2 = p_+ + \rho_+ u_+^2$$

$$h_- + \tfrac{1}{2} u_-^2 = h_+ + \tfrac{1}{2} u_+^2 \tag{3.5}$$

where h $(= \gamma p/(\gamma-1)\rho$ for a perfect gas) is the specific enthalpy. Combining these relations, we obtain an expression for the shock compression ratio $r \equiv u_-/u_+$

$$\frac{1}{r} = \frac{\gamma-1}{\gamma+1} + \frac{2}{(\gamma+1)M^2} \tag{3.6}$$

$$\text{where } M = \left[\frac{\rho_- u_-^2}{\gamma p_-}\right]^{1/2} \tag{3.7}$$

is the Mach number of the shock. Note that we have ignored possible magnetic and cosmic ray contributions to these conservation laws.

(v) Test particle acceleration

We consider a stationary collisionless shock and regard it as a discontinuity in the background fluid velocity, $\mathbf{u}$. Let us assume that there are scattering Alfvén waves ahead of and behind the shock and that the associated diffusion coefficient is $D(\mathbf{x},p)$. The cosmic ray distribution function ahead of and behind the shock must satisfy the equation

$$u\frac{\partial f}{\partial x} = \frac{\partial}{\partial x}D\frac{\partial f}{\partial x}. \qquad (3.8)$$

The solutions of this equation in the two regions are

$$f = f_- + (f_0 - f_-)e^{\int_0^x u dx/D} \quad ; \quad x < 0$$

$$f = f_+ ; \quad x > 0 \qquad (3.9)$$

where $f_0 = f(0_-, p)$, $f_+ = f(0_+, p)$, and $f_- = f(-\infty, p)$ is the distribution function for the cosmic rays a long way ahead of the shock. In a steady state, cosmic rays can only diffuse against the stream and therefore the distribution function must be constant behind the shock ($x > 0$).

These two solutions must be joined at the shock front. As usual with second order differential equations there are two junction conditions. The first junction condition is that the distribution function must be continuous (to $O(u/v)$) so that $f_0 = f_+$. The second condition is that the particle flux at a given momentum $F(\mathbf{x}, p)$ must be continuous. By integrating equation (3.2) across the discontinuity or more properly by direct evaluation we see that the conserved quantity is

$$F = -D\frac{\partial f}{\partial x} - \frac{u}{3}\frac{\partial f}{\partial \ln p} \qquad (3.10)$$

(The first term in equation (3.10) is the usual diffusive flux, the second term is the convected flux, corrected for the "Compton-Getting" factor.) this differs from the term uf which might have been expected because equation (3.10) measures the flux at a given energy measured in the inertial frame and the particle distribution is measured in the fluid frame.)

Conserving f, F at the shock front, we obtain

$$\frac{(u_+ - u_-)}{3}\frac{df_+}{d\ln p} = \left[D\frac{\partial f}{\partial x}\right]_{0_-} = u_-(f_+ - f_-) \qquad (3.11)$$

which can be solved to obtain

$$f_+ = qp^{-q}\int_0^p dp' f_-(p')p'^{(q-1)} \qquad (3.12)$$

where

$$q = \frac{3u_-}{u_- - u_+} = \frac{3r}{r-1}. \tag{3.13}$$

Equation (3.12) gives the Green's function solution for the distribution function transmitted by the shock f_+ when the incident distribution is f_-. If we set f_- to be a delta function at low energy, then f_+ is a power law of slope q. Substituting equation (3.6) for $\gamma = 5/3$ we obtain

$$q = \frac{4M^2}{M^2-1}. \tag{3.14}$$

Hence for a strong shock ($M \gg 1$) $q \sim 4$. The relativistic energy distribution function is then

$$N_E = 4\pi p E f(p)/c^2 \approx E^{-2}. \tag{3.15}$$

As the shock weakens, the transmitted distribution function will steepen.

What is happening is that a particle of speed V must cross the shock $O(V/u)$ times. However, on each double crossing it gets a secular momentum boost $\Delta p/p = O(u/V)$ giving total momentum boost $[\Delta p/p] = O(1)$. This is therefore a first order Fermi process. The transmitted spectral slope is determined by the ratio of the acceleration and escape times which is now determined solely by the kinematics of the shock front. In particular it is independent of the form of the diffusion coefficient (save that it be positive!) and the angles the magnetic field and the velocity make with the shock front (save that they not be very small).

(vi) Application to galactic cosmic rays

The steady state spectrum of galactic cosmic rays satisfies $N_E \approx E^{-2.6}$. However, as described in section 1(ii), the required cosmic ray injection spectrum is $S(E) \approx E^{-2.2}$. This is close to what the theory of the preceding subsection gives for acceleration by a strong, $\gamma = 5/3$ shock front. Provided that this theory remains approximately valid for efficiencies ~10% (allowing for some adiabatic decompression) then we conclude the supernova blast waves can account for the acceleration of galactic cosmic rays both energetically and spectrally.

Supernova blast waves cannot account for the acceleration of the highest energy cosmic rays. To see this we note that the scattering mean free path must certainly exceed the Larmor radius $\sim r_L c \sim 10^{12}(E/1\,\mathrm{GeV})$ cm for a typical interstellar field $\sim 3\mu G$. The concentration scale length ahead of the shock must be $\sim(c/u)$ mean free paths $\gtrsim 3\times10^{14}(E/1\,\mathrm{GeV})$ cm. If the theory outlined above is to be applicable then this scale length must not certainly exceed the radius of a strong supernova shock wave (~1 pc) for a typical young supernova remnant expanding with a speed ~1000 kms^{-1}. We conclude that supernova blast waves can only accelerate cosmic rays efficiently up to an energy $E_{\max} \sim 10^4$ GeV, somewhat below the "knee" in the cosmic ray spectrum.[35] Of course, we can appeal to acceleration at circumgalactic and extragalactic shock fronts to account for the remainder of the spectrum.

(vii) Some complications

The simple test particle theory can therefore explain semi-qualitatively quite a lot of what is known about cosmic ray acceleration. However, it is clearly not a complete description and we must now look at some necessary complications. The first of these concerns the influence of the cosmic ray pressure on the shock structure. If, as we require, shock waves accelerate cosmic rays with $\gtrsim$10 per cent efficiency, then the cosmic ray pressure is a significant fraction of the total momentum flux and ought therefore to be included in the Rankine-Hugoniot relations. (The magnetic contributions are also not always ignorable.) As the pressure of the accelerated particles builds up ahead of the shock, the incoming fluid will be decelerated. In fact, the cosmic ray precursor should be considered part of the total shock and the discontinuity in the background fluid velocity then becomes a subshock. All of this changes the transmitted cosmic ray spectrum in a manner which depends upon the energy dependence of the diffusion coefficient and which is still not properly understood. (In fact, it is possible to dispense with the subshock altogether and mediate the total shock with the cosmic rays so that the background fluid is just compressed adiabatically.[36] It is not known whether or not this can ever occur in nature.)

A further issue which must be addressed is the nature of the scattering Alfvén waves. We have already remarked that the interstellar medium must scatter the cosmic rays just like the interplanetary medium. However, a much larger intensity of Alfvén waves can be created in the vicinity of a shock by the accelerated particles themselves. These will be streaming with a speed $\sim u$ relative to the background fluid and as this speed exceeds the Alfvén speed, backward propagating Alfvén waves will grow at the resonant k-vector. The growth rate is

$$\gamma(k = eB/pc) \sim \Omega_i \left(\frac{u}{a}\right) \frac{n_{CR}}{n_i} \tag{3.16}$$

where $n_{CR} \simeq 4\pi p^3 f(p)$ is a measure of the number of resonant cosmic rays and n_i is the background ion density.[31,37,38] Non-resonant suprathermal particles may also create Alfvén waves.[39] This estimate is fine unless the cosmic ray pressure becomes so large that the Alfvén wave amplitude predicted by application of equation (3.16) becomes non-linear. (This occurs when $p_{CR} \gtrsim 0.1\rho_- u_- a_-$.) Under these circumstances non-linear saturation and damping (probably non-linear Landau damping) processes must be included if a self-consistent treatment of the diffusion coefficient is required. In practice, it is probably not possible to improve upon the guess $D_\parallel \approx 0.1 v^2/\Omega(p)$.

There is a further complication associated with the Alfvén waves. Most authors have been concerned with the creation of scattering waves ahead of the shock and have assumed that either they will be convected across the shock to provide the necessary downstream scattering or that these downstream waves will somehow be present in the turbulence spectrum produced by the thermalization of the bulk fluid (possibly as a consequence of a firehose instability). In the former case, we can compute the transmission coefficient for backward propagating Alfvén waves incident upon a shock front. It turns out that when the shock front is oblique with respect to the magnetic field the transmitted Alfvén modes also propagate obliquely and therefore have a compressive part. This means that they are subject to rapid transit-time damping (Landau damping by magnetized ions).[40] Unfortunately, it is no clear how important this damping will be for highly non-linear waves.

(viii) Interplanetary shock waves

We are still a long way from a complete, physical model of a collisionless shock. The complex non-linear plasma physics has exhausted our analytical capabilities. Fortunately, we can use direct observations of interplanetary shock fronts (under conditions quite similar to those encountered in the interstellar medium) to probe these shocks directly. Recent analyses[41] (see also M. Thomson, these proceedings) have reached the conclusion that there is a significant difference between the quasi-parallel shocks (in which the field makes an angle $\lesssim 45°$ with the shock normal) and the quasi-perpendicular interplanetary shocks. The quasi-parallel shocks contain a definite subshock in which the magnetic field changes over a few thermal ion Larmor radii. However, they also contain a prominent extended precursor containing a large pressure of supra-thermal ions and large amplitude Alfvén waves apparently generated by these streaming ions, just as predicted by the simple theory. The ions have a power-law distribution function that is compatible with the simple theory.[32] Intense, Langmuir, ion acoustic and whistler turbulence is also reported.

However, supra-thermal ions do not appear to be present in large numbers ahead of quasi-perpendicular interplanetary shocks. It is not yet clear whether or not this distinction is true for all Mach numbers or just the low Mach number shocks encountered in the interplanetary medium.

(ix) Numerical modeling

The other potential source of enlightenment is numerical simulation of collisionless shocks. Two complementary approaches have been followed. Firstly, as discussed above, a kinetic equation for the cosmic rays is solved in the presence of a background fluid and an ad hoc diffusion coefficient. This has been done both directly[42] or using Monte Carlo methods.[43] The shock structure is not treated self-consistently. Secondly, the shock structure is modeled using particle simulation (as described in these proceedings by Quest). The influence of the precursor ions is not included. Most attention has been devoted to perpendicular shocks (for obvious computational reasons). What is required is to marry particle simulations of quasi-parallel shocks to kinetic models of cosmic ray acceleration. If this allows us to compute adequate models of interplanetary shocks, then we should be able to proceed to the interstellar medium and beyond with considerable confidence. In view of the considerable effort devoted to extragalactic radio astronomy, cosmic ray measurements, and so on, it seems to be well worth pursuing this program.

ACKNOWLEDGEMENTS

I thank Richard Epstein and Doyle Evans for the opportunity to attend this workshop and several other participants for helpful comments. I thank an anonymous referee for a careful reading of the manuscript which led to an improvement in the presentation. Support under NSF grant AST82-13001 and the Alfred P. Sloan Foundation is gratefully acknowledged.

REFERENCES

[1] Axford, W. I. *Proc. Int. Conf. Cosmic Rays 17th* **12**, 155 (1981); Axford, W. I. *Origin of Cosmic Rays*, Proc. IAU Symposium No. 94, ed. Setti, Spada, and Wolfendale (Reidel, Dordrecht, Holland, 1981), p. 339; Axford, W. I. *Plasma Astrophysics*, ed. Guyenne and Levy (ESA, Paris, France, 1981c), p. 425.

[2] Longair, M. S., *High Energy Astrophysics* (Cambridge University Press, Cambridge, 1981).

[3] Hillas, A. M., Ann. Rev. Astron. Astrophys. **22**, 425 (1984).

[4] Meyer, P., *Origin of Cosmic Rays*, Proc. IAU Symposium No. 94, ed. Setti, Spada, and Wolfendale (Reidel, Dordrecht, Holland, 1981), p. 11.

[5] Bell, A. R., Gull, S. F., and Kenderdine, S., Nature **257**, 463 (1975).

[6] Phinney, E. S., unpublished thesis, University of Cambridge (1983).

[7] Swinbank, E. and Pooley, G. G., Mon. Not. R. astr. Soc. **186**, 776 (1979).

[8] Trimble, V., in *The Crab Nebula*, ed. R. D. Davies and F. G. Smith (Reidel, Dordrecht, Holland, 1971).

[9] Begelman, M. C., Blandford, R. D., and Rees, M. J., Rev. Mod. Phys. **56**, 255 (1984).

[10] Melrose, D. B., *Plasma Astrophysics* (Gordon and Breach, New York, 1980).

[11] Arons, J., Max, C. E., and McKee, C. F. (eds.) *Particle Acceleration Mechanisms in Astrophysics*, AIP Conference Proceedings No. 56 (AIP, New York, 1979).

[12] Guyenne, T. D. and Lévy, G. (eds.) *Plasma Astrophysics* (ESA, Paris).

[13] Lifshitz, E. M. and Pitaevskii, L. P., *Physical Kinetics*, (Pergamon, Oxford, 1981).

[14] Rybicki, G. and Lightman, A. P., *Radiative Processes in Astrophysics* (Interscience, New York, 1979).

[15] Ramaty, R., in *Particle Acceleration Mechanisms in Astrophysics*, AIP Conference Proceedings No. 56 (AIP, New York, 1979).

[16] Blandford, R. D. and Rees, M. J., Mon. Not. R. astr. Soc. **169**, 395 (1974).

[17] Achterberg, A., Astron. Astrophys. **76**, 276 (1979).

[18] Akheizer, A. I., Akheizer, I. A., Polovin, R. V., Sitenko, A. G., and Stepanov, K. N., *Plasma Electrodynamics* (Pergamon, Oxford).

[19] Kulsrud, R. M. and Ferrari, A., Ap. Sp. Sci. **12**, 302 (1971).

[20] Achterberg, A., Astron. Astrophys. **98**, 195 (1981).

[21] Achterberg, A., Astron. Astrophys. **98**, 161 (1981).

[22] Skilling, J., Mon. Not. R. astr. Soc. **177**, 557 (1975).

[23] Blandford, R. D., *Particle Acceleration Mechanisms in Astrophysics*, AIP Conference Proceedings No. 56 (AIP, New York, 1979).

[24] Blandford, R. D. and Scharlemann, Astrophys. and Space Sci. **36**, 303 (1975).

[25] Hones, A. G., *Proc. Chapman Conference on Magnetic Reconnection* (AGU Press, 1984).

[26] Schindler, K., Journ. Geophys. Res. **79**, 2803 (1974).

[27] Coroniti, F. V. and Kennel, C. F., *Particle Acceleration Mechanisms in Astrophysics*, ed. Arons, Max, and McKee, AIP Conference Proceedings, No. 56 (AIP, New York, 1979).

[28] Cavaliere, A. and Morrison, P., Astrophys. J. (Lett.) **238**, 63 (1980).

[29] Parker, E. N., Phys. Rev. **109**, 1328 (1958).

[30] Drury, L. O'C., Rep. Prog. Phys. **46**, 973 (1983).

[31] Blandford, R. D. and Eichler, D., Phys. Reports, to be submitted (1985).

[32] Lee, M. A., Journ. Geophys. Res. **88**, 6109 (1983).

[33] Eichler, D., Astrophys. J. **244**, 711 (1981).

[34] Landau, L. D. and Lifshitz, E. M., *Fluid Mechanics* (Pergamon, Oxford, 1959).

[35] Lagage, P. O. and Cesarsky, C. J., Astron. Astrophys. **118**, 223 (1983).

[36] Drury, L. O'C. and Völk, H. J., Astrophys. J. **248**, 344 (1981).

[37] Kulsrud, R. M. and Pearce, W., Astrophys. J. **156**, 459 (1969).

[38] Bell, A. R., Mon. Not. R. astr. Soc. **182**, 147 (1978).

[39] Gary, P. S., Astrophys. J. **288**, 342 (1985).

[40] Achterberg, A. and Blandford, R. D., Mon. Not. R. astr. Soc., to be submitted (1985).

[41] Kennel, C. F., Edmiston, J. P., and Hada, T., *Proc. of AGU Chapman Conference on Collisionless Shocks in the Heliosphere* (1984).

[42] Achterberg, A., *Radiation in Plasmas*, ed. McNamara (World Scientific, Singapore, 1984), vol. 11, p. 3.

[43] Ellison, D. C., Ph.D. thesis, Catholic University, Washington (1981).

HYDROMAGNETIC ASPECTS OF ACTIVE GALACTIC NUCLEI

R. D. Blandford
Theoretical Astrophysics 130-33, California Institute of Technology, Pasadena, CA 91125

ABSTRACT

The observed properties of active galactic nuclei are briefly summarized. The broad features of the observations are discussed in terms of a model incorporating a massive central black hole. The properties of the extended radio jets are also described and interpreted in terms of simple kinematical models. Models for the collimation of jets are outlined with particular attention being paid to hydromagnetic models. Recent work on the electromagnetic extraction of energy from a spinning black hole is mentioned.

1. PHENOMENOLOGY OF ACTIVE GALACTIC NUCLEI

Active galactic nuclei are a highly heterogeneous class of objects. They are observed from the lowest radio frequencies to the highest gamma ray energies, and their bolometric luminosities span the range from $\gtrsim 10^{47}\,\mathrm{erg\,s^{-1}}$ for the brightest distant quasars to $\lesssim 10^{39}\,\mathrm{erg\,s^{-1}}$ for the nucleus of a galaxy such as our own. (Recent research on bright, nearby galaxies indicates that most galactic nuclei can display activity at some level.[1])

In Fig. 1 we show schematically various observed manifestations of nuclear activity. Note that these have length scales which range from $\lesssim 10^{13}\,\mathrm{cm}$, the size of the postulated central black hole (and $c\times$ the shortest observed variability timescales) to $\gtrsim 10^{25}\,\mathrm{cm}$, the span of the giant double radio sources. It is as important to account for the distinctions between the various categories of active objects as it is to account for the union of their properties.

Quasars are recognizable optically by their bright starlike nuclei, broad emission lines and high redshifts. (In fact, most of the closer quasars appear to be surrounded by a faint nebulosity, which is almost certainly the host galaxy.) Roughly 10 per cent of quasars are "radio-loud." The remainder are "radio-quiet," though probably not radio-silent. Roughly 10 per cent of the radio-loud quasars are designated optically violent variable or OVV. They also exhibit large, rapidly variable linear polarization. The powerful extragalactic radio sources have traditionally been divided into two classes -- the compact sources in which most of the power at a frequency $\sim 1\,\mathrm{GHz}$ emerges from a region $\lesssim 1\,\mathrm{kpc}$ in size and the extended sources for which the converse is true and which generally exhibit a double structure. The radio emission is believed to be synchrotron radiation mainly because of its non-thermal power-law spectrum and linear polarization. The compact sources have flat spectra (spectral index $\alpha \equiv -d\ln S_\nu / d\ln\nu \sim 0$) whereas the extended sources generally have steep spectra ($\alpha \sim 0.7$). The flat spectra are believed to be due to the superposition of several self-absorbed components. Some compact sources exhibit superluminal expansion. That is to say, they contain features which appear to be moving faster than the speed of light. (Superluminal expansion is believed to be an illusion caused by having a radio-emitting region move at relativistic though sub-luminal speed V at a small angle θ to the observer direction. The observed transverse speed exceeds the speed of light if $V > c/(\sin\theta + \cos\theta)$; cf. Fig. 2.)

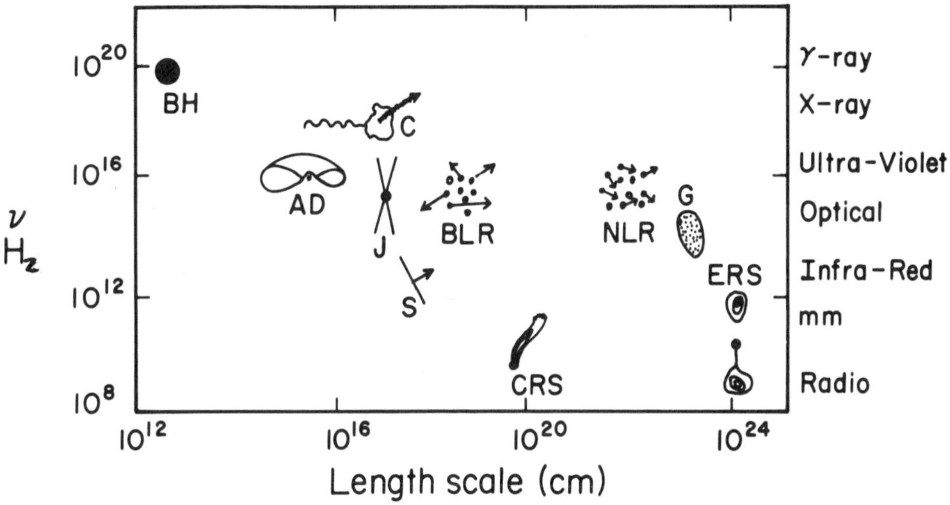

Fig. 1. Schematic illustration of types of structure inferred or directly observed in active galactic nuclei. Not all components are present in every object (and some components may be present in no objects!). This montage emphasizes that active galactic nuclei produce observable effects throughout the whole electromagnetic spectrum and over twelve decades of radius. BH is the central black hole whose presence is inferred from the observed rapid variability, stable radio source axes, and central light cusps. γ-rays may be created close to its event horizon. AD is a (possibly thick) accretion disk in orbit about the black hole. J represents the jets that we see on larger length scales and which may be collimated close to the black hole. C represents a cloud of hot plasma that can Comptonize (Compton upper scatter) soft photons to create a power law spectrum. S represents a shock front where relativistic electrons may be accelerated. These electrons can then radiate a non-thermal spectrum by the synchrotron mechanism or the inverse Compton process. CRS stands for compact radio sources as probed with VLBI. ERS stands for extended radio source as probed by the VLA and other similar interferometers. BLR is the broad optical emission line region and NLR is the narrow emission line region which are both believed to be photo-ionized by the central UV continuum source. *G* is the host galaxy.

Seyfert galaxies are like low luminosity optical quasars and are subdivided into types I and II depending upon whether or not they show broad optical emission lines. They are often weak radio sources as well.

Radio galaxies are powerful extended radio sources that are identified with elliptical galaxies (never spirals). They are subdivided into types I and II that correspond to the low and high radio power objects respectively. Type II radio galaxies are "edge-brightened" which means that most of the power comes from the extremities of the source, often from compact regions called hot spots (Fig. 3). Usually there is a compact radio source identified with the nucleus of the central galaxy. Type I radio galaxies are edge-darkened and the surface brightness typically fades with increasing distance from the central identification (Fig.

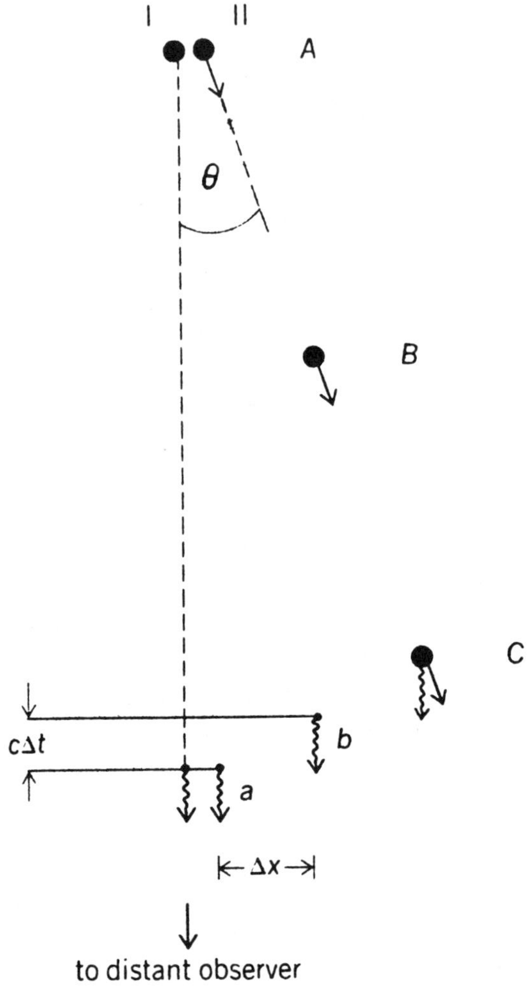

Fig. 2. Possible kinematic model for superluminal expansion. Suppose that we have a stationary emitting component I and a second component II, moving with speed V at an angle θ to the line of sight. As component II moves from A to B to C it sends radio waves to the observer. The apparent transverse expansion speed is $\Delta x/\Delta t = V\sin\theta/(1-(V/C)\cos\theta)$ which has a maximum value of $V/(1-V^2/C^2)^{1/2}$ when $\theta = \cos^{-1}(V/C)$.

4).

In recent years, it has been discovered that many radio sources contain jets -- linear features that connect the compact sources with the extended components. Jets are almost always found in the type I radio galaxies but are less frequently observed (and then are generally one-sided) in the case of the type II

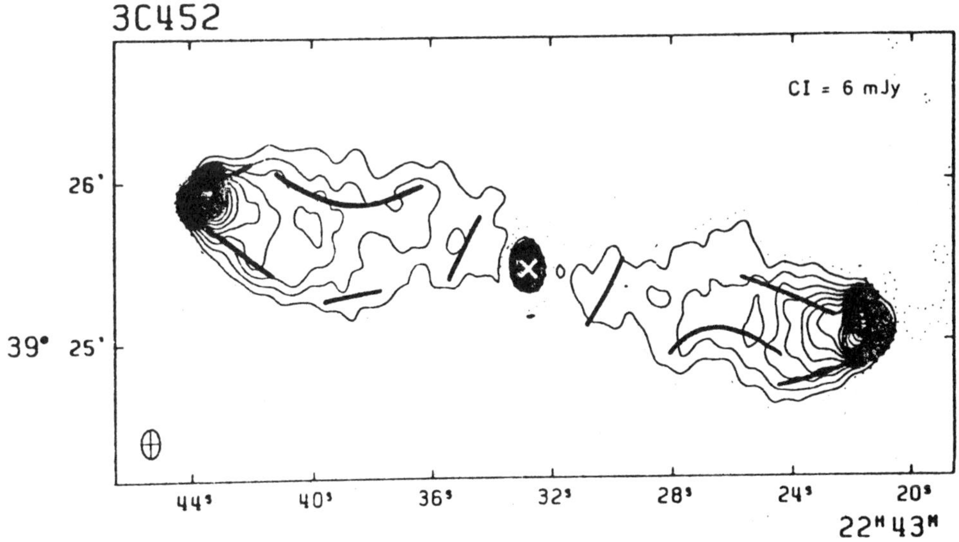

Fig. 3. 3C452, a typical type II (edge-brightened) radio galaxy from Högbom.[2] Intensity contours at 5 GHz are displayed. The inferred underlying magnetic field direction is shown with bold lines.

sources. Radio loud quasars of comparable power show one-sided jets more frequently. These jets are believed to be the ducts along which the "exhaust" gases from the nucleus are channeled into the extended radio components transporting mass, momentum, energy, and magnetic flux. Jets are not confined to extragalactic objects. Several galactic sources contain linear features, notably SS433,[4] SCOX-1,[5] and a variety of bipolar outflows associated with young stars.

Another class of radio galaxies is the BL Lac objects. These are related to the OVV quasars but are associated with elliptical galaxies. They are characterized optically by steep, featureless power law spectra and rapidly variable compact radio sources.

One of the first questions a theorist will ask about an astronomical object is in what form does it release the most power? Unfortunately, in the case of active galactic nuclei, we are not able to answer this with any certainty. In the preceding article on particle acceleration, we show a schematic spectra of the observed power per logarithmic bandwidth for different types of active objects. Note the large uncertainties in the far IR, the extreme UV, and the hard X-ray parts of the spectrum, where most of the bolometric power appears to be radiated. By contrast, in the powerful radio galaxies, the central bolometric luminosity of the nucleus appears to be much lower than the power carried essentially invisibly by the jets.

I shall not give many references in these pedagogic lectures because there exist several up to date reviews in which more detailed general discussions of active galactic nuclei can be found.[3,6–12]

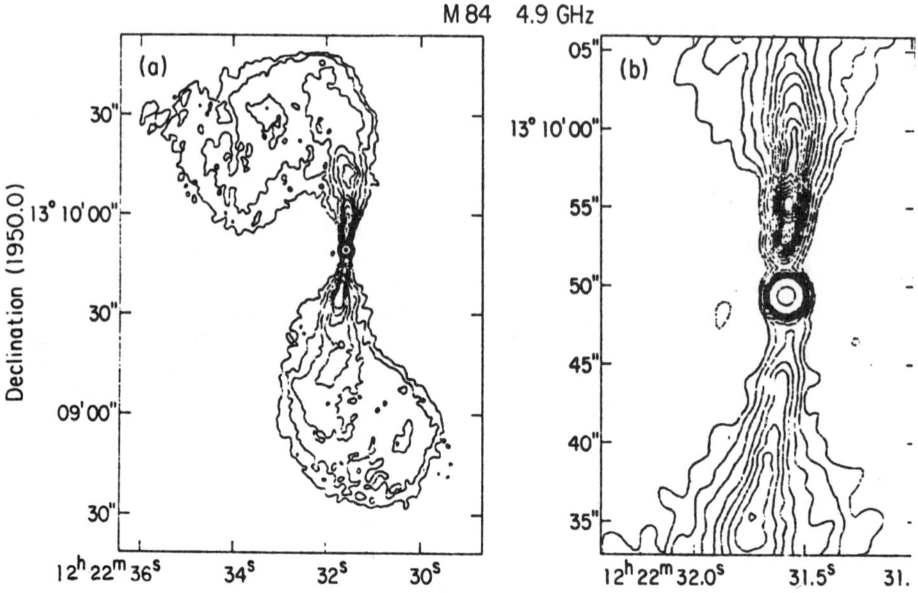

Fig. 4. M84 is a typical (edge-darkened) radio galaxy reproduced from Bridle and Perley.[3] (Data of R.A. Laing and A.H. Bridle in preparation.)

2. BLACK HOLE MODELS AND THE UNIFICATION OF ACTIVE GALACTIC NUCLEI

The notion that active galaxies are powered by massive black holes is almost as old as the discovery of quasars.[13–15] As is well known, the observational evidence in favor of this view is circumstantial and not even by the most lax standards of scientific proof can we claim to have demonstrated the existence of a black hole within the nucleus of any galaxy (or indeed anywhere else). The most direct observational argument for their existence probably follows from the discovery of 100-1000 s X-ray variability in a small number of quasars and Seyferts.[16] Central light cusps, the apparent long-term stability of radio source axes, superluminal expansion and observations of the galactic center provide further indirect evidence for spinning, relativistically deep gravitational potential wells. The most persuasive arguments for their existence are essentially theoretical. They include the requirement of a high radiative efficiency and the apparent inevitability of black hole formation during the evolution of a galactic nucleus.

If we adopt the black hole hypothesis, then we can associate some of the different types of object with different relative rates of mass accretion. The natural unit of mass accretion for a black hole of mass M is the Eddington rate: $\dot{M}_{Edd} = 4\pi GMm_p / C\sigma_T \simeq 10^{25}(M/10^8 M_\odot) g\ s^{-1}$. This is $c^{-2}\times$ the Eddington luminosity, the luminosity at which the pressure of the escaping radiation can oppose gravity and suppress accretion. When $(\dot{M}/\dot{M}_{Edd})$ is small, typically $\lesssim 0.1$, gas near the hole will be optically thin and unable to maintain thermal equilibrium. The energy will probably be released in a non-thermal form (i.e., relativistic particles, gamma rays, and large scale electromagnetic fields). We associate this case with the radio galaxies. Conversely, when the accretion rate is very large relative to the mass of the hole (i.e. $\dot{M} \gtrsim 10\dot{M}_{Edd}$) we expect that the gas near the hole will be quite optically thick and that most of the radiant energy release should occur at the black body temperature, typically in the ultraviolet. This case can be associated with the optical quasars (large mass holes) and Type I Seyfert galaxies (small mass holes). The radio quasars, which combine both characteristics can be identified with holes that are accreting at an intermediate rate.

A further parameter which may control the classification of an active nucleus is its orientation with respect to the observer direction. It has been suggested that the BL Lac objects and OVV quasars (sometimes referred to collectively as "blazars") are intermediate power radio galaxies and quasars, respectively, in which we happen to be looking into the outflowing jet. Non-thermal emission from the jet is Doppler boosted and dominates the emission. A similar boosting may occur in the compact radio sources.

This idea has some observational support. It has been found that compact sources frequently exhibit a core-jet morphology. However, the jet features in compact sources are often quite distorted and elongated along different position angles from the large scale radio structure in contrast to the extended radio galaxies where the alignment is generally quite good. This is what is expected if the compact sources are viewed along their jet directions.

There are further suggestions that the orientation of the observer may distinguish Type I from Type II Seyferts if, in the latter case the broad line region is hidden by a large obscuring disk in the nucleus of the galaxy. Of course, all of these associations between type of active nucleus and the character of black hole accretion are quite speculative at this stage. What is encouraging is that we are now in a position to start asking these questions seriously.

3. HYDROMAGNETIC MODELS OF RADIO JETS

(i) Fluid model

Most recent discussions of radio jets have treated the outflow in the fluid approximation. The reason for this is that it is believed that the effective particle mean free path is much smaller than the width of the jet. Even though this mean free path is not well defined, Larmor radii and Debye lengths are typically $\lesssim 10^{-9}$ of the jet widths and very small levels of turbulence should suffice to scatter the thermal electrons and protons effectively. Radio jets are therefore discussed in quite different terms from particle beams in the laboratory and in the atmosphere. An extension of this idea is that magnetohydrodynamics should be adequate to describe the large scale features of the flow. In fact, most of the serious simulations of jets so far have been gas dynamical rather than hydromagnetic. However, as this is a workshop on magnetospheric phenomena, I shall emphasize MHD ideas.

(ii) Relativistic beaming

As we have already remarked, the discovery of superluminal motion within some compact radio sources suggests that the outflow speed in a jet can be relativistic. We have also mentioned the prevalence of one-sided jets amongst quasars and intermediate power radio galaxies and how this one-sidedness frequently occurs within an associated compact radio source. This suggests that the speeds of these extended jets are also relativistic and that their one-sidedness is due to Doppler beaming. This effect can be quite large. The brightness ratio of two oppositely directed, but otherwise similar jets traveling with speed V and observed from an angle θ is $[(1+V\cos\theta/c)/(1-V\cos\theta/c)]^{2+\alpha}$. For example, a jet traveling with speed $V \sim 0.85c$ viewed from an angle $\theta \sim 30°$ will be $\sim$100 times brighter than the counter jet. In VLBI observations this will render the counter jet invisible.

There are some difficulties with the relativistic model. It is clearly inappropriate to the type I radio galaxies in which two comparably bright jets are usually seen. However we believe that the flow speeds in these low power sources is generally much less than c. Subrelativistic flows should be less efficient at particle acceleration and magnetic field amplification than relativistic flows. However, some of the one-sided jets show large bends around which the observed surface brightness does not change as dramatically as might be expected if the flow speed were relativistic. A second difficulty that has been noted is that the projected sizes of many of the one-sided sources are already quite large and if they have to be deprojected then they become unusually large.[17]

For these, and other reasons, many radio astronomers (not including this reviewer) have begun to take seriously an alternative model for one-sided jets in which the nuclear source sends out an intrinsically one-sided jet whose direction flip-flops on a time long compared with the transit time through the whole jet but short compared with the radiative cooling time of the two-sided extended radio lobes. An obvious prediction of this model is that faint counter-jets will not be seen as more high dynamic range observations are carried out with the VLA.

If relativistic beaming is important, then it can alter the deduced physical conditions within the jets. In what follows, we shall, for simplicity, assume that the flow speeds are no more than mildly relativistic.

(iii) Magnetic confinement

Jets in general appear to be confined rather than freely expanding because the angles that they subtend at the nucleus decrease with distance from the nucleus. To understand how this confinement occurs consider the internal and external pressure in a jet. Synchrotron radiation theory tells us that when we observed a surface brightness I_ν from a source of depth l, then the minimum total (relativistic electron plus magnetic) pressure within the source region is attained when there is approximate equipartition between the two components its value, $p_{\min} \propto (I_\nu/l)^{4/7}$. We can use this argument to place a lower bound on the internal pressure within a jet. Relativistic proton, thermal, and turbulent contributions will only increase this pressure.

We can use X-ray observations to put an upper bound on the thermal pressure of the external gas. (The free-free emissivity for a given gas temperature is proportional to the square of the pressure.) There are still some ambiguities (e.g. the possibility that the high pressure regions are transients attributable to internal shocks) but if we ignore these then there appear to be several examples of sources in which the maximum external pressure is less than the minimum jet pressure. The next most reasonable source of confinement to thermal gas is a toroidal magnetic field that can pinch the jet (see Fig. 5).

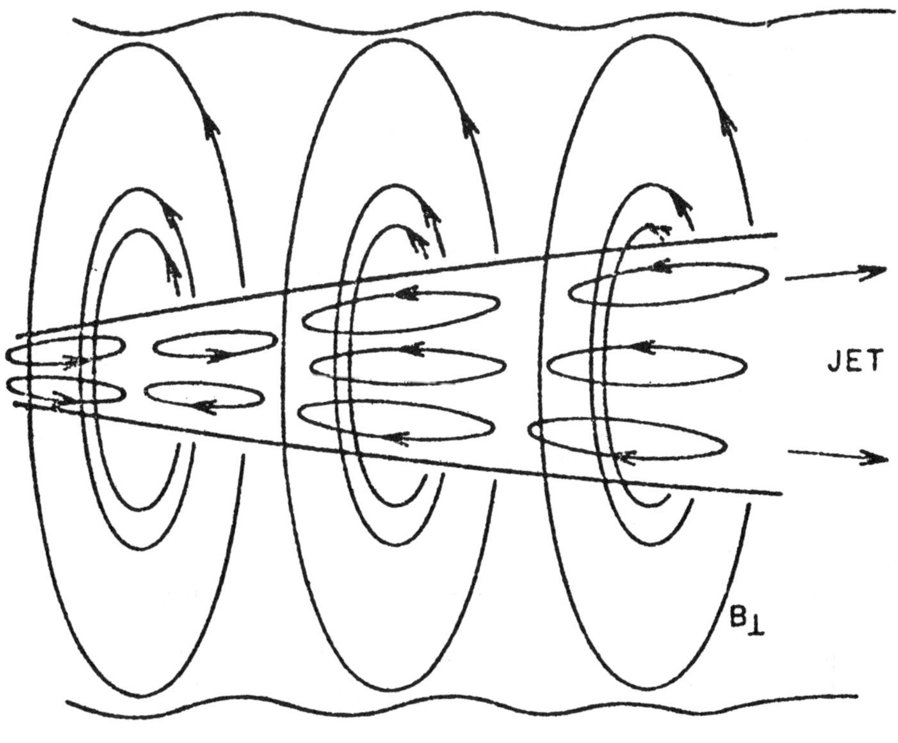

Fig. 5. Schematic illustration of a magnetically confined jet (after Ref. 6).

In this case it is helpful to think about the current distribution. There will be a current in the jet which acts as the source for the toroidal field within the jet. If the return current flows at some large distance r_{ret} much greater than the jet radius, r_{jet}, then the toroidal field will decrease inversely with radius, $B_\varphi \propto r^{-1}$ for $r_{jet} < r < r_{ret}$. The magnetic stress decreases $\propto r^{-2}$ and can be balanced by a thermal pressure smaller by a factor $\sim(r_{jet}/r_{ret})^2$ than the pressure within the jet. This idealized model demonstrates that it is possible to confine a jet transversely, with a much lower gas pressure through the intermediary of a magnetic field. In fact, the toroidal field should be built up naturally if a small quantity of magnetic flux is convected by the jet and spun off around the end of the jet.

(iv) Polarization and magnetic geometry

The direction of linear polarization in synchrotron radiation is perpendicular to the projected magnetic field. We can therefore use polarization observations to learn about the magnetic geometry of jets. A very clear pattern has emerged. The powerful or type II jets generally exhibit parallel magnetic fields although these may become predominantly perpendicular at large distances from the nucleus. By contrast, the lower power jets have perpendicular fields. Sometimes, these perpendicularly magnetized jets have a parallel magnetic field at their sides.

This pattern has a natural hydromagnetic interpretation (Fig. 6). The velocity profile of a supersonic jet is likely to be fairly centrally peaked close to the origin. In this case there will be a strong velocity shear and loops of magnetic field frozen into the flow will be strung out along the jet. However, viscous stresses will gradually flatten the velocity profile (this will happen sooner in the lower power jets) and the sideways expansion will dominate the transverse gradient in the parallel velocity. Small loops of field will be elongated perpendicular to the jets in this case. However, there would also be a fairly thin boundary layer in which the strong shear should again create a parallel field.

In this view, there is a fairly disorganized field within the jet confined by an organized field in the stationary external medium. Alternatively, the jet field may well be highly organized. However, it seems very difficult for the parallel field to be unidirectional, because if this were the case and the magnetic flux were conserved along the jet, then the magnetic pressure ($\propto r_{jet}^{-4}$) would be prohibitively large in the inner jet.

A further possibility is that the jet fluid becomes magnetically dominated. This seems reasonable, because magnetic pressure will decline no faster than $\propto r_{jet}^{-2}$, whereas, for example, thermal gas pressure would decline $\propto r_{jet}^{-10/3}$ if the expansion were adiabatic. In fact, we know from the slow decline in surface brightness measured in the observed jets that there must be some dissipation during the outflow, but this does not alter the tendency of the jet to achieve a low beta state. (Beta is the ratio of gas pressure to magnetic pressure.) When the magnetic stresses dominate the internal structure of the jet will probably rearrange itself to become force-free (see Aly's article in these proceedings), so that the current density flows along the direction of the magnetic field. More formally

$$\nabla \times \mathbf{B} = \alpha \mathbf{B}$$

$$\text{with } \mathbf{B} \cdot \nabla \alpha = 0 \,. \tag{3.1}$$

A variety of force-free configurations have been explored in fusion, solar, and astronomical contexts. An important subset of these are those in which α is spatially constant. These are appropriate when the field is in some sense ergodic as for example is believed to occur in a reversed field pinch.[18] Königl and Choudhuri[19] have argued that many jets should contain constant α force-free fields and have attempted to fit the fine observations of NGC6251.[20] They argue that the lowest energy magnetic configurations should be non-axisymmetric. Note that if a constant α magnetized jet has no net magnetic flux threading its

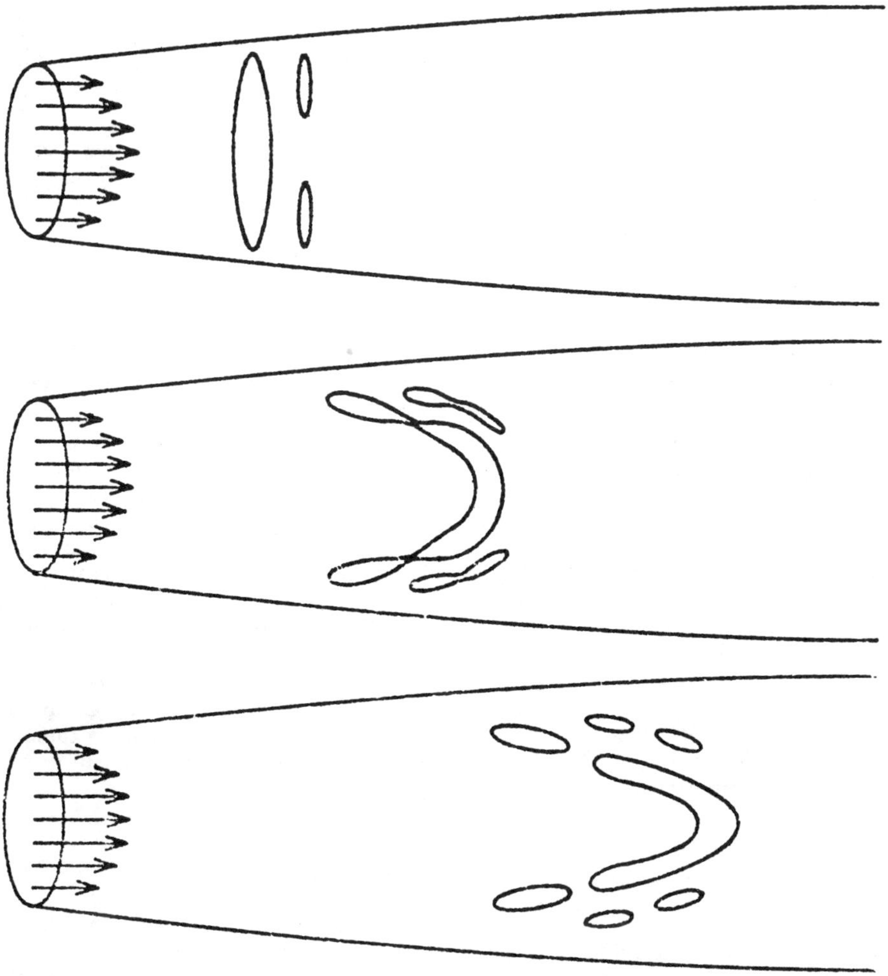

Fig. 6. Diagram illustrating the production of loops of magnetic field elongated along the jet axis in a jet with velocity shear (after Ref. 6).

cross-sectional area as argued above then it must also carry no net current and therefore cannot be confined by an external toroidal field. These attempts to model the underlying magnetic geometry are particularly important as many more jets will be mapped with comparable resolution and sensitivity to the observations of NGC6251 in the near future.

(v) Stability

Another issue raised by the possibility of magnetic collimation is that of stability. Aerodynamicists and fusion plasma physicists are keenly aware that jets and confined plasmas respectively are prone to destructive instability. Yet many of the more powerful radio jets are impressively straight as they expand through more than five decades of radius.

There has been much recent work on the linear stability of supersonic jets. If we just consider a plane vortex sheet separating two similar compressible fluids moving relative to each other, we can find propagating sound wave solutions in the two fluids. If we match the pressure perturbations and the displacement in the usual manner at the interface, we can derive a dispersion relation for the frequency which has complex roots, some of which correspond to temporally growing modes. (In fact, for the problem we have just described, there are only growing modes with **k** parallel to the relative velocity when the Mach number $M \leq 8^{1/2}$. However, it is possible to find modes with oblique values of **k** for all Mach numbers and even some modes parallel to the velocity when the shear layer is resolved. See Aly, Lovelace, these proceedings.)

The modes propagating in a cylindrical jet of different density from its surroundings are rather more complicated. They can be labeled by azimuthal and radial quantum numbers, m and n respectively, and we can again solve the dispersion relation for the complex value of ω for a given value of the k-vector resolved along the jet ($k_{\parallel}$). The $m = 0$ modes correspond to the "pinching" modes and the $m = 1$ modes to the "kink" modes. The mode number n corresponds to the number of radial modes in the pressure perturbation. In a growing mode, the jet acts as an acoustical waveguide with sound waves crossing the jet from side to side gaining energy from each reflection.

Most calculations have given the complex frequency for real values of the k-vector whereas the converse specification may be more appropriate for a radio jet. However, as long as the growth rate is small, complex variable theory can be used to estimate a spatial growth rate from a temporal growth rate using the Cauchy-Riemann equations

$$\frac{\omega_i(k_r, k_i=0)}{k_i(\omega_r(k_r), \omega_i=0)} \simeq -\left(\frac{\partial \omega_i}{\partial k_i}\right)_{k_r} = -\left(\frac{\partial \omega_r}{\partial k_r}\right)_{k_i} . \tag{3.2}$$

That is to say, the two growth rates are related by the group velocity and not the phase velocity. This approximation, which ignores the change in the real part of k_r on changing from temporal to spatial growth appears to be quite good for slowly growing modes.[21,22]

Unfortunately, these growth rates can only be computed in the linear regime, whereas any unstable jet that can be mapped is almost certainly nonlinear. Further, if the angular deflection of a supersonic jet exceeds the reciprocal of the Mach number, then internal shocks will develop within the jet. This means that numerical computations are necessary. Two dimensional gas dynamical computations have been carried out by Norman, Smarr, Winkler, and colleagues (Hawley and Smarr, these proceedings). These have exhibited a variety of structures depending upon the Mach number and the ratio of the densities of the jet and the ambient medium. It is hoped to extend this work to encompass hydromagnetic jet production in the near future.

(vi) Large Scale Structure

It is conceivable that magnetic fields have a big influence on the large scale shape of jets. A particular case in point is 3C75.[23]

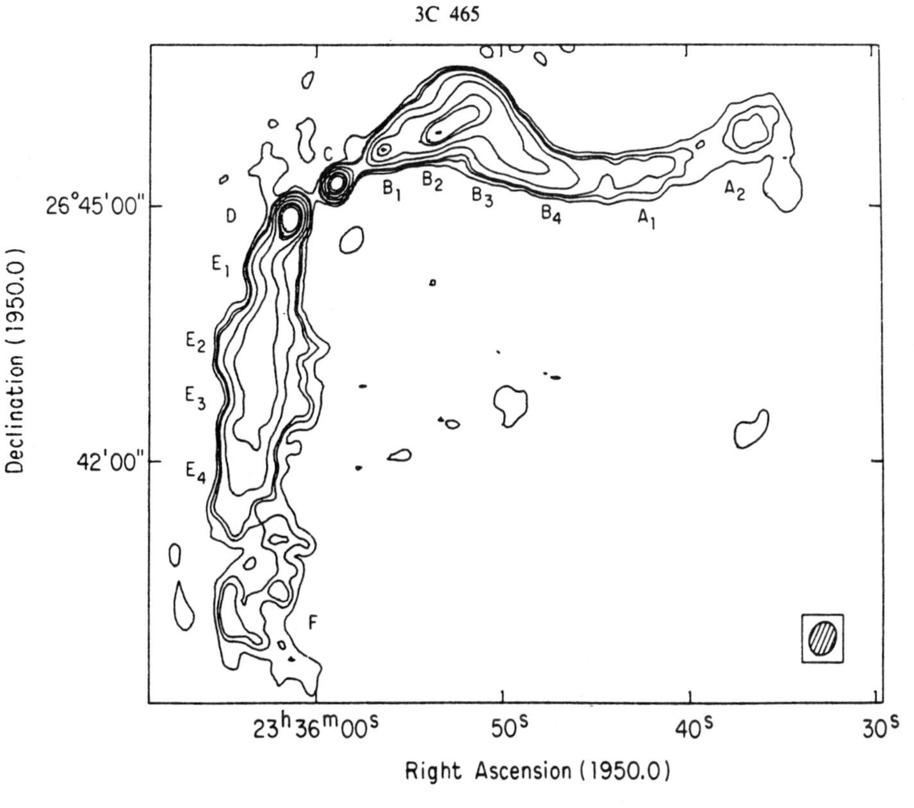

Fig. 7. The wide angle tail radio 3C465.[24]

Here we see an example of the "wide angle tail" morphology. However, what is curious is that there are two pairs of jets each pair emerging from a component of a cD galaxy. (A cD galaxy is a dominant elliptical galaxy with an extended envelope found within a cluster of galaxies; cD galaxies frequently have multiple nuclei.) Both pairs of jets appear to bend together. This could be coincidence. It might also be purely hydrodynamical response to the intergalactic meteorological conditions or the motion of the cD.[24] However, a third possibility is that the individual jets are confined magnetically as described above and that consequently electrical currents flow along their lengths. If these currents are parallel then there will be an attractive force between the two jets which could account for their apparent linkage. Indeed this force might extend across the whole radio source and be responsible for the C-type symmetry of the wide angle tails in the first place.[24] This type of interaction can only be effective when the jet speed is neither highly supersonic nor highly super-Alfvenic as is believed to be true of the lower power class I radio sources.

A second possible manifestation of large scale fields is exemplified by a recent map of Hercules A.[25] This shows large loop-like structures that have been observed in several other strong radio sources. These have been associated with bubbles or blast waves created by the advancing jet. However it is just possible that we are instead seeing illuminated flux tubes of toroidal field.

The best way to see if these explanations that are predicated upon the existence of large scale toroidal field are correct is to measure the Faraday rotation across the source. If this changes sign (after subtracting the foreground rotation) then it would certainly be consistent with having a large scale toroidal field.

4. MAGNETIC COLLIMATION OF JETS

(i) Location of Collimation

The above discussion of the behavior of jets is mostly independent of the mechanism by which they are created in the active nucleus. There is no widely accepted model for jet collimation as yet although several distinct mechanisms have been advanced. Indeed, the observed heterogeneity of jets leads one to suspect that several different mechanisms are at work in different types of sources. Any model of jet collimation must account for the production of large amounts of apparently relativistic fluid and its channeling into two antiparallel jets that can maintain their direction in space for times in excess of several million years and, in the case of several radio galaxies, over length scales from pcs to Mpcs. Black holes are natural candidates for accomplishing both of these tasks because, as we shall see, their spin energy can be extracted in the form of a relativistic hydromagnetic wind and they are excellent gyroscopes. (In fact some of the large scale inversion symmetric structure that has been seen in sources like 3C315 has been attributed to precession or alignment of the black hole through torques due to the accreted gas or a companion black hole.)

However it is not necessary that the collimation actually occur very close to the black hole and some models like the "twin exhaust" mechanism attribute it to the pressure of a spinning gas cloud some tens of parsecs in size. In this case the jet axis need bear no relation to the axis of a central black hole. The discovery that many VLBI sources show compact features elongated along several different position angles relative to the large scale radio structure also suggests that the job is not completed until the jet has escaped the nucleus of the galaxy. Further evidence that jets are still being collimated as they climb out of the galaxy's potential well is furnished by the observation that jets often appear to be focussed as they propagate away from the nucleus.

(ii) "Twin Exhaust" Model.

Most of the models combine a gravitating mass, angular momentum, a source of high entropy gas and possibly magnetic flux in order to produce jets. In the "twin exhaust" model, two anti-parallel converging-diverging nozzles are opened up along the symmetry axis of a flattened, spinning gas cloud (Fig. 8).

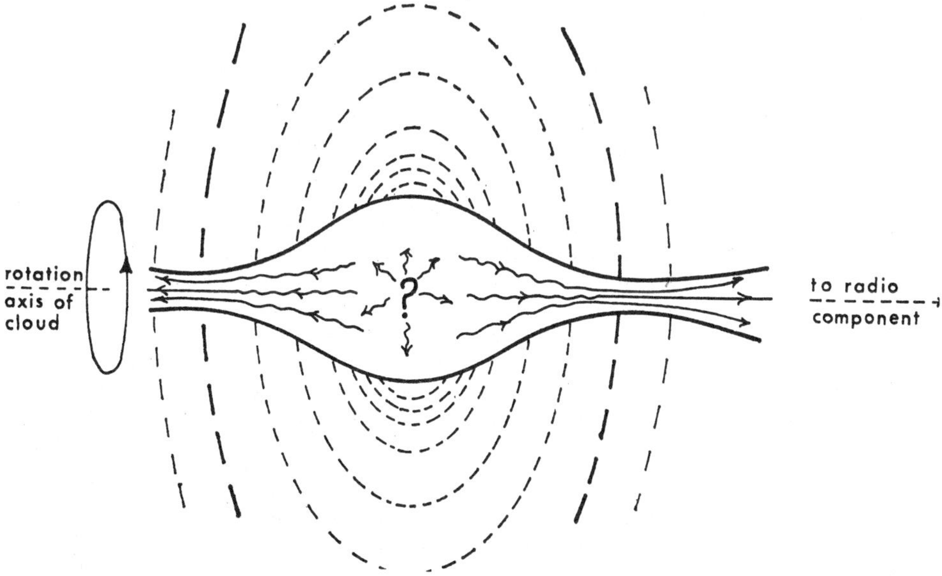

Fig. 8. The twin exhaust model.

The flow speed increases from being subsonic to becoming supersonic as the cross-section of the flow passes through a minimum. As the density of the jet decreases, the internal energy of the fluid is converted into bulk kinetic energy. If the jet can be regarded as thin, the internal pressure (assuming adiabatic conditions for simplicity) must adjust to balance the external pressure and falls to roughly half its stagnation value at the trans-sonic point. Thereafter, if the external pressure p scales with radius r as $p \propto r^{-n}$, where $n \sim 2$, then the density $\rho \propto r^{-3n/5}$ assuming that the specific heat ratio of the jet fluid is 5/3. If the jet velocity is supersonic, then it can be approximated as roughly constant implying in turn from the equation of mass conservation that the jet diameter $d \propto r^{3n/10}$. This simple example demonstrates that a jet can actually be focussed by a pressure gradient. Unfortunately, there is no guarantee that the jet flow will be adiabatic. If the jet were able to radiate away its internal energy, then collimation would be even easier to bring about. However, by monitoring the decay of surface brightness with distance, it is possible to measure the fall off of equipartition pressure with distance. If the equipartition pressure is a fair measure of actual pressure then we infer that $p \propto \rho^{0.8}$, approximately. The jet fluid therefore gets <u>hotter</u> as it expands, presumably as a consequence of ongoing dissipation of some of the bulk kinetic energy. In this case the jet will be collimated only if

$n < 1.6$.

The "twin exhaust" model describes the simplest type of critical point flow. If we include the direct influence of the gravitational field or magnetic field on the flow then we generate a much richer variety of possible flows.[26] Every time the flow speed equals a wave phase velocity, then a critical point will occur.

(iii) Funnels

A smaller scale version of the twin-exhaust mechanism can occur within the centrifugally-supported vortices created by gas swirling around a spinning black hole. Two types of funnel have been proposed. Firstly, as we described in section 2 above, a radiation-dominated torus will be created when the accretion rate is large enough. Unfortunately, this is unlikely to be present in the radio galaxies because the central nuclear luminosities are far below the Eddington limit for a black hole massive enough to have supplied the minimum energy requirements of the extended radio components. Nevertheless it could be relevant to the radio-loud quasars. However, in this case the influence of radiative viscosity and the strong interaction of any outflowing jet with the funnel walls makes it extremely unlikely that a powerful supersonic outflow can be created. Furthermore, the stability of the underlying torus is in question.[27]

The second possibility is that a funnel supported by ion pressure be sustained at low mass accretion rates. In this case we anticipate that there will be a strong electromagnetic coupling between the funnel walls and any outflowing plasma again invalidating the adiabatic approximation. Nevertheless several exploratory calculations of flow in the vicinity of a black hole have been carried out and, at the very least, they demonstrate the richness of the solutions of the relevant stationary fluid equations. An alternative approach to this problem is to integrate the time dependent equations numerically (Hawley and Smarr, these proceedings).

(iv) Radiation-driven Winds from Disks

Another distinguishable class of models involves thin flat disks confined to the equatorial plane of the black hole. Even if the total luminosity of the disk is less than the Eddington-limit, it can still drive a wind away from its surface by radiation pressure. There are three ways in which this can occur. Firstly, as the effective temperature of the inner disk will lie in the ultra-violet it will behave somewhat like a flattened O-star and accelerate gas through resonance-line scattering. The gas will tend to flow roughly normal to the disk plane although the collimation will be at best poor and must be supplemented by additional focusing at large distances from the disk.[28]

A second way to increase the specific opacity of the gas over and above the standard Thomson opacity and thereby allow radiation pressure to overcome the gravitational stress is to create a large number of electron-positron pairs most plausibly through gamma ray - gamma ray interactions. The mean molecular weight per electron is then reduced by a factor of a thousand within the electron-positron region.[29]

A third method is indirect and relies upon creating a hard X-ray spectrum in the central parts of the disk and illuminating the outer parts of the disk.[30] The gas there will be heated by Compton recoil to an equilibrium temperature

$$T_e = \frac{h\bar{\nu}}{4k}$$

where $\bar{\nu}$ is the intensity-weighted mean frequency of the illuminating radiation field. If this temperature is large enough, the gas will be able to escape the gravitational potential well and again may produce a poorly -collimated wind.

(v) Centrifugally-driven winds

Of more interest in the present context is a mechanism that Payne and I have discussed in detail.[31] In this model, a thin accretion disk is supposed to be threaded by poloidal magnetic field which is convected around with the Keplerian angular velocity. If the magnetic field has a sufficiently large radial component, then plasma in the disk will be flung out along the field by the centrifugal force. However the plasma will still be tied to the field lines. When it has traveled some radial distance out from the foot of the field lines its inertia will become important and the field lines will be bent backwards, thus developing a toroidal component. This toroidal field will eventually dominate the poloidal field component and cause a collimated jet to form just as we described in section 3 above. (c.f also Lovelace, these proceedings.)

This mechanism may provide an answer to the question of where does the angular momentum in an accretion disk end up? It is generally assumed that in the case of an accreting compact object in a binary system the disk angular momentum is absorbed by the orbit of the two stars, through the action of tidal torques. However, there is no such natural repository in the case of a galactic nucleus, (although a central star cluster might conceivably perform this function). It seems reasonable that an essentially axisymmetric system like this will lose angular momentum in the same way as the sun and similar stars are supposed to have lost their original spins; that is to say through the action of magnetic torques.

We can be more quantitative if we introduce a simple equivalent mechanical model of the plasma flowing out from the disk along the field lines. Suppose that we have a long straight wire that moves with the Keplerian angular velocity ω appropriate to its point of intersection with an equatorial disk. Now suppose that a bead moves without friction along this wire. Let the radial excursion of the bead from the point of intersection (at radius r) with the disk be x and the angle that the wire makes with the radial direction be θ. The potential energy Φ of the particle for small displacements x is

$$\Phi=[\omega^2 rx+\tfrac{1}{2}\omega^2 x^2(\sec^2\theta-3)]-[\omega^2 rx+\tfrac{1}{2}\omega^2 x^2]$$

The first term is the gravitational energy and the second term is the centrifugal potential energy. The terms linear in x cancel as they should for we are perturbing about an equilibrium. The quadratic terms can be combined to give the condition for this equilibrium to be unstable. This is that $\cos\theta>\tfrac{1}{2}$. So if the wire makes an angle of less than 60° with the radial direction, the bead will be flung outwards (or indeed inwards) along the wire. In other words if magnetic field lines emerge from the surface of an accretion disk they can drive a centrifugal wind.

The analytical models of this mechanism that have been constructed so far are all rather artificial and we must await the development of a robust MHD code to study this problem further. In particular it is extremely improbable that the flow will adopt the self-similar form assumed by Blandford and Payne.[31] It is far more likely that the field adjust in such a way that most of the angular momentum is carried in a sheath at large radius whereas most of the energy which is liberated near the black hole be carried in a central jet confined by the magnetic fields that are wrapped around it.

5. BLACK HOLE MAGNETOSPHERES

(i) Black Holes

It is convenient to measure the mass $10^8 M_8 M_\odot$ of a black hole in geometrical units $G=c=1$ as a length (half its Schwarzschild radius)

$$m = M_8 \, \mathrm{AU}$$

(Unfortunately there are not many further points of contact with space physics!) The generic black hole will be spinning and so can be described by a second parameter a which measures its angular momentum per unit mass again in geometrical units as a length so the total spin angular momentum in physical units is amc^3/G. The spacetime around a spinning black hole is described by the Kerr metric.[32] We are ignoring the gravitational influence of any surrounding matter. The "surface" of the black hole is called the event horizon and has a radius

$$r_+ = m + (m^2 + a^2)^{1/2}$$

The angular velocity of the hole is defined kinematically to be

$$\Omega_H = \frac{a}{a^2 + r_+^2}$$

It is possible to designate some of the total rest mass energy of the hole as spin energy and in principle to extract it. The fraction of the total mass that is extractable is

$$f = 1 - \tfrac{1}{2}[1 + (1 - (a/m)^2)^{1/2}]^{1/2}$$

f can be as large as 29 per cent if a has its maximum allowed value of m. This is of interest because the nuclei of radio galaxies do not seem to be accreting rapidly at present and yet they appear to be supplying large amounts of power to the radio lobes. The most energetic radio sources have energies of order $10^{61} erg$. This is one third the spin energy of a maximally rotating black hole of mass $10^8 M_\odot$ and so it is quite plausible that the radio galaxies are energized by the rotational energy of a black hole spun up in an earlier phase of rapid accretion when the radio source was presumably a quasar.

(ii) Electromagnetic Extraction of Spin Energy

The problem is then to find a way of extracting the spin energy in a suitably relativistic and not excessively radiative form. One thought experiment devised by Penrose[33] goes as follows. Sufficiently close to the event horizon, within a region known as the ergosphere, particles can move along orbits that have negative total energy (including rest mass). These orbits obviously do not extend to infinity but a collision between two particles within the ergosphere can place one particle on a negative energy orbit and allow it cross the event horizon thereby extracting some of the spin energy. The other particle will have positive energy and can be said to have acquired this at the expense of the hole. Useful as this thought experiment has been, it does not appear to be practical to realize it in a self-consistent model. The phase space associated with the negative energy orbits is very small unless the spin of the hole is unreasonably close to its maximal value.[29]

The spin energy can also be extracted using electromagnetic torques.[29,34,35]

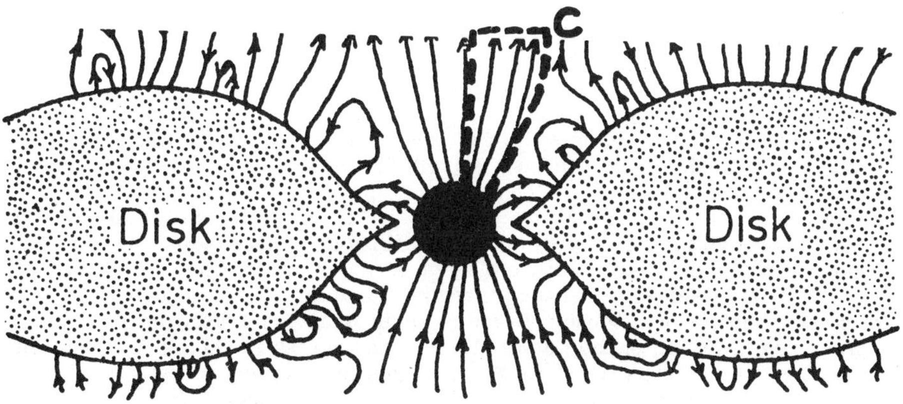

Fig. 9. Electromagnetic extraction of energy from a Kerr black hole.

These large scale magnetic fields must be supported by electrical currents external to the black hole (e.g. in an accretion disk) and in this way a black hole magnetosphere is different from a planetary magnetosphere. The reason for this is that although a black hole can be regarded as an electrical conductor, it is not a very good conductor. A helpful way to understand this is to note that a field that threads the hole and which is not supported by electrical currents will decay away in a few light crossing times. Equating this decay time to the standard estimate $4\pi\sigma m^2/c^2$ gives an estimate for the surface conductivity of a black hole horizon of $(120\pi)^{-1} ohm^{-1}$. This analogy can be put on a formal basis.[36] It also implies that the field lines threading the hole are likely to be rather well-ordered.

A spinning black hole can therefore act as a dissipative unipolar inductor. It makes a battery with an internal resistance of $\sim 100 ohm$. The emf generated by the battery is given by using the relativistic version of Faraday's law on a closed curve, C in the curved spacetime (see Fig. 9)

$$\oint_C \mathbf{E}\cdot d\mathbf{l} = -\frac{1}{c}\frac{d}{dt}\Phi - \oint_C (\boldsymbol{\gamma}\times\mathbf{B})\cdot d\mathbf{l}$$

where γ is a gravitomagnetic potential and Φ is the magnetic flux bounded by C.[35] The fields **E** and **B** should be regarded as defined by the Lorentz force law for a charge q, $d\mathbf{p}/dt = q(\mathbf{E} + \mathbf{v}\times\mathbf{B}/c)$, where t is a universal time.

Now choose for the curve C the dashed line in Fig. 9, which connects the horizon of the hole with the "acceleration region" where charged particles can cross the field lines. This curve is a path along which current can flow. (Current which enters the horizon can be regarded as flowing along the horizon until it exits again.[36] If the magnetic field is stationary and axisymmetric, the EMF is produced by the gravitomagnetic potential γ. Numerically it is $V \sim aB \sim 10^{20} B_4 M_8\, Volts$. If the electromagnetic conditions well outside the horizon are relativistic, then the load impedance will be essentially that of free space and roughly half the power generated will be dissipated in or beyond the magnetosphere, with the other half being dissipated within the horizon. The total power will be $\sim V^2/100 ohm \sim B^2 a^2 c$. If the field strength is $10^4 G$ and the black hole mass is $\sim 10^8 M_\odot$, then the power generated is $\lesssim 10^{45} ergs_{-1}$ which is adequate to account for even the most powerful radio sources. All this presupposes that electrical current can flow through the magnetosphere. This poses an interesting problem. The dissipation at large distances from the hole probably causes acceleration of an outflowing relativistic wind. However we also know that plasma must flow inwards across the horizon of the hole. Therefore we need to have an ongoing source of fresh plasma within the magnetosphere. Cross field diffusion from the disk is unlikely to be fast enough to accomplish this as the outflow speeds are of order the speed of light. However, there are likely to be sufficient electron positron pairs to carry the necessary current created through gamma ray -gamma ray collisions. These gamma rays may be radiated by the high temperature plasma in the vicinity of the hole or may just as well attributable to breakdown of the vacuum in the magnetosphere by processes analogous to those believed to occur in a radio pulsar magnetosphere. Only a small fraction of the potential difference across the hole need be tapped to accelerate electrons to relativistic energies so that they can scatter photons with energies in excess of the threshold for pair production $\sim 1 MeV$.

The minimum power that need be carried mechanically by the charged particles through the magnetosphere need only be a small fraction of the total Poynting flux carried by the electromagnetic fields. To order of magnitude this fraction is the ratio of the specific angular momentum a, (still regarded as a length) to the electron Larmor radius (typically a few cm). For this reason, the force-free approximation, (essentially that $\rho\mathbf{E}+\mathbf{j}\times\mathbf{B}$ vanishes) was made in some of the first treatments of black hole magnetospheres. However, this is insufficient to specify the electromagnetic solution for a given poloidal field distribution in the magnetosphere and a hydromagnetic treatment is required.[29] The constraint that the flow pass through Alfvenic and magnetosonic critical points ensures that the efficiency of energy extraction (i.e. the fraction of the spin energy lost that is dissipated at large distances from the hole) is close to the maximum value of $\sim$ 50 per cent.

The magnetosphere has several interesting properties that contrast with those conventionally attributed to pulsar or planetary magnetospheres. For example, there are both inner and outer light surfaces that play the role of the usual light cylinder in a radio pulsar. That is to say any plasma that corotates with the field within the inner light surface (according to the relativistic generalization of Ferraro's law of isorotation) must be moving inward towards the hole. Similarly, any plasma tied to the magnetic field beyond the outer light surface must move outwards.

As we mentioned above, the most powerful radio galaxies seem to produce most of their power invisibly without a large associated ultraviolet luminosity. This requires that the electromagnetic torques acting within the disk be much less than those acting on the hole. It is not clear whether or not this is self-consistent. It is also not clear whether or not this mechanism can also work in a radio quasar where the large UV luminosity is present. Nevertheless, it does provide a framework for discussing the origin of radio jets in physical terms.

ACKNOWLEDGEMENTS

I thank Richard Epstein and Doyle Evans for the opportunity to attend this workshop and several other participants for helpful comments. Support under NSF grant AST82-13001 and the Alfred P. Sloan Foundation is gratefully acknowledged.

REFERENCES

[1] A. V. Filippenko and W. L. W. Sargent, Astrophys. J. Suppl. **57** (1985).

[2] J. A. Högbom, Astron. Astrophys. Suppl. **36**, 173 (1979).

[3] A. H. Bridle and R. A. Perley, Ann. Rev. Astron. Astrophys. **22**, 319 (1984).

[4] B. Margon, Ann. Rev. Astron. Astrophys. **22**, 507 (1984).

[5] E. B. Fomalont, B. J. Geldzahler, R. M. Hjellming, and C. M. Wade, Astrophys. J. **275**, 802 (1983).

[6] M. C. Begelman, R. D. Blandford, and M. J. Rees, Rev. Mod. Phys. **56**, 255 (1984).

[7] R. D. Blandford, Ann. NY Acad. Sci. **422**, 303 (1983).

[8] M. J. Rees, Ann. Rev. Astron. Astrophys. **22**, 471 (1984).

[9] P. Wiita, Physics Reports (1985), in press.

[10] J. E. Dyson, ed., *Active Galactic Nuclei*, Proceedings of Workshop held at Manchester University (1985), in press.

[11] A. H. Bridle and J. Eilek, Proceedings of Greenbank Workshop on Radio Jets (1985), in press.

[12] J. Miller, ed., Proceedings of Santa Cruz Workshop on Astrophysics of Active Galaxies and Quasi-Stellar Objects (1985).

[13] Ya. B. Zel'dovich and I. D. Novikov, Doklo Acad. Nauk. SSSR **158**, 811 (1964).

[14] E. E. Salpeter, Astrophys. J. **140**, 796 (1964).

[15] D. Lynden-Bell, Nature **233**, 690 (1969).

[16] A. F. Tennant, R. F. Mushotzky, E. A. Boldt, J. M. Swank, Astrophys. J. **251**, 15 (1981).

[17] R. T. Schilizzi and A. G. deBruyn, Nature **303**, 26 (1983).

[18] J. B. Taylor, Phys. Rev. Lett. **33**, 1139 (1974).

[19] A. Königl and A. R. Choudhuri, Astrophys. J. **289**, 173 (1985).

[20] R. A. Perley, A. M. Bridle, and A. G. Willis, Astrophys. J. Supp. Ser. **54**, 291 (1984).

[21] D. G. Payne and H. Cohn, preprint (1984).

[22] M. Birkinshaw, Mon. Not. R. astr. Soc. **208**, 887 (1984).

[23] Owen, O'Dea, and Inoue, in preparation (1984).

[24] J. A. Eilek, J. O. Burns, C. P. O'Dea, and F. N. Owen, Astrophys. J. **278**, 37 (1984).

[25] J. Dreher and E. D. Feigelson, Nature **308**, 43 (1984).

[26] A. Ferrari, S. R. Mabbul, R. Rosner, and K. Tsingouros, Astrophys. J. Lett. **277**, L35 (1984).

[27] J. C. B. Papaloizou and J. E. Pringle, Mon. Not. R. astr. Soc. **208**, 721 (1984).

[28] V. Icke, Astron. J. **85**, 329 (1980).

[29] E. S. Phinney, unpublished thesis, University of Cambridge (1983).

[30] M. C. Begelman, C. F. McKee, and G. Shields, Astrophys. J. (1984), in press.

[31] R.D. Blandford and D. Payne, Mon. Not. R. astr. Soc. **199**, 883 (1982).

[32] C. W. Misner, K. S. Thorne, and J. A. Wheeler, *Gravitation*, (Freeman, New York, 1973).

[33] R. Penrose, Nuovo Cimento **1**, 252 (1969).

[34] K. S. Thorne and R. D. Blandford, in *Extragalactic Radio Sources*, Proceedings of IAU Symposium No. 97, ed. Heeschen and Wade (Reidel, Dordrecht, Holland, 1981).

[35] D. Macdonald and K. S. Thorne, Mon. Not. R. astr. Soc. **198**, 345 (1982).

[36] R. Znajek, Mon. Not. R. astr. Soc. **185**, 833 (1978).

SOME TOPICS IN THE MAGNETOHYDRODYNAMICS OF ACCRETING MAGNETIC COMPACT OBJECTS*

J. J. Aly
University of Illinois at Urbana-Champaign
Department of Physics
1110 W. Green Street
Urbana, IL 61801
and
Service d'Astrophysique - CEN Saclay
91191 Gif-sur-Yvette Cedex - France

ABSTRACT

Magnetic compact objects (neutron stars or white dwarfs) are currently thought to be present in many accreting systems that are releasing large amounts of energy. The magnetic field of the compact star may interact strongly with the accretion flow and play an essential role in the physics of these systems. Some magnetohydrodynamic (MHD) problems that are likely to be relevant in building up self-consistent models of the interaction between the accreting plasma and the star's magnetosphere are addressed in this series of lectures. The basic principles of MHD are first introduced and some important MHD mechanisms (Rayleigh-Taylor and Kelvin-Helmholtz instabilities; reconnection) are discussed, with particular reference to their role in allowing the infalling matter to penetrate the magnetosphere and mix with the field. The structure of a force-free magnetosphere and the possibility of quasistatic momentum and energy transfer between regions linked by field-aligned currents are then studied in some detail. Finally, the structure of axisymmetric accretion flows onto magnetic compact objects is considered.

CONTENTS

*These lecture notes were prepared with the assistance of Dr. Steve McMillan, University of Illinois.

2. Fluid approximations

   A. Closure of the system of moment equations
   B. Constitutive relations for collision-dominated plasmas
   C. Constitutive relations for collisionless plasmas
   D. Radiative effects

3. Ideal MHD

   A. Equations
   B. Flux freezing
   C. MHD waves
   D. Discontinuities
   E. Stability

4. Accreting magnetic compact objects

   A. Overview
   B. Theoretical problems
   C. The MHD approach
   D. Force-free magnetospheres

II. Some basic MHD processes

1. The Rayleigh-Taylor instability (RTI)

   A. Physical mechanisms and conditions for instability
   B. Non-linear development of the RTI

2. The Kelvin-Helmholtz instability (KHI): Physical mechanisms

   A. Incompressible fluids
   B. Compressible fluids
   C. Magnetic effects
   D. Other effects

3. The KHI in an accreting magnetosphere

   A. Linear theory
   B. Non-linear development
   C. KHI at the interface between an accretion disk and the magnetosphere

4. Reconnection of magnetic field lines

   A. Tearing instability
   B. Forced stationary reconnection
   C. Reconnection in an accreting magnetosphere

LECTURE I
THE MAGNETOHYDRODYNAMIC APPROACH TO THE THEORY OF ACCRETING MAGNETIC COMPACT OBJECTS

In this first lecture, we introduce the basic ideas of magnetohydrodynamics (MHD) that will be used in the following lectures to describe some aspects of the physics of accretion by magnetic compact objects. We first recall (Section 1) the macroscopic equations which determine the evolution of a plasma. In Section 2, we describe one-fluid approximations to these equations and discuss their domains of validity. Some classical topics in ideal MHD, which is defined by the simplest set of equations among those introduced in Section 2, are presented in Section 3. Finally, in Section 4, we give a short overview of the systems in which we are interested. We outline some of the problems we will consider in more detail later, and present justification for approaching them through MHD techniques.

1. Macroscopic equations for a plasma

The formalism developed in this section and the two following ones is quite classical and may be found in many textbooks. Useful references are Krall and Trivelpiece,[1] Roberts,[2] and Rossi and Olbert.[3]

A. Boltzmann equations

A plasma consisting of n species of particles may often be described statistically by n "single-particle distribution functions," $f_\alpha(\mathbf{r}, \mathbf{u}, t)$ ($\alpha = 1,\ldots,n$), which depend on position, $\mathbf{r}$, velocity, $\mathbf{u}$, and time, t. $f_\alpha(\mathbf{r}, \mathbf{u}, t)\, d\mathbf{r}\, d\mathbf{u}$ represents the number of particles of type α contained at time t in the volume $d\mathbf{r}\, d\mathbf{u}$ around the point $(\mathbf{r}, \mathbf{u})$ in phase space. If the particles are non-relativistic ($u/c \ll 1$), the f_α obey the Boltzmann equation,

$$\frac{\partial f_\alpha}{\partial t} + \mathbf{u} \cdot \nabla_{\mathbf{r}} f_\alpha + \left\{\frac{q_\alpha}{m_\alpha}\left(\mathbf{E} + \frac{\mathbf{u}\times\mathbf{B}}{c}\right) + \mathbf{g}\right\} \cdot \nabla_{\mathbf{u}} f_\alpha = \left.\frac{\partial f_\alpha}{\partial t}\right|_{coll} , \qquad (1)$$

where m_α and q_α are the mass and charge of a particle of type α, $\mathbf{g}$ is the gravitational field, and the "average" electric field, $\mathbf{E}$, and magnetic field, $\mathbf{B}$, are related to each other, and to the current $\mathbf{j}$ and charge density ρ_q, by Maxwell's equations,

$$\nabla \cdot \mathbf{B} = 0 , \qquad (2)$$

$$\nabla\times\underset{\sim}{B} - \frac{1}{c}\frac{\partial}{\partial t}\underset{\sim}{E} = \frac{4\pi}{c}\underset{\sim}{j} = \frac{4\pi}{c}\sum_{\alpha} q_{\alpha} \int f_{\alpha}\,\underset{\sim}{u}\,d\underset{\sim}{u} \quad , \tag{3}$$

$$\nabla \cdot \underset{\sim}{E} = 4\pi\,\rho_q = 4\pi \sum_{\alpha} q_{\alpha} \int f_{\alpha}\,d\underset{\sim}{u} \quad , \tag{4}$$

$$\nabla \times \underset{\sim}{E} + \frac{1}{c}\frac{\partial}{\partial t}\underset{\sim}{B} = 0 \quad . \tag{5}$$

The average force, $q_{\alpha}\,(\underset{\sim}{E} + \underset{\sim}{u} \times \underset{\sim}{B}/c)$, acting on a particle is due to a large number of distant particles. The effect of nearby particles ("collisions") is contained in the RHS of Eq. (1). Actually, this term cannot be computed from the f_{α} alone. A knowledge of the so-called two particle distribution functions, $f^{(2)}_{\alpha\beta}(\underset{\sim}{r}_1, \underset{\sim}{u}_1, \underset{\sim}{r}_2, \underset{\sim}{u}_2, t)$, which contain information about the correlations between the positions and velocities of pairs of particles, is required. Fortunately, the collision term may often be evaluated approximately using simple models.

B. Macroscopic variables and moments of the Boltzmann equation

A description equivalent to that given by Eq. (1) is obtained by multiplying these equations by tensorial powers, $\otimes^{(p)}\underset{\sim}{u}$, of $\underset{\sim}{u}$ ($p = 0,2,\ldots,\infty$) and integrating over $\underset{\sim}{u}$ to obtain a series of equations relating the "moments" of f_{α}. These moments can be combined to form "one-fluid" quantities that obey equations easily derived from the p-moment equations.

From now on, we will assume that the plasma is composed of only two species: electrons ($\alpha = e$, $q_e = -e$) and protons ($\alpha = i$, $q_i = e$). The one-fluid quantities one obtains are

the mass density,

$$\rho = \sum_{\alpha} \rho_{\alpha} = \sum_{\alpha} m_{\alpha} \int f_{\alpha}\,d\underset{\sim}{u} \quad , \tag{6}$$

the center of mass velocity,

$$\underset{\sim}{v} = \rho^{-1} \sum_{\alpha} \rho_{\alpha}\,\underset{\sim}{v}_{\alpha} = \rho^{-1} \sum_{\alpha} \rho_{\alpha} \int f_{\alpha}\,\underset{\sim}{u}\,d\underset{\sim}{u} \quad , \tag{7}$$

the pressure tensor,

$$\underset{\approx}{t} = \sum_{\alpha} \underset{\approx}{t}_{\alpha} = \sum_{\alpha} m_{\alpha} \int f_{\alpha} (\underset{\sim}{u} - \underset{\sim}{v}) \otimes (\underset{\sim}{u} - \underset{\sim}{v})\,d\underset{\sim}{u} \quad , \tag{8}$$

the internal energy density,

$$\rho e = \sum_\alpha \rho_\alpha e_\alpha = \frac{1}{2} \sum_\alpha m_\alpha \int (\mathbf{u} - \mathbf{v})^2 f_\alpha \, d\mathbf{u} \tag{9}$$

and the heat flux,

$$\mathbf{q} = \sum_\alpha \mathbf{q}_\alpha = \sum_\alpha \frac{1}{2} m_\alpha \int (\mathbf{u} - \mathbf{v}) (\mathbf{u} - \mathbf{v})^2 f_\alpha \, d\mathbf{u} \quad , \tag{10}$$

and they obey the following equations:

<u>continuity equation</u>

$$\frac{\partial \rho}{\partial t} + \nabla \cdot \rho \mathbf{v} = 0 \quad , \tag{11}$$

<u>momentum equation</u>

$$\rho \frac{d\mathbf{v}}{dt} = \rho \left(\frac{\partial \mathbf{v}}{\partial t} + \mathbf{v} \cdot \nabla \mathbf{v}\right) = \rho_q \mathbf{E} + \frac{\mathbf{j} \times \mathbf{B}}{c} + \rho \mathbf{g} - \nabla \cdot \mathbf{t} \quad , \tag{12}$$

<u>generalized Ohm's law</u>

$$\frac{\partial \mathbf{j}}{\partial t} + \nabla \cdot (\mathbf{v} \otimes \mathbf{j} + \mathbf{j} \otimes \mathbf{v} - \rho_q \mathbf{v} \otimes \mathbf{v}) =$$

$$\left(\sum_\alpha \frac{\rho_\alpha e^2}{m_\alpha^2}\right) \mathbf{E} + \frac{e^2}{m_e m_i} \rho \frac{\mathbf{v} \times \mathbf{B}}{c} - e \frac{m_i - m_e}{m_e m_i} \frac{\mathbf{j} \otimes \mathbf{B}}{c}$$

$$- e \nabla \cdot \left(\frac{\mathbf{t}_i}{m_i} - \frac{\mathbf{t}_e}{m_e}\right) + \sum_\alpha q_\alpha \int \mathbf{u} \left.\frac{\partial f_\alpha}{\partial t}\right|_{coll} d\mathbf{u} \quad , \tag{13}$$

<u>energy equation</u>

$$\rho \frac{de}{dt} = - \nabla \cdot \mathbf{q} - \mathbf{t} : \nabla \mathbf{v} - (\mathbf{j} - \rho_q \mathbf{v}) \cdot (\mathbf{E} + \mathbf{v} \times \mathbf{B}/c) \quad . \tag{14}$$

The evolution of $\mathbf{t}$ is determined by an equation (of which Eq. (14) is the contraction) in which there appears a new quantity, $Q^{(3)}$ (a third rank tensor), whose evolution equation introduces another new quantity (a fourth rank tensor), and so on.

2. Fluid approximations

A. Closure of the system of moment equations

The moment equations derived from the Boltzmann equation constitute an infinite system and are practically impossible to handle. Under some circumstances, however, it is possible to derive "plasma models," in which only a few equations of the chain, complemented by some "constitutive relations", are used. In the models introduced below, the only necessary moment equations are Eqs. (10)-(14); these must be supplemented by relations expressing $\underset{\approx}{t}_{e,i}$, $\underset{\sim}{q}$ and the collision term in Eq. (13) as functions of the other macroscopic variables present in the equations.

Before giving these relations, we make the following, generally well justified, approximation: hereafter, the plasma is assumed to be quasi-neutral (i.e. $\rho_q \simeq 0$, or $n_e \simeq n_i \simeq n$, where n_e and n_i are the number densities of the electrons and protons, respectively); therefore, we may neglect the terms $\rho_q \underset{\sim}{E}$ (electric force) in Eq. (11), $\rho_q \underset{\sim}{v} \otimes \underset{\sim}{v}$ in Eq. (12) and $\rho_q \underset{\sim}{v}$ in Eq. (13). A condition for this approximation to be valid is that the length scale, ℓ, of the phenomenon under consideration be much larger than the so-called Debye length, λ_D, of the plasma, i.e.

$$\ell^2 \gg {c_e}^2/{\omega_p}^2 = {\lambda_D}^2 \quad , \tag{15}$$

where ${c_e}^2 = kT/m_e$ (T is the temperature) and $\omega_p = (4\pi n e^2/m_e)^{1/2}$. It should be noted that it is not possible to set $\rho_q = 0$ in Maxwell equation (4) also, as this would result in the suppression of a variable (ρ_q) without the accompanying suppression of an equation. Rather, Eq. (4) may be used, when the problem has been solved and $\underset{\sim}{E}$ is known, to compute ρ_q and thus to check a posteriori the consistency of the quasi-neutrality assumption.

B. Constitutive relations for collision-dominated plasmas

If ℓ and τ (the characteristic length- and time-scales of the phenomenon considered) are such that

$$\lambda_\alpha \ll \ell \quad \text{and} \quad \nu_{c\alpha}^{-1} \ll \tau \quad , \tag{16}$$

where λ_α is the mean free path of particles of type α and ν_α their collision frequency, the plasma is said to be collision-dominated. In that case, the distribution functions f_α are nearly isotropic at all times, and it may be supposed that, to zeroth order in λ_α/ℓ, the stress tensors have the form $\underset{\approx}{t}_\alpha = p_\alpha \underset{\approx}{I} = (p/2)\underset{\approx}{I}$ ($p = p_e + p_i$ and one assumes equal temperatures for the electrons and the protons) and the heat flow vector is null. In this approximation, the momentum equation reduces to

$$\rho \frac{d\underset{\sim}{v}}{dt} + \nabla p = \rho \underset{\sim}{g} + \frac{\underset{\sim}{j} \times \underset{\sim}{B}}{c} \quad , \tag{17}$$

while the energy equation becomes

$$\frac{d}{dt} \frac{p}{\rho^{5/3}} = \frac{2}{3\rho^{5/3}} \underset{\sim}{j} \cdot \left(\underset{\sim}{E} + \frac{\underset{\sim}{v}}{c} \times \underset{\sim}{B}\right) \quad , \tag{18}$$

where we have used $3p = \mathrm{tr}\ \underset{\approx}{t} = 2\rho e$ and the continuity equation, (11). When its RHS (which contains, in particular, the Joule heating term) is zero, Eq. (18) gives the usual adiabatic law ($p/\rho^{5/3}$ = constant for each element of plasma). To the next order (first order in λ_α/ℓ), one obtains deviations of $\underset{\approx}{t}_\alpha$ from isotropy. If the cyclotron frequency of a particle of species α, $\omega_{c\alpha} = eB/m_\alpha c$, satisfies $\omega_{c\alpha} \ll \nu_\alpha$, these can be treated as viscous effects,

$$\underset{\approx}{t}_\alpha - p_\alpha \underset{\approx}{I} = \underset{\approx}{\pi}_\alpha = -\eta_\alpha(\nabla\underset{\sim}{v} + {}^T\nabla\underset{\sim}{v} - 2/3(\nabla \cdot \underset{\sim}{v})\ \underset{\approx}{I})$$

$$+ \xi_\alpha(\nabla \cdot \underset{\sim}{v})\ \underset{\approx}{I} \quad , \tag{19}$$

where the superscript "T" denotes the transposed matrix. The quantities η_α and ζ_α (which depend on T and ρ), are called the coefficients of shear and bulk viscosity, respectively. For $\omega_{c\alpha} \gtrsim \nu_\alpha$, the viscosity is anisotropic, and tensorial coefficients must be introduced. It is worth noting that $\eta_i \gg \eta_e$ and $\zeta_i \gg \zeta_e$, so $\underset{\approx}{\pi} = \underset{\approx}{\pi}_e + \underset{\approx}{\pi}_i \simeq \underset{\approx}{\pi}_i$.

To the same order, there also appear heat flows ($\underset{\sim}{q}_\alpha \neq 0$); again, these are linear functions of gradients (in particular of the temperature gradient), with anisotropic tensorial coefficients when $\omega_{c\alpha} \gtrsim \nu_\alpha$ (in which case the heat flow normal to $\underset{\sim}{B}$ is substantially reduced). We refer the reader to Rossi and Olbert[3] for expressions for the conductivity (and viscosity) tensors. Let us just note that, in contrast to the viscosity, the heat transport is mainly due to electrons (at least if the magnetic field is not too strong).

Let us now consider Ohm's law. For a collisional plasma, the collision term in Eq. (13) may be written as $-\nu_e \underset{\sim}{j}$ (where ν_e is actually the electron-proton collision frequency), if one neglects the slight anisotropy (due to $\underset{\sim}{B}$) of the coefficient. Defining the resistivity, η, of the plasma, and its conductivity, σ, by

$$\eta = \frac{1}{\sigma} = \frac{\nu_e m_e}{ne^2} = \frac{4\pi\nu_e}{\omega_p^2} \quad , \tag{20}$$

and neglecting, for simplicity, viscous effects and terms of order $m_e/m_i \ll 1$, one obtains, from Eq. (13),

$$\underset{\sim}{E}_* = \underset{\sim}{E} + \frac{\underset{\sim}{v}}{c} \times \underset{\sim}{B} = \eta \underset{\sim}{j} + \frac{1}{nec} \underset{\sim}{j} \times \underset{\sim}{B} - \frac{1}{2ne} \nabla p + \frac{m_e}{ne^2} \{\frac{\partial \underset{\sim}{j}}{\partial t} + \nabla \cdot (\underset{\sim}{j} \otimes \underset{\sim}{v} + \underset{\sim}{v} \otimes \underset{\sim}{j})\} \quad , \qquad (21)$$

where we have introduced $\underset{\sim}{E}_*$, the electric field in a frame moving locally with the plasma. This expression is still quite complicated, and other simplifications may be sought. Let us compare the magnitude of the terms in the RHS of Eq. (21) to $|\underset{\sim}{v} \times \underset{\sim}{B}/c|$.

(i) The ohmic resistive term, $\eta \underset{\sim}{j}$, may be neglected if

$$\frac{vB}{c\eta j} \simeq \frac{4\pi v\ell}{c^2\eta} = \frac{v\ell}{D_B} \equiv R_m \gg 1 \quad , \qquad (22)$$

where $D_B = c^2\eta/4\pi = c^2/4\pi\sigma$ is the coefficient of magnetic diffusion and R_m is the magnetic Reynolds number.

(ii) The term $\underset{\sim}{j} \times \underset{\sim}{B}/nec$, which represents the "Hall effect", is of order (ω_{ce}/ν_e) times the ohmic term, and is negligible if

$$\frac{vne}{j} = 4\pi \frac{vne\ell}{B} = 4\pi \frac{v}{c} \frac{\omega_p^2 \ell}{\omega_{ce}} \gg 1 \quad . \qquad (23)$$

(iii) The term $\nabla p/2ne$, which represents the "ambipolar electric field," may be neglected if

$$\frac{vBe\ell}{ckT} = \frac{vc}{c_e^2} \frac{\ell}{r_{ce}} \gg 1 \quad , \qquad (24)$$

where $r_{ce} = c/\omega_{ce}$.

(iv) The last term, which concerns the inertia of the electrons, is negligible if

$$\frac{vBne^2}{m_e c} \frac{4\pi\ell\tau}{cB} = \left(\frac{\ell}{c/\omega_p}\right)^2 \gg 1 \qquad (25)$$

In many astrophysical situations, conditions (22)-(25) are satisfied and Ohm's law reduces to its simplest form,

$$\underset{\sim}{E}_* = \underset{\sim}{E} + \frac{\underset{\sim}{v}}{c} \times \underset{\sim}{B} = 0 \quad . \qquad (26)$$

Thus, for the collision-dominated case, we have found a set of equations which, when supplemented by the three Maxwell equations (2), (3) and (5), is obviously complete. This system may be simplified when some simple criteria are satisfied. A further approximation involves Maxwell equation (3), in which the displacement current, $c^{-1}\ \partial \underset{\sim}{E}/\partial t$, may be neglected for sufficiently slow processes.

The main assumption supporting the development of this section is the weak anisotropy of the distribution functions, ensured by the dominance of coulomb collisions. In many astrophysical situations, the mean free paths of the particles are quite large. However, it is often still possible to apply MHD theory if isotropy can be established by other processes, e.g. collisions of the particles with turbulent plasma waves, or small-scale magnetic irregularities (the role of the mean free path being played in that case by the correlation length, ℓ_B, of the field, assuming $m_e c/eB \equiv r_c \lesssim \ell_B$). Of course, in such situations, the transport coefficients depend on the mechanism involved.

C. <u>Constitutive relations for collisionless plasmas</u>

It is interesting to note that collisionless plasmas in a magnetic field may also be described by a fluid theory if

$$\ell \gg r_c \text{ and } \omega_c \tau \gg 1 \quad , \tag{27}$$

where r_c and ω_c are the gyroradius and gyrofrequency of the particles. In that case, however, the stress tensor is anisotropic. To zeroth order in r_c/ℓ and $(\omega_c \tau)^{-1}$, assuming negligible heat flow along the lines,

$$\underset{\approx}{t}^{\alpha} = p_{\parallel}^{\alpha}\ \underset{\sim}{B} \otimes \underset{\sim}{B}/B^2 + p_{\perp}^{\alpha}\ (\underset{\approx}{I} - \underset{\sim}{B} \otimes \underset{\sim}{B}/B^2) \quad , \tag{28}$$

where the pressures $p_{\parallel}^{\alpha}$ and $p_{\perp}^{\alpha}$, parallel and perpendicular to the field, obey the adiabatic relations

$$\frac{d}{dt}\, \frac{p_{\perp}^2\, p_{\parallel}}{\rho^5} = 0 \tag{29}$$

and

$$\frac{d}{dt}\, \frac{p_{\perp}}{\rho B} = 0 \quad . \tag{30}$$

On the other hand, under restricted conditions, the "perfect" Ohm's law is satisfied. The model resting on these relations is called the "double-adiabatic theory."

The expansion may be continued to first order, as in the case of the collisional plasma considered above.[4] The tensor $\underset{\approx}{t}$ acquires off-diagonal terms proportional to the Larmor radius, which are similar (to within a numerical factor) to the off-diagonal terms of the viscous-stress tensor of Subsection B in the limit $\omega_c \nu_{coll} \to \infty$. These terms may be important in stabilizing some instabilities when the wavelength is very small ("finite Larmor radius effects"; see Lecture II). Of course, they do not produce dissipation. The heat flow perpendicular to the lines may also be computed, as well as a new form of Ohm's law, which is similar to Eq. (21) with $\eta = 0$ and $m_e = 0$.

The intent of this subsection was to show that MHD (with an appropriate form for the stress tensor) may be used to represent quite a large class of phenomena. The form of MHD theory with an anisotropic pressure, which could be appropriate in describing some phenomena in the underdense parts of some magnetospheres (see below, Section 4-D), has recently been used for the first time in the theory of accreting magnetic compact objects (R. Lovelace, these proceedings).

D. Radiative effects

In almost all astrophysical situations, radiation plays a very important role, and must be taken into account. The simplest way of introducing radiation effects into the MHD equations is to add "by hand" terms which represent the exchange of momentum and energy between the plasma and the radiation field. The explicit form of these terms depends, in general, on quantum processes.

In the momentum equation, radiation adds a force

$$\underset{\sim}{F}_r = \sigma \, n \, \underset{\sim}{J}/c \quad , \qquad (31)$$

where $\underset{\sim}{J}$ is the flux of radiation energy and σ some average interaction cross-section. In accreting systems, the radiation comes mainly from near the compact object and so opposes the gravitational force, $\underset{\sim}{F}_g$. If the radiation is emitted with spherical symmetry, $\underset{\sim}{F}_r$ may be combined with $\underset{\sim}{F}_g$ to give

$$\underset{\sim}{F} = - \frac{GM \rho}{r^2} \left(1 - \frac{L}{L_{edd}}\right) \hat{r} \quad , \qquad (32)$$

where L is the luminosity of the object, L_{edd} (the Eddington luminosity) is defined by

$$L_{edd} = \frac{4\pi \; GM \; m_p c}{\sigma} \quad , \qquad (33)$$

and the plasma has been assumed to be optically thin to the radiation. Clearly, L_{edd} represents a limit to the luminosity which can be released as a consequence of the accretion process, because accretion itself is driven by the force (32), and would stop if this force were directed outward. Some restrictions have been imposed to establish Eq. (32); however, the maximum luminosity cannot greatly exceed L_{edd} even if they are relaxed.

Of course, the energy equation also has to be completed by a term, ε_r, representing the interaction of matter and radiation, and equations for the radiation field must be added to close the system. As they will not be used here, we refer the reader to the abundant literature on this subject (see, for example, Pomraning[5]).

3. Ideal MHD

A. Equations

Ideal MHD is obtained when the pressure tensor reduces to diagonal form, $\underset{\approx}{t} = p\,\underset{\approx}{I}$, adiabatic evolution ($\underset{\sim}{q} = 0$) is assumed (with an adiabatic index γ), the simplest form of Ohm's law (Eq. (26)) is used and the displacement current is neglected in Eq. (3). In that case, the system of equations reduces to

$$\frac{\partial \rho}{\partial t} + \nabla \cdot \rho \underset{\sim}{v} = 0 \ , \tag{34}$$

$$\rho \frac{d\underset{\sim}{v}}{dt} + \nabla p = \rho \underset{\sim}{g} + \frac{(\nabla \times \underset{\sim}{B}) \times \underset{\sim}{B}}{4\pi} \ , \tag{35}$$

$$\frac{d}{dt}(p\,\rho^{-\gamma}) = 0 \ , \tag{36}$$

$$\frac{\partial \underset{\sim}{B}}{\partial t} = \nabla \times (\underset{\sim}{v} \times \underset{\sim}{B}) \ , \tag{37}$$

and

$$\nabla \cdot \underset{\sim}{B} = 0 \ , \tag{38}$$

where Eq. (37) is obtained by combining Ohm's law (26) with Eq. (5) and Eq. (35) results from the combination of Eqs. (3) and (17). Other useful expressions for the magnetic force are

$$\frac{\underset{\sim}{j} \times \underset{\sim}{B}}{c} = \nabla \cdot \left\{ -\frac{B^2}{8\pi}\underset{\approx}{I} + \underset{\sim}{B} \otimes \underset{\sim}{B} \right\} = -\nabla \frac{B^2}{8\pi} + \kappa \frac{B^2}{4\pi}\hat{n} \ , \tag{39}$$

where $\hat{n}$ is the principal normal to a magnetic field line and κ is its curvature. This expression clearly shows the anisotropic character of the Lorentz force, which may be decomposed into a pressure force (isotropic) and a tension force acting along the lines, tending to straighten them.

B. Flux freezing

An elementary consequence of Eq. (37) is the conservation of the magnetic flux through any closed contour moving with the plasma. Hence, if $\underset{\sim}{B} = 0$ initially in some region, V_o, it will be zero at any time t in V_t, the volume occupied at that time by the particles originally in V_o.

Another consequence is the following: if, at t = 0, a set of particles is on the same field line, C_o, then, at time t, they will be on the same line, C_t. We can physically identify the lines C_t: they are the positions at different times of the same line, whose motion is tied to that of the plasma. This property is often referred to as the "frozen-in law." It implies the conservation of the topology of the field lines as the system evolves.

C. MHD waves

Consider a uniform plasma filling all space and threaded by a uniform magnetic field, $\underset{\sim}{B}$. Any small perturbation of this state may propagate as a linear wave as a result of the restoring forces associated with the pressure (thermal + magnetic) and the tension of the field lines.

Linearizing the perfect MHD equations about the uniform state and seeking plane-wave solutions, in which any physical quantity, f, varies as

$$f(\underset{\sim}{r}, t) = \tilde{f}\, e^{i(\underset{\sim}{k} \cdot \underset{\sim}{r} - \omega t)}, \qquad (40)$$

one finds that the frequency, ω, and wave number, $\underset{\sim}{k}$, must satisfy

$$\omega = \pm\, \underset{\sim}{k} \cdot \underset{\sim}{c}_A \qquad (41)$$

or

$$\omega = \pm \left(\tfrac{1}{2} \{ k^2 (c_s^2 + c_A^2) \pm [k^4 (c_s^2 + c_A^2)^2 - 4 c_s^2 (\underset{\sim}{k} \cdot \underset{\sim}{c}_A)^2]^{1/2} \} \right)^{1/2}, \qquad (42)$$

where

$$\underset{\sim}{c}_A = \frac{\underset{\sim}{B}}{\sqrt{4\pi\rho}} \qquad (43)$$

is the so-called Alfvén velocity and

$$c_s = (\gamma p/\rho)^{1/2} \tag{44}$$

is the adiabatic sound speed.

The waves corresponding to Eq. (41) are called Alfvén waves. They are supported by the tension of the lines and there is no compression of the plasma associated with them. They carry energy at the group velocity, $\underset{\sim}{c}_g = \pm \underset{\sim}{c}_A$. The waves corresponding to Eq. (42) are called fast (+) and slow (-) magnetosonic waves. In contrast to Alfvén waves, they are compressive. The phase velocities, $c_{ph} = \omega/k$, of all three waves are plotted in Fig. 1. For a given $\underset{\sim}{k}$, one always has the ordering $c_{ph}^{slow} \leqslant c_{ph}^{Alfven} \leqslant c_{ph}^{fast}$. On the other hand, one has $\max(c_s^2, c_A^2) \leqslant (c_{ph}^{fast})^2 \leqslant c_s^2 + c_A^2$ and $0 \leqslant (c_{ph}^{slow})^2 \leqslant \min(c_s^2, c_A^2)$.

It should be noted that the diagrams in Fig. (1) do not represent the wave front which would be produced at time t = 1 by an isotropic pulse emitted at t = 0 at the origin. Rather, the length of $\underset{\sim}{c}_{ph}(\theta)$ is the distance travelled by the plane phase surface normal to $\underset{\sim}{c}_{ph}$.[6] The wave front is the envelope of these planes. Two examples are shown in Fig. 2. These considerations are of some importance in understanding some aspects of the theory of accretion flows (see Lecture IV).

D. Discontinuities

MHD equations may admit surfaces across which the physical variables ρ, p, $\underset{\sim}{v}$, $\underset{\sim}{B}$ change discontinuously. However, the changes cannot be arbitrary, but must satisfy conditions expressing the conservation of momentum and energy that are easily derived from Eqs. (34)-(38). In a frame moving with the discontinuity, the jump conditions are:

$$[\rho u_n] = 0 \quad , \tag{45}$$

$$\left[\rho v_n \underset{\sim}{v} + \frac{B^2}{8\pi}\hat{n} - \frac{B_n \underset{\sim}{B}}{4\pi} + p\hat{n}\right] = 0 \quad , \tag{46}$$

$$[B_n] = 0 \quad , \tag{47}$$

and

$$[B_n \underset{\sim}{v}_t - \underset{\sim}{B}_t v_n] = 0 \quad , \tag{48}$$

where indices n and t refer to components normal and tangential to the discontinuity, respectively, and [f] denotes the jump in the quantity f. Discontinuities may be classified according to the

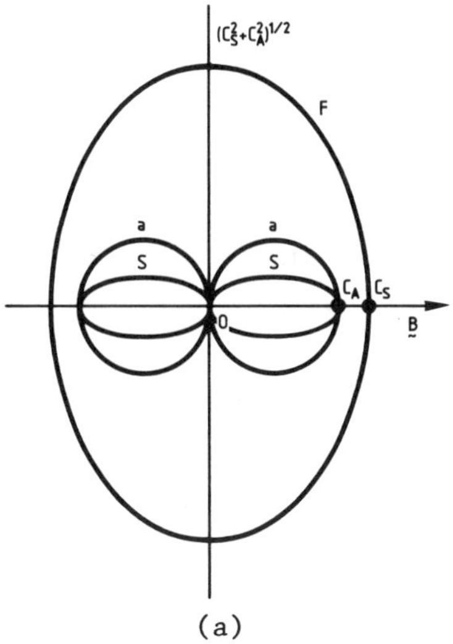

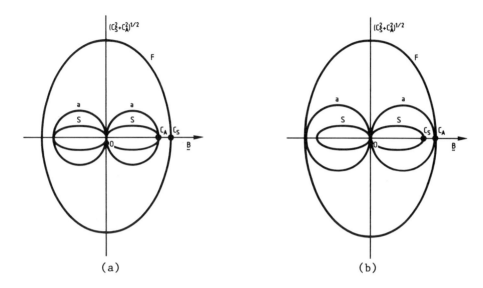

Figure 1. Phase velocity diagrams for MHD waves.
This diagram shows the variation of the phase velocities of the three MHD modes (fast F, Alfvén a, slow s) as a function of the angle θ with $\underset{\sim}{B} = B\,\hat{x}$. In 1a, $C_A^2 < C_s^2$, while, in 1b, $C_s^2 < C_A^2$.

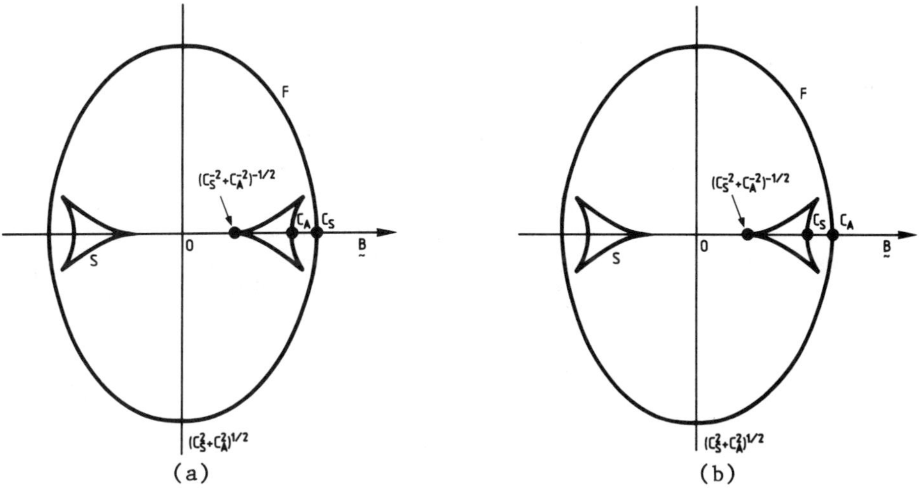

Figure 2. Friedrichs diagrams for MHD waves.
The curves represents the shape at time t = 1 of the fast (F), Alfvén (A) and slow (s) wavefronts emanating from a fixed point 0. In 2a, $C_A^2 < C_s^2$, while, in 2b, $C_s^2 < C_A^2$.

following scheme:

$v_n = 0$	$B_n \neq 0$	contact discontinuity
"	$B_n = 0$	tangential discontinuity
$v_n \neq 0$	$B_t = 0$	parallel shock
"	$B_n = 0$	perpendicular shock
"	$B_t\ B_n \neq 0$	oblique shock

They are considered in more detail in R. Blandford's lectures.

E. Stability

To linear order, the displacement, $\underset{\sim}{\xi}(\underset{\sim}{r}, t)$, of a plasma element with respect to its position in some equilibrium configuration satisfies

$$\rho \frac{\partial^2 \underset{\sim}{\xi}}{\partial t^2} = \underset{\sim}{F}(\underset{\sim}{r}; \underset{\sim}{\xi}) \quad , \tag{49}$$

where the linear operator F is self-adjoint (this results from energy conservation). F does not depend explicitly on time, so one may look for solutions to Eq. (49) of the form $\tilde{\underset{\sim}{\xi}}(r)e^{-i\omega t}$. Equation (49) then becomes

$$-\omega^2 \rho\, \tilde{\underset{\sim}{\xi}} = F(\underset{\sim}{r}; \tilde{\underset{\sim}{\xi}}) \quad . \tag{50}$$

Together with appropriate boundary conditions, Eq. (50) defines an eigenvalue problem; an eigensolution, $\tilde{\underset{\sim}{\xi}}_n$, corresponding to ω_n^2 (necessarily real) is called a proper mode. If $\tilde{\omega}_n^2 > 0$, the $\tilde{\underset{\sim}{\xi}}_n$ are stable, while if $\omega_n^2 < 0$, they are unstable and grow exponentially. The ω_n may be determined using a variational method (Rayleigh principle). One looks for functions which are solutions of $\delta(\omega^2) = 0$, where $\omega^2[\underset{\sim}{\xi}]$ is the functional

$$\omega^2[\underset{\sim}{\xi}] = 2\delta W[\underset{\sim}{\xi}] \;/ \int \rho_o |\underset{\sim}{\xi}|^2 d\underset{\sim}{r} \tag{51}$$

and $\delta W[\underset{\sim}{\xi}] \equiv -\frac{1}{2}\int \underset{\sim}{\xi}\cdot\underset{\sim}{F}(\underset{\sim}{r};\underset{\sim}{\xi})d\underset{\sim}{r}$.

The equilibrium is stable (unstable) if $\min_{\xi} \omega^2[\underset{\sim}{\xi}] > 0$ (<0). In general, the problem of minimizing $\omega^2[\underset{\sim}{\xi}]$ (and hence solving the eigenvalue problem) is too difficult; however, one may still decide the stability simply by looking at the sign of $\delta W[\underset{\sim}{\xi}]$: if $\delta W[\underset{\sim}{\xi}] > 0$ for <u>all</u> possible $\underset{\sim}{\xi}$, one has stability, while if $\delta W[\underset{\sim}{\xi}] < 0$ for at least <u>one</u> $\underset{\sim}{\xi}$, the equilibrium is unstable. This is the well-known

"Energy Principle" of Bernstein et al.[7] Physically, $\delta W[\underset{\sim}{\xi}]$ represents the potential energy associated with the perturbation. To say that $\delta W[\underset{\sim}{\xi}] < 0$ for some $\underset{\sim}{\xi}$ means that there are states in a neighborhood of the equilibrium which have less energy than the equilibrium. Explicit expressions for $\delta W[\underset{\sim}{\xi}]$ will be given when needed (see Sections II-1-A and III-2-C).

4. Accreting magnetic compact objects

In this section, a short overview of the physics of accretion onto magnetic compact objects is presented. Much more detail may be found in F. K. Lamb's lectures and in some recent review papers,[8-12] in which an account is given of the numerous developments which have followed the pioneering works of the early 1970's.[13-15]

A. Overview

To date, accretion by magnetic compact objects has been considered mainly in the two limiting cases where the flow at large distances from the accreting star is either spherically symmetric or forms a geometrically thin viscous disk. In understanding these cases, it is helpful to define some characteristic length scales. The stellar radius is R_s. The accretion radius, R_a, is the radius at which the flow starts to be influenced by the gravitational field of the star and the magnetospheric radius, R_m, is the radius at which the flow starts to be influenced by the stellar magnetic field. We will assume hereafter that $R_m >> R_s$.

The first case occurs if the angular momentum carried by the accreted matter is negligible compared to its Keplerian angular momentum with respect to the star and if $R_a >> R_m$. The flow will then be roughly spherically symmetric between R_a and R_m, and these radii may be estimated by writing $GM/R_a \sim (v^2 + c_s^2)(R_a)$ and $B^2(R_m)/8\pi \sim \rho(v^2 + c_s^2)(R_m)/2$, respectively. This type of flow might be produced, for example, when the star is located in the interstellar medium or in the wind of a companion.

In spherically symmetric accretion, the field of the compact object, which has been pushed in by the conducting infalling plasma, is confined to a cavity of average radius of order R_m. As a result of processes discussed in more detail below, the matter suspended above the field may be able to enter the magnetospheric cavity and flow towards the surface of the star. There is now an extensive literature on this problem.[6,17-28]

A thin accretion disk is formed in a binary system when a companion filling its Roche lobe transfers matter through the inner Lagrangian point. The matter is captured at once by the compact object, but, because of the large amount of angular momentum it carries, it is forced to orbit the object, forming a Keplerian disk. The angular momentum is removed by viscosity and similar processes, and the matter slowly spirals inward. The extracted angular momentum is eventually carried away by ejected matter or by tidal transfer to the binary motion. A disk (which is not necessarily thin) may also be formed in wind accretion when the infalling matter has acquired a large angular momentum from the binary motion.

In disk accretion, as in spherical accretion, the field is pushed inward by the infalling conducting plasma. In turn, the field exerts outward-directed forces on the disk, compressing it. At some radius the interaction between the disk and the magnetosphere becomes strong enough for the plasma to become turbulently mixed with the field. The magnetic field lines of the central object are then sheared by the disk motion, resulting in a braking of the plasma, which may eventually fall towards the star surface along the field lines.

Theories (or elements of a theory) for this type of accretion flow have been worked out by Ichimaru,[29] Scharleman,[30] Ghosh and Lamb,[31,32] Anzer and Börner,[33,34] Aly,[35] Kundt and Robnik,[36] Riffert,[37] Lipounov,[38,39] Lipounov and Shakura,[40,41] Lipounov et al.,[42] Horiuchi et al.,[43,44] Horiuchi.[45]

In some binary systems containing a disk, the magnetic field may be strong enough for direct interaction between the compact object and its companion to have non-negligible effects. In this case, the interaction between the two stars may arise "at a distance", or through field-aligned currents flowing in the magnetosphere. Different models for the magnetic interaction in such a system have been considered by Joss et al.,[46] Lamb et al.,[47] Campbell,[48] Chanmugam and Dulk.[49]

Transfer of matter through a thin disk and spherically symmetric inflow represent two extreme modes between which a wide range of accretion flows is possible. But, because of their relative simplicity, they have received most attention from theoreticians. The situation where both types of accretion occur simultaneously has also been investigated.[38,39]

B. Theoretical problems

The space around an accreting magnetic compact object may be roughly divided into three regions of interest:

- an outer accretion flow,
- an inner magnetospheric accretion flow,
- a magnetospheric quiescent (?) region.

The last two regions constitute the magnetosphere, i.e., the region where the magnetic field dominates the motion of the plasma.

(i) To explain the high luminosity of objects like the x-ray pulsars, and the existence of a pulsation in their radiation, it is necessary to assume that the accreted matter enters the magnetosphere at a high rate and gets threaded by the magnetic field. The matter may then fall along the field lines toward the magnetic poles; pulses result because of the localized character of the emission region and the rotation of the star. The main problem here is that an initially field-free region of plasma can get threaded by the stellar lines only by microscopic diffusion (Section 3-B), and this process, by itself, is inadequate to account for the observed luminosity. The large length scale, ℓ, of the plasma regions implies a very long diffusion time, $\tau_D = \ell^2/D_B$, where $D_B = \eta c^2/4\pi$ is the coefficient of diffusion of the magnetic field.

Plasma entry into the magnetosphere may be aided by the development of a large-scale MHD instability. Two such instabilities have been proposed in this context: the Rayleigh-Taylor instability (RTI), which arises when a dense fluid lies above a lighter one (or a magnetic field) in a gravitational field, and the Kelvin-Helmholtz instability (KHI) which can arise at the interface between two fluids in relative motion. It must be noted, however, that these instabilities cannot, by themselves, produce a threading of the plasma by the field. However, by increasing the surface of contact between the plasma and the magnetosphere by breaking the plasma into smaller pieces, they may greatly enhance the rate at which diffusive mixing occurs.

Another mechanism which allows plasma to enter the magnetosphere and become threaded by the stellar field (actually, it does both at once) is magnetic reconnection, a process whereby two regions with initially topologically different field lines can become magnetically connected in a characteristic time much shorter than the diffusion time. Threading may also be favored if the microscopic diffusion is assisted by some form of turbulent diffusion.

These processes will be considered in detail in the next lecture. The regions where they play a role in the interaction between the magnetosphere and the outer accretion flow are shown in Fig. 3.

(ii) Let us now suppose that we have two distinct regions of plasma: the star, V_1, and a second region, V_2, which may be part of a disk, the companion, the boundary of a closed magnetospheric cavity, etc. Consider what happens if the matter in V_2 is threaded by the stellar field lines, but is not able to fall toward the star along these lines. Because of the motion of the plasma in V_1 and V_2, the field in the magnetosphere evolves. If there is some plasma on the lines, currents are induced, and, if the density is not too large, these currents flow along the field lines between V_1 and V_2. The currents may close across the lines within V_1 and V_2, exerting Lorentz forces on these regions. This type of situation, in which two regions are linked by a "force-free" magnetosphere and exchange momentum and energy by means of field-aligned currents, is the subject of Lecture III.

(iii) A much more complicated problem arises if plasma is allowed to flow in the magnetosphere between the outer accretion flow region and the star. In that case, energy and angular momentum are transported not only by magnetic stresses, but also by the flow of the plasma itself. Near the compact object, we expect the magnetic field to dominate the plasma motion completely (so the field is nearly force-free). However, near the region where the inner magnetospheric flow meets the outer flow, the energies in the plasma and in the field may be comparable, and the existence of the flow has to be taken into account in determining the structure of the field. One expects to have transitions between sub-c_k and super-c_k flows in that region, where c_k is any velocity of

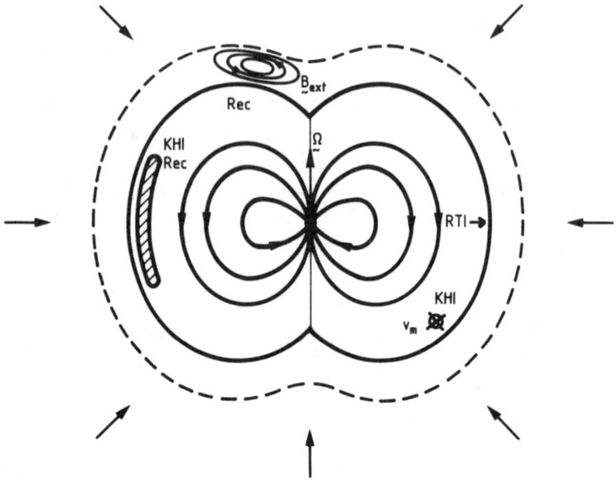

(3a)

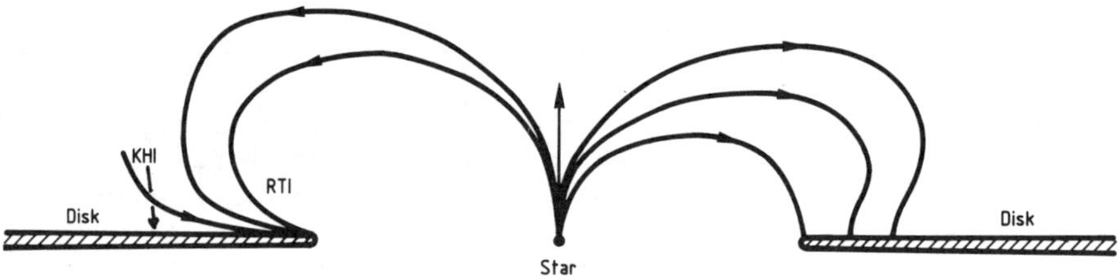

(3b)

<u>Figure 3. Possible magnetospheric configurations around accreting magnetic compact objects (not to scale!).</u>
Figure (3a) shows the case of spherical accretion, in which a closed cavity confining the magnetic field is formed by the infalling plasma. A filament of matter which has passed through the boundary is shown on the left. Some places where Kelvin-Helmholtz (KHI) or Rayleigh-Taylor (RTI) instabilities, and reconnection (Rec) may play a role are indicated.
Figure (3b) shows the case of disk accretion. Lines which do not thread the disk are shown on the left, while threading is assumed on the right. The magnetic field is shown to extend up to the companion, with which it may interact significantly in some cases. KHI, RTI, Rec, are as above.

propagation of a perturbation (see Section 3-C), and these transitions may generate shocks and instabilities. All of these problems will be examined, if not solved, in Lecture IV.

Before considering these particular topics, it is necessary to check that the tool we intend to use for this exploration, i.e., the MHD theory introduced in Sections 1-3, is adequate for our purposes.

C. The MHD approach

The applicability of an MHD description of a phenomenon depends on the ratios, ℓ/ℓ_{pl} and τ/τ_{pl}, of the typical length- and time-scales, ℓ and τ, of this phenomenon to the characteristic dimensions, ℓ_{pl} and τ_{pl}, of the region of plasma in question. We cannot consider here all the possible values of ℓ, τ, ℓ_{pl} and τ_{pl} that might be encountered in the study of accreting magnetic compact objects, so we will just discuss a few situations which may occur in the magnetosphere of a strongly magnetized ($B_{surface} \approx 10^{12}$ G) one-solar-mass neutron star accreting at a rate of around 10^{17}g s^{-1} (typical of X-ray pulsars); see Table I.

As a first example, we consider, in a system undergoing spherical accretion, the region between the magnetopause (surface of separation between the magnetosphere and the outer accretion flow) and the accretion shock a distance δ above it.[16,17,19,20] The values given for the parameters are just illustrative; the actual values in a specific system may differ by as much as one order of magnitude. The typical time scale is the free-fall time. The plasma parameters involving the magnetic field have been computed assuming an internal field equal to α times the external magnetospheric field. The temperature is some average between the ion and electron temperatures, which may differ.

In the second example, typical values corresponding to the inner part of an accretion disk in the magnetosphere of the neutron star are shown. The disk has a thickness δ, a Keplerian velocity v_K, and the typical time is the Keplerian time. The field inside the disk is αB_m ($\alpha = 1$ if the disk is threaded by the stellar field lines).

In the last example, we assume that plasma is outflowing at the sound speed from a ring of width δ at the inner edge of the previous accretion disk. The parameters correspond to the flow just above the surface of the disk. The typical value of δ is taken to be of the order of the value given by Ghosh and Lamb's[31] model. The plasma is threaded by the stellar field and B has the the magnetospheric value.

It is clear from the table that all the microscopic lengths (except for the mean free paths of the particles, in some cases) are small compared to the characteristic macroscopic dimensions, and the microscopic times are small compared to the macroscopic evolutionary time scale. A fluid theory of the large-scale phenomena is thus well justified. Phenomena involving smaller length- and time-scales, such as short-wavelength instabilities, may also be considered within the framework of MHD, although it may

Table I. Characteristic parameters in x-ray pulsar magnetospheres (cgs units).

Parameter	Definition	Magnetopause	Disk	Accretion Flows
distance to the star	r (cm)	10^8	10^8	10^8
thickness	δ (cm)	10^7	10^6	10^7
shortest typical time	τ (s)	$4.\ 10^{-2}$	$2.8\ 10^{-2}$	$\sim 2.8\ 10^{-2}$
density	n (cm^{-3})	10^{15}	10^{20}	10^{16}
temperature	T (K)	10^9	10^6	10^6
magnetospheric field	B (G)	10^6	10^6	10^6
Debye length	$\lambda_o = (3kT/4\pi ne^2)^{1/2}$ (cm)	$1.2\ 10^{-2}$	$1.2\ 10^{-6}$	$1.2\ 10^{-4}$
Λ	$\Lambda = 24\ \pi n\lambda_D^3$	$1.3\ 10^{11}$	$1.3\ 10^4$	$1.3\ 10^6$
thermal e-gyroradius	$r_{ce}^t = m_e c c_e^t/eB$ (cm)	$3.2\ 10^{-4}\ \alpha^{-1}$	$3.8\ 10^5\ \alpha^{-1}$	$3.8\ 10^{-5}$
thermal p-gyroradius	$r_{cp}^t = m_p c c_p^t/eB$ (cm)	$5.2\ 10^{-2}\ \alpha^{-10}$	$1.6\ 10^{-3}\ \alpha^{-1}$	$1.6\ 10^{-3}$
e-p-mean free path	$\lambda_{ee} = \lambda_{pp} = c_p^t/\nu_{pp}$ (cm)	10^7	$3.9\ 10^{-4}$	3

Table I (continued).

Parameter	Definition		Magnetopause	Disk	Accretion Flows
plasma frequency	$\omega_p = (4\pi ne^2/m_e)^{1/2}$	(s^{-1})	$1.8\ 10^{11}$	$5.6\ 10^{14}$	$5.6\ 10^{12}$
e-cyclotron frequency	$\omega_{ce} = eB/m_e c$	(s^{-1})	$1.8\ 10^{13}\ \alpha$	$1.8\ 10^{13}\ \alpha$	$1.8\ 10^{13}$
p-cyclotron frequency	$\omega_{cp} = eB/m_p c$	(s^{-1})	$9.6\ 10^{9}\ \alpha$	$9.6\ 10^{9}\ \alpha$	$9.6\ 10^{9}$
e-p collision frequency	$\nu_{ep} = 6\omega_p(\log\Lambda/\Lambda)$	(s^{-1})	$2.1\ 10^{3}$	$2.5\ 10^{12}$	$3.7\ 10^{8}$
p-p collision frequency	$\nu_{pp} = \nu_{ep}(m_e/2m_p)$	(s^{-1})	$3.5\ 10$	$4\ 10^{10}$	$6\ 10^{6}$
electric conductivity	$\sigma = \omega_p^2/(4\pi\nu_{ep})$	(cgs)	$1.2\ 10^{18}$	$1.\ 10^{16}$	$7.7\ 10^{15}$
diffusion time	$\tau_B = (4\pi\sigma\delta^2/c^2)$	(s)	$1.7\ 10^{12}$	$1.4\ 10^{8}$	$1.1\ 10^{10}$
kinematic viscosity	$\nu = (3kT)^{5/2}(e^4 m_p^{1/2} \mathrm{Log}\Lambda)$	(cgs)	$1.5\ 10^{15}$	$1.3\ 10^{3}$	$0.9\ 10^{7}$
viscous time	$\tau_v = \delta^2/r$	(s)	$7\ 10^{-2}$	$8\ 10^{8}$	$1.1\ 10^{7}$

The values given for the conductivity and the viscosity correspond to B=0; when $\omega_{c\alpha}/r_\alpha >> 1$, these quantity are anisotropic, and some of the components of the corresponding tensor are much less than the values given here.

be necessary in some situations (in particular at the magnetopause) to appeal to the astrophysicist's usual justification of invoking turbulence or small-scale magnetic fields to isotropize the distribution functions.

From the last two lines of Table I it is evident that dissipative effects may play a role only on very small scales. In most situations, they can be neglected, along with the non-resistive terms in the RHS of Ohm's law (26), and ideal MHD applies. In some cases (e.g. accretion disks), the dissipative terms must be retained, but they should be reinterpreted as being due not to microscopic processes, but to macroscopic turbulence. These terms are very important, as they allow for the existence of evolving disks (viscous terms) and may control the interaction of disks with the magnetospheres of compact objects (see F. K. Lamb's lectures).

D. Force-free magnetospheres

From the inferred values of the accretion rate and the field of the compact star, it has been possible to obtain estimates for the plasma characteristics in the outer and inner (magnetospheric) parts of the accretion flow. Similar estimates are much more difficult to establish in other parts of the magnetosphere, where the source of plasma is rather uncertain.

These regions cannot be pure vacuum. If they were, it is well known[50] that huge electric fields would develop, with non-zero components, $E_\parallel$, along the field lines, and electric charges would be electrostatically pulled out from the surface of the star and the plasma region threaded by the field lines. (For an instance in which this region is a disk, see Michel and Dessler.[51]) Charge would be sucked into the magnetosphere until the charge density became high enough to cancel $E_\parallel$ and make $\tilde{E} + \tilde{v} \times \tilde{B}/c = 0$. Using Eq. (4), this gives a minimum density in the rotating magnetosphere equal to (Goldreich and Julian[50])

$$n_{GJ} \approx \frac{\tilde{\Omega} \cdot \tilde{B}}{4\pi ce} \approx 3 \times 10^5 \ \Omega \ B_6 \ \mathrm{cm}^{-3} \quad , \qquad (52)$$

where $B_6 = B/10^6$G and Ω is the angular frequency of the plasma on a line. If the system is stationary, Ω must be equal to the angular frequency of the star, Ω_s; at a point on a line undergoing differential shearing, Ω may be between Ω_s and the Keplerian frequency, Ω_K, of the matter in the disk. This minimum density corresponds to a charge-separated magnetosphere and is quite low. Even if we assume that all the particles are moving at the speed of light, the maximum value of the magnetic perturbation δB due to their current is of the order of $\delta B/B \sim r\,\Omega/c \ll 1$.

Actually, non charge-separated plasma may be in equilibrium only in the part of the magnetosphere where the centrifugal force acting on a particle corotating with the star exceeds the gravitational force. In that case, plasma trapped in the magnetosphere, with the centrifugal force balancing the tension of the lines, may reach densities up to the critical value

$$n_{cr} \sim B^2/(4\pi m_p r^2 \Omega^2) \sim (c/r\Omega)^3 (r/r_{cp})\, n_{GJ} \qquad (53)$$

($r_{cp} = m_p c^2/e\,B \approx 10^{-3}/B_6$ cm), for which the kinetic energy density of the plasma is comparable to the magnetic energy density. If plasma is injected continuously, the field lines open when n reaches n_{cr} and a wind is produced (see Fig. 4). In the part of the magnetosphere which is dominated by gravity, no large-scale equilibrium is possible, gravity acts as a vacuum cleaner, and plasma must be continuously replenished by some source.

The details of all the processes which could populate the magnetosphere have never been worked out, and, therefore, we can only conjecture about the density of plasma in order to compute the magnetospheric structure. We will assume here (and in Lecture III, which is completely devoted to this problem) that the density of plasma is larger than the n_{GJ} $(c/r\Omega)$ necessary to support currents able to create $\delta B/B \sim 1$. This seems fairly reasonable as, in an accreting system, there is likely to be a large amount of diffuse plasma, some part of which may penetrate the magnetosphere; also, plasma may be ejected from the main stream of accreting matter into the magnetosphere or be produced inside the magnetosphere by the interaction of γ-rays with the magnetic field. We will assume that its density is not large enough for the matter stresses to distort

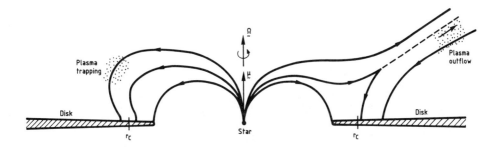

Figure 4. Plasma in the corotating magnetosphere of a disk accreting compact object.

For densities smaller than n_{cr} given by Eq. (52), plasma trapping (left side of the figure) is possible in the outer part of the magnetosphere, where centrifugal forces dominate gravitational ones. If too much plasma is injected into the magnetosphere, magnetic stresses are no longer able to balance the centrifugal force and some plasma must be ejected. This implies an intermittent opening of the field lines (right side).

the stellar field appreciably. In the case of disk accretion, this requires that there be no appreciable outflow from the surface of the disk, as in Anzer and Börner,[33,34] and Arons, McKee and Pudritz (work quoted in Arons et al.[12]) (see Section II-3-C), or at least that this outflow, which opens the field lines, is quite intermittent. In these circumstances, the magnetosphere will assume a force-free configuration, i.e. the distortion of the field relative to the potential configuration will be created primarily by field-aligned currents generated by shearing motions in the regions where the field lines are anchored.

LECTURE II
SOME BASIC MHD PROCESSES

We saw in the previous lecture that large inflows of plasma into a magnetosphere, as well as high-rate mixing of the plasma with the magnetic field, require large-scale MHD processes. In this lecture, we study those MHD processes which are most important to accreting compact objects. In Section 1, we consider the Rayleigh-Taylor instability, in Sections 2 and 3 the Kelvin-Helmholtz instability, and in Section 4 the resistive reconnection between the fields carried by adjoining regions of plasma.

1. The Rayleigh-Taylor instability (RTI)

The RTI is an instability which develops when a heavy fluid rests on top of a lighter one in a gravitational field. A similar instability arises if the light fluid is replaced by a magnetic field. This is the situation we will consider here. A great deal of work on RTI may be found in Chandrasekhar.[53]

A. Physical mechanisms and conditions for instability

(a) Let us first consider the simple situation in which a plasma occupying the half-space $\{z>0\}$ is supported against the action of a gravitational field, $\mathbf{g} = -g\hat{z}$, by a magnetic field, $\mathbf{B} = B\hat{y}$, which is uniform in $\{z<0\}$ and zero outside (Fig. 5a). From the discontinuity relation (I-42), $p(0) = B^2/8\pi$, where $p(z)$ is the

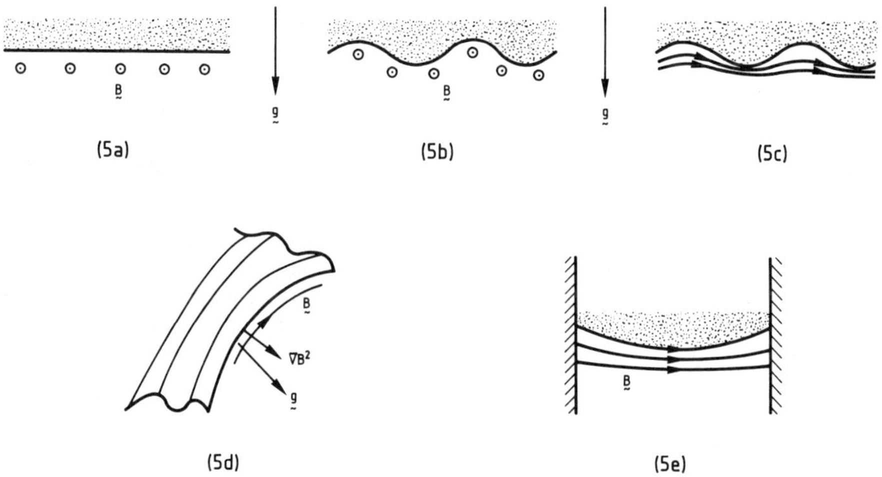

Figure 5. Various situations considered in the physical analysis of RTI in Section 1-A.
Plane situation: unperturbed (a), perturbed with $\mathbf{k} \cdot \mathbf{B} = 0$ (b), perturbed with $\mathbf{k} \cdot \mathbf{B} \neq 0$ (c); perturbed curved interface (d); perturbed plane line-tied situation (e) (in this case, there is necessarily a bending of the lines).

pressure of the plasma.

Let us now suppose that the boundary {z=0} is slightly deformed into a sinusoidal surface, $z = \eta \sin kx$ (Fig. 5b), and assume that the plasma is incompressible (this is a good approximation if the wavelength, $\lambda = 2\pi/k$, and frequency, ω, of the perturbation are much smaller than the scale height of the plasma and kc_s, respectively) and a perfect fluid. In the perturbed configuration, the magnetic energy is the same as in the initial one (the field lines are still straight), while the potential energy is lower; therefore there are states in a neighborhood of the equilibrium state which have lower energy and the equilibrium is unstable.

Another way of looking at the instability is to compute the force acting on a cylindrical volume element bounded by the interface and a parallel surface at $z=\varepsilon$. From the equations for an incompressible fluid ($\nabla \cdot \underset{\sim}{v} = 0$; $\rho\, \partial \underset{\sim}{v}/\partial t + \nabla \delta p = 0$, one finds that the pressure perturbation, δp, is a harmonic function, $\Delta \delta p=0$, and is given by $\delta p = \delta p(0)e^{-kz}\sin kx$ (one imposes boundedness on δp), where $\delta p(0) = \eta\rho g$ and ρ is the density of the medium. The resulting "buoyancy" force, $-k\eta\rho g\varepsilon \sin kx \neq 0$, produces the instability, which grows as $e^{\omega t}$, with $\omega = \sqrt{kg}$.

The buoyancy force may be balanced if the perturbation has a component of the wave number, $\underset{\sim}{k}$, parallel to $\underset{\sim}{B}$. In that case, there is a force, $\varepsilon\eta \sin(\underset{\sim}{k}\cdot\underset{\sim}{x})\, (\underset{\sim}{k}\cdot\underset{\sim}{B})^2/4\pi$, resulting from the tension in the lines (Fig. 5c) and the condition for instability is

$$k\rho g > \frac{(\underset{\sim}{k}\cdot\underset{\sim}{B})^2}{4\pi} . \tag{1}$$

Thus, short-wavelength perturbations can be stabilized if $\underset{\sim}{k}$ is not exactly normal to $\underset{\sim}{B}$.

(b) We now consider the case in which the boundary between the plasma and the field is curved. For a perturbation with $\underset{\sim}{k}$ normal to $\underset{\sim}{B}$ (Fig. 5d), an element which moves into the field must do work against the increasing magnetic pressure if the boundary is convex towards the plasma. This has a stabilizing effect that overcomes the effects of gravity when

$$\xi\hat{n} \cdot \rho \underset{\sim}{g} > \xi\hat{n} \cdot \nabla \frac{B^2}{8\pi} = \xi\, \kappa \frac{B^2}{4\pi} , \tag{2}$$

where $\xi\hat{n}$ ($\xi>0$) is the displacement of the boundary, κ is the curvature of a line on the boundary (the last equality comes from Eq. (39)) and $\hat{n}$ is the normal to the field. If the boundary is concave towards the plasma, the magnetic term is destabilizing and its effect adds to that of the gravitational field.

The above argument may be made more rigorous by using the "energy principle" of Section I-3-D. In the situation considered here, δW is the sum of three terms: a plasma region term (δW_{pl}), a field region term (δW_v) and a term that includes the effect of the interface between the two regions (δW_s). δW_{pl} is positive if $\underset{\sim}{g} \cdot \nabla s < 0$ (s is the specific entropy), which is the usual condition for the stability of an atmosphere against convection (but this possible source of instability is independent of the magnetic field and is not of interest here). δW_v is always ≥ 0, while δW_s can be written as

$$\delta W_s = -\frac{1}{2}\int_{\text{interface}} (\hat{n} \cdot \underset{\sim}{\xi})^2 \, \hat{n} \cdot [\nabla p + \nabla B^2/8\pi] \, d\sigma \; . \tag{3}$$

The analysis of δW yields a result similar to that obtained with our first simple argument (Eq.(2)). Using $\nabla p = \rho \underset{\sim}{g}$ and $\nabla B^2 = 2\kappa B^2 \hat{n}$, one obtains the following condition for instability:

$$\hat{n} \cdot \nabla p = \rho \, \hat{n} \cdot \underset{\sim}{g} > \hat{n} \cdot \nabla \frac{B^2}{8\pi} = \frac{\kappa B^2}{4\pi} \; . \tag{4}$$

(c) One effect which was ignored in the previous analysis is "line-tying". In the situation in which we are interested, the "feet" of the lines of the magnetospheric field are firmly anchored in the heavy conducting matter of the stars and cannot move freely. Because the magnetosphere is also conducting, some perturbations cannot be realized without bending the lines. This contributes a term to δW_v, which then may be >0, leading to a stabilizing effect.

The magnitude of the line-tying effect may be estimated quite simply by considering the model used in (a) and adding two conducting heavy plates, located a distance L apart in two planes normal to $\underset{\sim}{B}$ (Fig. 5e). In this case, it is no longer possible to effect the perturbation introduced at the beginning (a), because we must now have $\underset{\sim}{\xi} = 0$ on the plates. To satisfy this condition, the lines must be bent, and stability may result if

$$\frac{\pi B^2}{4L^2} > \rho g k \; , \tag{5}$$

i.e. for long wavelengths (this is the same relation as (1)).

(d) So far, we have neglected all possible non-ideal MHD effects. Let us now briefly consider the consequences of two of them: viscosity and finite Larmor radius effects (for details, see Elsner[19]). Viscosity does not change the instability criteria, but it affects the growth of perturbations with wave number larger than $k_* = (g/\nu_c^2)^{1/3}$ (ν_c is the kinematic viscosity), reducing the rate by a factor of $(k_*/k)^{3/2}/2$ under the conditions of interest here. Note that ν_c can be greatly reduced by the field itself. In contrast, finite Larmor radius effects may stabilize the RTI if there is an embedded field and if $k > k_{FLR} \approx g^{1/2} r_{cp}^{-4/3} \omega_{cp}^{2/3}$, where r_{cp} is the ion Larmor radius and ω_{cp} is the ion Larmor frequency (based on the embedded magnetic field).

B. Non-linear development of the RTI

The linear theory presented above predicts the conditions which allow the RTI to grow from small initial fluctuations, and permits a description of its development so long as the amplitude of the perturbation remains small (i.e. so long as $k\eta \ll 1$). When this is no longer the case, non-linear terms play an important role, and the problem in general becomes analytically intractable. One then has to turn to physical analysis and intuitive guesses, or to extensive numerical simulations.

The first attempts to describe the non-linear stage of the RTI in the context of accreting magnetic compact objects were proposed by Arons and Lea,[16] who used analogies with some well-studied laboratory situations. In Arons and Lea's picture, the evolution of an initial perturbation with $\underset{\sim}{k} \cdot \underset{\sim}{B} = 0$ results in the formation of plasma "spikes" which are accelerated downward, while magnetic "bubbles" rise. The ascending motion of the latter is stopped by the weight of the slowly falling plasma above, and a magnetic piston effect develops, cutting the neck of the spikes, and allowing them to fall freely through the magnetosphere as "melon seed" diamagnetic filaments.

More quantitative investigations of this problem, based on 2-D numerical calculations, have been reported recently by Wang and Nepveu[54] and Wang, Nepveu and Robertson.[55] The calculations presented in the former paper deal with the evolution of a symmetric perturbation at the interface between a non-magnetic and a magnetic or non-magnetic medium. The infalling plasma is found to have a tendency to swirl out and form elongated trailing loops. The swirling motions are conjectured to lead to large-scale mixing between the accreting plasma and the magnetospheric field, which could favor the latter threading the former by diffusive effects.

In the second paper, the evolution of non-symmetric perturbations at the interface between two field-free regions is considered. The condition $\underset{\sim}{B} = 0$ is used to simplify the calculations; it is argued that the results should not be very different when $\underset{\sim}{B} \neq 0$ in one of the media, insofar as magnetic tension does not come into play. It is shown that mushroom-shaped structures of rising light fluid, or of falling heavy fluid, are formed by complementary circulatory motions and become extremely elongated after several linear growth times (Fig. 6). Short-wavelength modes are found to dominate the long ones, which otherwise could transport mass more efficiently; the latter seem to have difficulty developing in a medium strongly perturbed by the short modes. It appears that the size of the pockets of falling matter is controlled by effective viscous damping. (See, however, the discussion of this problem by F. K. Lamb in these proceedings, where the opposite conclusion is reached.)

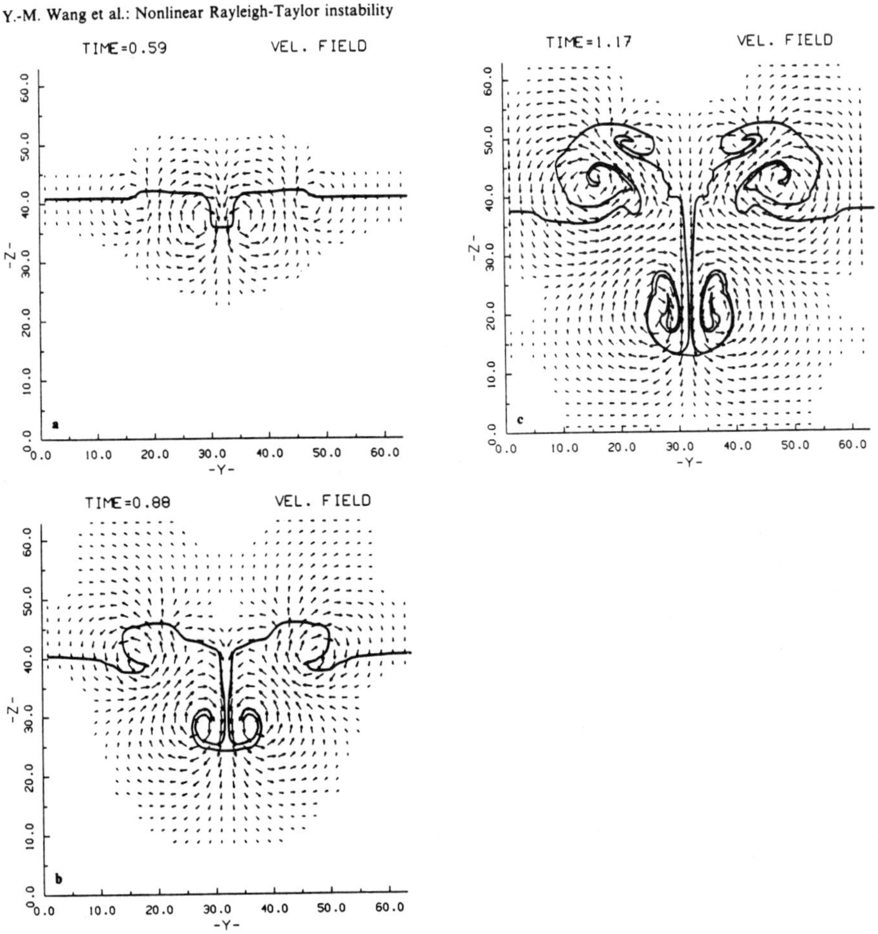

Figure 6. Non-linear development of the RTI.
Non-linear development of the RTI according to the numerical simulations by Wang, Nepveu and Robertson[55] (reprinted by kind permission of the authors and "Astronomy and Astrophysics").

2. The Kelvin-Helmholtz instability (KHI): Physical Mechanisms

The Kelvin-Helmholtz instability may develop at the interface between two fluids in relative motion (see, for example, Gerwin,[56] and references therein). It can be understood by considering the simple situation in which the two fluids are incompressible and field-free ($\mathbf{B} = 0$). We will consider this case first; then the compressibility of the plasma, the presence of magnetic fields, and some non-ideal MHD effects will be taken into account.

A. Incompressible fluids ($\mathbf{B} = 0$) (e.g. Chandrasekhar[53])

Consider the following equilibrium situation. Fluid 1 (2), of density $\rho_1(\rho_2)$, fills the half-space {z>0} ({z<0}) and flows with velocity $\mathbf{v}_1$ ($\mathbf{v}_2$). Both fluids are incompressible and we assume $\mathbf{B} = 0$ and $\mathbf{g} = 0$. Suppose now that a disturbance of wave number $\mathbf{k} = k\hat{x}$ is propagating along the "vortex sheet" {z=0}, making its shape sinusoidal, with $z = \eta \sin kx$. In the rest frame of this wave, consider the force acting on an infinitesimal volume element bounded by two surfaces parallel to the perturbed boundary and located at a distance ε on either side of it. As remarked in Section 1-A, the pressure is harmonic ($\Delta p = 0$) inside each fluid and, in addition, is continuous on the separating surface. Because $\mathbf{g} = 0$, the pressures are equal on either side of the element. There is thus no net pressure force on the cylinder, but there is a centrifugal force, which acts in the same direction on either side of the boundary, making the wave grow.[57] The configuration is thus unstable.

Another possible way of looking at the instability is as follows. Again consider what happens in the frame of a wave, but now suppose that the perturbed boundary is rigid, and that the motions on either side are stationary and satisfy Bernoulli's law, $v_i'^2/2 + p_i/\rho_i = c_i^{te}$, where v_i' is the velocity in the wave frame. By conservation of matter, the velocity near the boundary must be larger than average above the crests and smaller above the troughs. The pressure is therefore increased over a trough and decreased over a crest; the net force is thus in the direction to drive the wave to greater amplitude.

The effect of a gravity field, $\mathbf{g} = -g\hat{z}$ ($g>0$), is easily included in the argument. There is a buoyancy force, $\varepsilon(\rho_1 - \rho_2)\, kg\eta \sin kx$, acting on the element and the centrifugal force is $\varepsilon[\rho_1(\mathbf{k}\cdot\mathbf{v}_1')^2 + \rho_2(k\cdot v_2')^2]\eta \sin kx$. One sees easily that there is a wave velocity, $\mathbf{v}_0$, for which balance between the two forces, and hence stability, is possible if $\rho_2 > \rho_1$ and

$$\rho_1\rho_2[(\mathbf{v}_2 - \mathbf{v}_1)\cdot\mathbf{k}]^2 \leqslant (\rho_1 + \rho_2)(\rho_2 - \rho_1)kg \quad . \qquad (6)$$

Thus, for $\rho_2 > \rho_1$, the long-wavelength modes are stabilized by gravity. For $\rho_2 < \rho_1$, the RTI augments the KHI.

The above argument furnishes the conditions under which the KHI may appear. To obtain the characteristic timescale, t_{KH}, of its development, it is necessary to solve the equations of motion. The method is classical. One looks for solutions of the form $f(\underset{\sim}{r}_\perp, z, t) = \tilde{f}(z)\exp\{i(\underset{\sim}{k}\cdot\underset{\sim}{r}_\perp - \omega t)\}$ (where $\underset{\sim}{r}_\perp = (x,y)$ and f is any physical variable) in each region, and imposes continuity of pressure and normal displacement at the interface {z=0} and a "radiation condition" at infinity (all perturbations are produced at the interface {z=0}). One finds that (i) $\tilde{f}(z) = e^{-k|z|}$, so a perturbation of wave number $\underset{\sim}{k}$ is confined to a layer of thickness k^{-1} on either side of {z=0}; and (ii) non-trivial solutions may exist only if ω and $\underset{\sim}{k}$ satisfy the dispersion relation:

$$\omega = \frac{\rho_1(\underset{\sim}{k}\cdot\underset{\sim}{v}_1) + \rho_2(\underset{\sim}{k}\cdot\underset{\sim}{v}_2)}{\rho_1 + \rho_2} \pm i\left[\frac{\rho_1\rho_2}{\rho_1 + \rho_2}[(\underset{\sim}{v}_2 - \underset{\sim}{v}_1)\cdot\underset{\sim}{k}]^2 - kg\,\frac{\rho_2 - \rho_1}{\rho_2 + \rho_1}\right]^{1/2} . \tag{7}$$

Note that, for $\rho_1 << \rho_2$, the unstable wave is nearly comoving with fluid (2) and its growth rate is quite small.

B. <u>Compressible fluids</u> ($\underset{\sim}{B} = 0$; $\underset{\sim}{g} = 0$)

Now let the two plasmas, 1 and 2, be compressible and consider, as above, their flows in the frame of the perturbation.[58] Assuming stationarity and rigidity of the wavy boundary (equation $z = \eta \sin kx$, again), one obtains from the linearized equations of motion ($\rho_i \nabla\cdot\delta v'_i + \underset{\sim}{v}'_i\cdot\nabla\rho_i = 0$; $\rho_i(\underset{\sim}{v}'_i\cdot\nabla)\delta\underset{\sim}{v}'_i + c^2_{si}\nabla\delta\rho_i = 0$, where δp_i, $\delta\rho_i$ and $\delta\underset{\sim}{v}'_i$ are the perturbations of the pressure, density and velocity of fluid i in the wave frame and $\delta p_i = c^2_{si}\,\delta\rho_i$),

$$\left[(1-M'_i)^2\frac{\partial^2}{\partial x^2} + \frac{\partial^2}{\partial z^2}\right]\delta p_i = 0 , \tag{8}$$

where $M'_i = |\underset{\sim}{k}\cdot\underset{\sim}{v}'_i|/kc_{si}$ and $\underset{\sim}{k} = k\hat{x}$. The solution to Eq. (8) is

$$\delta p_i = -\eta k\rho_i\, v'^2_{ix}\,(1 - M'^2_i)^{-1/2}\sin kx\; e^{-kz(1-M'^2_i)^{1/2}} \qquad (M'_i < 1) , \tag{9}$$

or

$$\delta p_i = \eta k\rho_i\, v'^2_{ix}\,(M'^2_i - 1)^{-1/2}\cos[k\{x - z(M'^2_i - 1)^{1/2}\}] \quad (M'_i > 1) . \tag{10}$$

The equations of motion have been used to determine the boundary condition on δp_i at {z=0} from the shape of the boundary, and the solutions have been required to be bounded ($M'_i < 1$) or to travel downstream (for $M'_i > 1$; this is a "radiation condition"). There is a clear difference between the subsonic and the supersonic cases. In the subsonic case ($M'_i < 1$), the pressure at the boundary is greatest on a trough and least at a crest while, in the supersonic case ($M'_i > 1$), it is zero at the crests and troughs, and alternately maximum and minimum at the zeros of z (Fig. 7).

Clearly, pressure balance at the interface is impossible if one of the flows is subsonic, but it may be possible if both flows are supersonic. A frame exists such that $\delta p_1 = \delta p_2$ at $\{z=0\}$ if (taking $\gamma_1 = \gamma_2$ for simplicity)

$$|\hat{x} \cdot (\underset{\sim}{v}_2 - \underset{\sim}{v}_1)| > (c_{s1}^{2/3} + c_{s2}^{2/3})^{3/2} \quad . \tag{11}$$

When $|\underset{\sim}{v}_2 - \underset{\sim}{v}_1| > (c_{s1}^{2/3} + c_{s2}^{2/3})^{3/2}$, compressibility can stabilize all perturbations whose angle, θ, with the velocity satisfies $\cos^2\theta > (c_{s1}^{2/3} + c_{s2}^{2/3})^{3/2}/|\underset{\sim}{v}_2 - \underset{\sim}{v}_1|$. Nevertheless, there are always unstable perturbations (those propagating at large angles to the velocity).

C. Magnetic effects

If magnetic fields, $\underset{\sim}{B}_1$ and $\underset{\sim}{B}_2$, are present in the fluids and are parallel to $\{z=0\}$ [B_z must be zero as a consequence of the discontinuity relations I (45) - (48)], new effects are introduced:

(i) stabilization by the magnetic tension,
(ii) modification of the compressibility of the plasma,
(iii) an increase in the inertia of the media.

The first effect is readily shown by considering the incompressible situation of Subsection A and the force acting on the small element of volume introduced therein (Fig. 8). The expression for that force must now contain a magnetic tension term, $\varepsilon\eta \sin kx\, [(\underset{\sim}{k} \cdot \underset{\sim}{B}_1)^2 + (\underset{\sim}{k} \cdot \underset{\sim}{B}_2)^2]/4\pi$. The force may be zero in one frame (i.e., one has stability) if[57]

$$4\pi\, \rho_1\rho_2\, [\hat{k}\cdot(\underset{\sim}{v}_2 - \underset{\sim}{v}_1)]^2 < (\rho_1 + \rho_2)\, [(\hat{x} \cdot \underset{\sim}{B}_1)^2 + (\hat{x} \cdot \underset{\sim}{B}_2)^2] \quad , \tag{12}$$

where we have taken $\underset{\sim}{g} = 0$ to simplify matters.

When the plasma is compressible, the situation becomes much more complicated. (The dispersion relation is a tenth-order equation.) Nevertheless, one expects relations of type (11) to hold, with c_s replaced by combinations of c_s and c_A (Section I-3-C). The analysis of this situation by Southwood[59] indicates that there is no configuration which is stable with respect to all possible perturbations.

To understand the origin of the last effect [(iii) above], consider the Poynting vector, $\underset{\sim}{P}$, in a frame moving with velocity

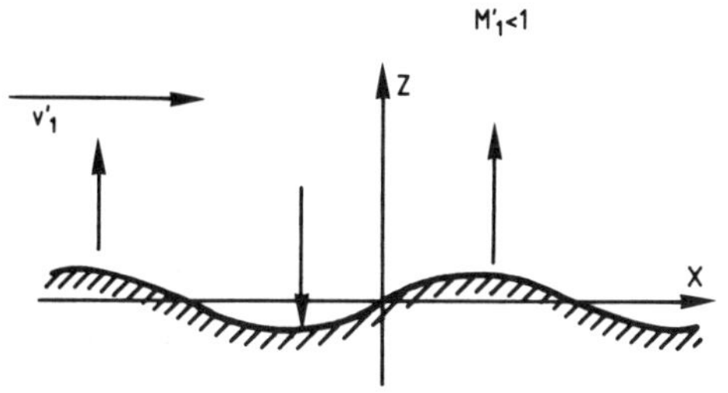

(7a)

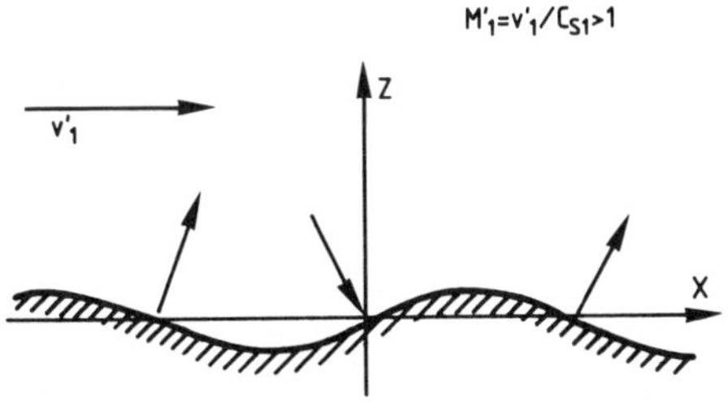

(7b)

Figure 7. Direction of the forces exerted by a compressible fluid on a wavy boundary.
Fig. 7a corresponds to a subsonic motion in the boundary frame, while Fig. 7b corresponds to a supersonic one.

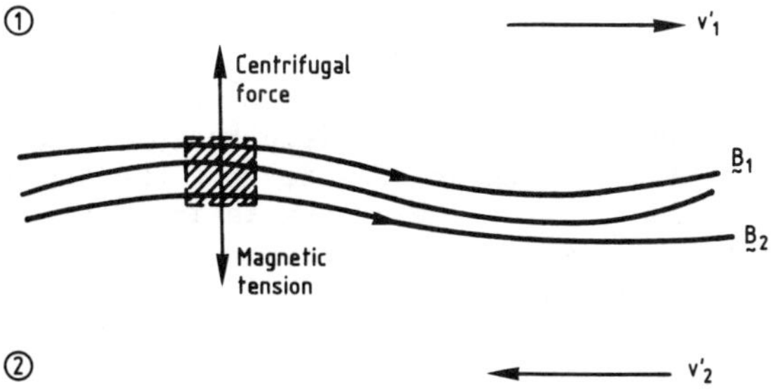

Figure 8. KHI in incompressible MHD.
By considering the force balance on the small element of volume, one may show the instability condition (12). KHI is driven by the centrifugal force acting on the element. Magnetic tension clearly has a stabilizing effect.

$\underset{\sim}{v}$ relative to the frame in which $\underset{\sim}{E} = 0$: $\underset{\sim}{P} = (B^2/4\pi c^2)\underset{\sim}{v}$. This expression is formally the relation between the momentum $\underset{\sim}{P}$, and the velocity, $\underset{\sim}{v}$, of a fluid of mass $B^2/4\pi c^2$, and may dominate the inertia of the medium if $B^2 \gg 4\pi\rho c^2$. This condition may be satisfied in an accreting magnetosphere; such a situation will be considered in detail in Section 3 below.

D. Other effects

The above models are very idealized ones. In particular, taking a sharp separation between the two plasmas may be too strong a hypothesis. If the vortex sheet is replaced by a transition layer of thickness δ, one finds that, as might be expected, the evolution of perturbations with $k\delta \lesssim 1$ is about the same as when $\delta=0$, but that perturbations with $k\delta \gg 1$ are stabilized (see, e.g. Chandrasekhar[53] for the incompressible case).

In the calculations just presented, the plasmas in both regions extend to infinity and the interface is plane. For actual regions of typical dimension ℓ, this finite size imposes an upper limit, of order ℓ^{-1}, on k.

So far, we have ignored possible non-ideal MHD effects. Viscosity may impede the development of short-wavelength perturbations for which ω and k in the ideal theory satisfy $k^2 \nu \gtrsim \omega$ (ν is the kinematic viscosity). In particular, if a medium is turbulent, eddies of size smaller than k^{-1} may provide an efficient viscosity mechanism. Finite Larmor radius effects may also stabilize those short-wavelength modes for which the ideal theory predicts $k^2 c_s r_{cp} \gtrsim \omega$.

3. The KHI in an accreting magnetosphere

Let us now consider the KHI at the interface between an accretion flow (disk, cavity boundary, falling plasmoid, etc.) and the low-density part of a magnetosphere. Our main assumption will be that

$$\frac{B^2}{4\pi\rho c^2} \gg 1 \tag{13}$$

in the magnetosphere; the plasma therein has no dynamical role, but is assumed to be dense enough to ensure that the magnetosphere corotates with the star when the system is in a stationary state. The analysis that follows in Subsections A and C is taken mainly from Aly, Ghosh and Lamb.[60]

A. Linear theory

a. First model: The system consists of two uniform parts, the magnetosphere (1) and the accretion flow (2), occupying the half-spaces $\{z>0\}$ and $\{z<0\}$, respectively. The velocity, $\underset{\sim}{v}_1$ ($\underset{\sim}{v}_2 = 0$), and the fields, $\underset{\sim}{B}_1$ and $\underset{\sim}{B}_2$, are parallel to the plane $\{z=0\}$. The plasma in (2) is taken to be incompressible and describable by ideal MHD and there is a gravitational field, $\underset{\sim}{g}_2 = -g\hat{z}$. We will define $a^2 = B_1^2/4\pi\rho_2$ and consider the case $a^2/c^2 \ll 1$. From the pressure balance condition, one has

$$a^2 = c_{A2}^2 + \frac{2}{\gamma} c_{s2}^2 \tag{14}$$

and $c_{A2} \lesssim c_{s2} \ll c^2$ (the first condition is necessary for the field $\underset{\sim}{B}_2$ not to be ejected by buoyancy forces.[61]

The ideal MHD linearized equations in the flow region and the magnetosphere can be solved if one assumes that any quantity f varies as $f(\underset{\sim}{r}_\perp, z, t) = \tilde{f}(z)\exp\{i(\underset{\sim}{k}\cdot\underset{\sim}{r}_\perp - \omega t)\}$ $[\underset{\sim}{r}_\perp = (x,y)]$. After imposing continuity of the pressure (thermal + magnetic) and the normal displacement on the boundary and the boundedness of the solution, one finds that (i) $\tilde{f}(z) = e^{-k|z|}$ and (ii) ω and $\underset{\sim}{k}$ must satisfy a dispersion relation:

$$\omega = \underset{\sim}{k} \cdot \underset{\sim}{v}_1 \frac{a^2}{c^2} \pm \{- (\underset{\sim}{k} \cdot \underset{\sim}{v}_1)^2 \frac{a^2}{c^2} + kg + \frac{1}{4\pi\rho_2} [(\underset{\sim}{k} \cdot \underset{\sim}{B}_1)^2 + (\underset{\sim}{k} \cdot \underset{\sim}{B}_2)^2]\}^{1/2} \quad . \tag{15}$$

For a given value of the real vector $\underset{\sim}{k}$, one has an unstable solution with Im $\omega > 0$ if

$$(\underset{\sim}{k} \cdot \underset{\sim}{v}_1)^2 \frac{a^2}{c^2} > kg + \frac{1}{4\pi\rho_2} [(\underset{\sim}{k} \cdot \underset{\sim}{B}_1)^2 + (\underset{\sim}{k} \cdot \underset{\sim}{B}_2)^2] \quad . \tag{16}$$

If $g > 0$, one obtains a first necessary condition for instability:

$$k > k_g = gc^2/a^2 v_1{}^2 \approx h^{-1}(c/v_1)^2 \quad , \tag{17}$$

where h is the scale height of the plasma in (2) near {z=0}. Because of the stabilizing effect of gravity, only short-wavelength perturbations are unstable. The necessity of overcoming the magnetic tension provides a further stabilizing effect and, in the worse case, where $\underset{\sim}{v}_1 \cdot \underset{\sim}{B}_1 = 0$, and $\underset{\sim}{B}_2 \cdot \underset{\sim}{B}_1 = 0$, one finds

$$\frac{v_1}{c} > \frac{B_2}{B_1} \tag{18}$$

for instability (see also Wang and Welter[62]). The perturbation develops in a characteristic time τ_{KH}. When the stabilizing terms are small, τ_{KH} is given by

$$\tau_{KH} \approx \frac{2\pi}{\underset{\sim}{k}\cdot\underset{\sim}{v}_1} \frac{c}{a} \quad . \tag{19}$$

In this simple model, we have neglected the compressibility of the plasma both in the determination of the stationary states and in the computation of the evolution of a perturbation. The first approximation limits the applicability of our results to perturbations having kh >> 1. This condition is satisfied for all the unstable modes we found [see Eq. (17)]. The second approximation turns out, a posteriori, to be well justified (see Subsection A-b), because the modes we obtained all satisfy $|\omega| \ll k(c_{s2}^2 + c_{A2}^2)^{1/2}$ in the frame of the plasma. Consideration of the complete dispersion relation shows that no other unstable modes exist.[62] [The first analysis of the KHI at the interface between a streaming plasma and a vacuum field was given by Northrop.[63] The effect of gravity was taken into account by Ghosh and Lamb[31] and Aly.[35] Arons and Lea[18] considered a nonzero ρ_1 and gave a relation which reduces to Northrop's when $B_1^2 \gg 4\pi\rho_1 c^2$ and to the usual incompressible MHD relation in the opposite limit; they assumed $\underset{\sim}{B}_2 = 0$, $\underset{\sim}{g} = 0$. The corresponding relation, including the effects of $\underset{\sim}{g}$ and $\underset{\sim}{B}_2$, was worked out by Aly, Ghosh and Lamb,[60] and (with $\rho_1 = 0$) by Wang and Welter.[62]]

b. *Second model*: We now consider a layer of plasma (2), of thickness $2h$, sandwiched between two parts, $\{z > h\}$ and $\{z < h\}$, of a magnetosphere (1). As in (a), we assume that $\underset{\sim}{v}_1$, $\underset{\sim}{B}_1$, and $\underset{\sim}{B}_2$ are parallel to $\{z=0\}$, $\underset{\sim}{v}_2 = 0$ and the plasma in (2) is incompressible and non-dissipative. However, we take $\underset{\sim}{g} = 0$ ($\underset{\sim}{g} \neq 0$ is easily included, but has no effect in the circumstances under which the model applies).

In this case, the solution of the dispersion relation is

$$\omega = r(kh) \frac{a^2}{c^2} (\underset{\sim}{k} \cdot \underset{\sim}{v}_1) \pm \left\{ -r(kh) \frac{a^2}{c^2} (\underset{\sim}{k} \cdot \underset{\sim}{v}_1)^2 \right.$$

$$\left. + \frac{1}{4\pi\rho_2} [r(kh) (\underset{\sim}{k} \cdot \underset{\sim}{B}_1)^2 + (\underset{\sim}{k} \cdot \underset{\sim}{B}_2)^2] \right\}^{1/2} , \quad (20)$$

where

$$r(kh) = \begin{cases} \coth(kh) & \text{for even modes, } f(z) = f(-z) \\ \tanh(kh) & \text{for odd modes, } f(z) = -f(-z) \end{cases} \quad (21)$$

and $a^2 \coth(kh) \ll c^2$ has been assumed for the even modes. There is instability if

$$\left(\frac{\underset{\sim}{k} \cdot \underset{\sim}{v}_1}{c}\right)^2 > (\underset{\sim}{k} \cdot \hat{B}_1)^2 + \left(\underset{\sim}{k} \cdot \frac{\underset{\sim}{B}_2}{B_1}\right)^2 \frac{1}{r(kh)} . \quad (22)$$

For short wavelengths ($kh \gg 1$), $r(kh) \approx 1$ and we have the same result as in (a), as expected. For long wavelengths ($kh \ll 1$), $r(kh) \simeq (kh)^{-1}$ for even modes and the effect of $\underset{\sim}{B}_2$ is diminished. For $\underset{\sim}{v}_1 \cdot \underset{\sim}{B}_1 = 0$ and $\underset{\sim}{B}_1 \cdot \underset{\sim}{B}_2 = 0$, the instability condition is

$$\frac{v_1}{c} > \frac{B_2}{B_1} (kh)^{1/2} . \quad (23)$$

The effect of $\underset{\sim}{B}_2$ is reinforced for odd modes.

B. Non-linear development

A description of the non-linear development of the KHI in the conditions considered here has been given by Arons and Lea.[18] This description is based on observations of the evolution of perturbations at the interface between a liquid and air (for example, at the surface of a falling drop of water, or at the surface of the ocean in a strong shear region).[64] In that case, a thin turbulent layer exists at the interface, and turbulent cells are entrained by the wind, forming a fine spray.

Arons and Lea argue that a similar phenomenon must occur at the unstable interface between an accretion flow and a magnetosphere. Fine drops of plasma are produced at the surface and become entrained by the magnetosphere. The rate at which plasma is removed from the flow is determined by the time necessary to break a KH wave. It can be described by a "mixing velocity" which, when $B_1^2 \gg 4\pi\rho_1 c^2$ and the stabilizing term in Eq. (15) is negligible, is given by

$$v_{mix} \approx \eta_{KH} \frac{\mathrm{Im}\,\omega_{KH}(k)}{k} \approx \eta_{KH} v_1 \frac{a}{c} \quad , \tag{24}$$

where η_{KH} is a phenomenological parameter, taken to be of order 0.1. The scale of the drops is of the order of the shortest possible unstable wavelength, which in most cases is likely to be determined by finite Larmor radius effects (which were neglected in our simple linear models). Since this dimension is small, the drops may quickly be threaded by the field.

This analysis presupposes that the plasma which evaporates from the flow can be rapidly removed from the neighborhood of the interface and that the field is not greatly distorted by the entrained matter. The validity of these assumptions can be verified for RTI-formed plasmoids falling between the lines of the inner magnetosphere of an accreting object (the Arons and Lea model[18] was initially devised to describe this situation) and for the interaction of a rotating closed magnetosphere with the plasma constituting the boundary. In these cases, the matter is removed by gravity and falls towards the star once it is mixed with the magnetic field.[28]

It is worth noting that, in quite a different context (the interaction of the solar wind with ionosphere of Venus), Wolff, Goldstein and Yeates[65] have introduced another scheme for the evolution of the KHI. In their description, the development of a KHI leads to the formation of magnetic flux ropes inside the dense region and of bubbles of dense plasma in the magnetic region (such flux ropes have been observed in Venus's ionosphere). The possibility of a similar effect in an accreting magnetosphere should be investigated.

C. KHI at the interface between an accretion disk and the magnetosphere

Consider an accretion disk around a magnetic dipole whose orientation is taken, for simplicity, to be perpendicular to the disk plane. We assume that the outer part of the disk, outside some radius r_{th}, is not threaded by the field of the central object. This part of the disk is effectively diamagnetic and the lines near it are parallel to its surface. The structure of the disk for $r > r_{th}$ may be determined if the magnetic field is known;[60] for instance, if $r_{th} \simeq r_i$, the internal radius of the disk, we can use the solution computed by Aly[35] (also Kundt and Robnik;[36] Riffert[37]) for the field around an infinitesimally thin perfectly diamagnetic disk.

In the determination of the structure of the diamagnetic portion of the disk, a characteristic radius is found to be important: it is the pressure-balance radius, r_{PB}, at which $p_c - p_s \approx B^2/8\pi$, where p_c and p_s are the plasma pressure at the mid-plane and the surface of the disk, respectively (of course, $p_s = B^2/8\pi$ for $r \geqslant r_{th}$). For $r \gtrsim \max(r_{th}, r_{PB})$, the influence of the magnetic field is negligible while, for $r_{th} \leqslant r \leqslant r_{PB}$ (if this region exists), it has a major effect: the disk, which was vertically confined essentially by the z-component of the star's gravitational field for $r > r_{PB}$, is confined instead by the external magnetic pressure. For $r \lesssim r_{PB}$, $p_c - p_s \ll B^2/8\pi$ and the pressure is roughly constant in the z-direction. However, an inward gradient of plasma pressure, $\partial p/\partial r \approx \partial(B^2/8\pi)/\partial r$, balances the centrifugal force and gravity. In the region $r_{th} \leqslant r \leqslant r_{PB}$, one thus has $\Omega < \Omega_K = (GM/r^3)^{1/2}$. This analysis shows that there is no reason for a diamagnetic disk to be destroyed at r_{PB}, as assumed by several authors,[30,33] who took $r_i = r_{PB}$. The minimum radius, at which the existence of the disk becomes impossible is, in fact, the radius where the magnetically induced pressure gradient balances the gravitational attraction. It can be deduced from the relations given by Aly et al.[60] for an α-disk.

Let us now consider in turn the stability of the two regions of the diamagnetic part of a disk. In the "classical" part (which is gravitationally confined in the z-direction), the interface may be subject to the KHI. We will use our first model (with $g = g_z$, $v_1 = r(\Omega_s - \Omega_K)$ and $\underset{\sim}{B}_1 \perp \underset{\sim}{v}_1$) to see if this is indeed the case. We must distinguish two situations, depending on the strength of the field, B_2, frozen in the disk plasma (we will take this field to be random, with correlation length ℓ_B). If B_2 is too strong [i.e., if $B_2 \gtrsim 10^{-2} - 10^{-1} B_1$ for $r = 10^8 - 10^9$ cm; see Eq. (18)], the KHI may develop only in those places where $\underset{\sim}{B}_1$ and $\underset{\sim}{B}_2$ are roughly parallel; the unstable waves have $\underset{\sim}{k} \cdot \underset{\sim}{B}_i \approx 0$ and $k\ell_B \gg 1$. The area which is involved in the KHI is relatively small, and this instability cannot cause an efficient mixing of the disk plasma with the magnetosphere.

If B_2 is small enough, the surface is everywhere KH unstable. However, due to the stabilizing effect of gravity, only short-wavelength perturbations, with $kh > k_g h \gg 1$ [Eq. (17)], can develop. A turbulent shear layer, of thickness $\delta \lesssim k_g^{-1}$, is expected to form at the surface of the disk. Within this layer, the plasma is threaded by the field as described in Subsection B.

The analysis hereafter is taken from Anzer and Börner,[34] and Arons et al.[12] The evolution of the layer depends mainly on its density, ρ. If ρ is of the same order as the density at the surface of the disk before the mixing, then $\rho v^2 \approx \rho v_K^2 \gg \rho c_s^2 \approx B^2/8\pi$, and the field is sheared by the differential rotation of the plasma. As a result of the azimuthal magnetic field created by the shearing, the layer expands vertically to remain in pressure balance with the overlying magnetosphere,[34] and angular momentum is transferred from the inner part of the layer to the outer part. The latter part thus flows outwards while the former flows inwards. The KHI is likely to be stabilized by the presence of the shear layer,[66] and can restart only after its disappearance.

If, on the other hand, ρ is small enough for the ordering $\rho v^2 \lesssim B^2/8\pi$ to hold, the plasma may continuously evaporate from the disk and be brought into corotation with the magnetosphere. The part of the layer beyond the corotation radius, r_c, may be centrifugally driven outwards, while the part within r_c is accreted. In this case, which seems to be favored by Arons, Mckee and Pudritz (see Arons et al.[12]), and which is used by Anzer and Börner[34] to describe what happens at the top of the expanding layer considered above, the excess angular momentum $\int_{\text{layer}} 2\pi\rho\delta r^3 (\Omega_s - \Omega_K)\, dr$ in the layer is gained or lost from the star.

Thus, the KHI in the outer part of the disk ($r > r_{PB}$) may be responsible for the creation of a wind. The wind is time-dependent, as it entails intervals during which the field lines open and the star and the disk are no longer connected, so the instability stops working. Simple estimates show that the wind may involve only a small fraction of the matter in the disk, most of which is unaffected by the KHI and does not get threaded by the magnetic field.

Let us now consider the stability of the magnetically confined region of the diamagnetic part of the disk. If the disk is completely diamagnetic ($r_{th} = r_i$), then it is likely to be subject to the RTI near its inner edge. The plasma is partially supported against the effective gravity (gravity minus centrifugal force; see above) by the external magnetic pressure, and the curvature of the boundary, at least near the inner edge, has the "wrong" sign for stability (Section 1-A). If the disk were in that state, it might break down into plasmoids separated by stellar magnetic flux tubes. These plasmoids, which might retain a large amount of angular momentum, could then be destroyed by the KHI and become threaded by the stellar field. If $r_{th} > r_i$ (e.g. as a result of the previous process), the RTI cannot work because the lines are anchored in a dense plasma for $r_i \leqslant r \leqslant r_{th}$.

In both situations ($r_i = r_{th}$ or $r_i < r_{th}$), the KHI may render the interface unstable (competing with the RTI when $r_i = r_{th}$). For a magnetically confined disk, gravity cannot stabilize the instability. The second model of Subsection A, which can be used here, and shows that long-wavelength perturbations, having $kh \lesssim 1$, may develop, completely destroying the disk, at least if B_2 is not too large. The effect of B_2 may be diminished by taking very small wavenumbers (although we must keep $\underset{\sim}{k} \cdot (\underset{\sim}{r}_{PB} - \underset{\sim}{r}_{th}) \gtrsim 1$), but the further action of the KHI in breaking down the large turbulent cells thus produced would be impeded.

It must be noted that the above analysis considers an over-idealized model, involving pure KHI, without the superposition of phenomena which characterizes an actual situation. In a real disk, the development of the KHI interacts, for example, with the development of the reconnection of $\tilde{B}_2$ and $\tilde{B}_1$, or with the turbulent diffusion of $\tilde{B}_1$ into the disk, and a more sophisticated analysis is necessary to handle the problem.

4. Reconnection of magnetic field lines

Magnetic reconnection may be defined as "a process whereby plasma flows across a surface that separates regions containing topologically different magnetic field lines."[67] Such a surface, called a separatrix, contains, by definition, a singular line (neutral line) along which two magnetic surfaces intersect. Reconnection cannot occur in ideal MHD, because, in that case, flux surfaces are frozen into the flow and the normal velocity of the plasma with respect to any surface is zero. Reconnection becomes possible when departures from the perfect Ohm's law (I-26) are introduced. These departures may be due, for example, to a non-zero resistivity, η, of the plasma in the MHD case, or to collective effects (electron or ion Landau damping) in the case of a collisionless plasma.[68] Two main types of reconnection processes have been considered in the literature: a "spontaneous" process (tearing instability) in which neutral lines, not present initially, are produced on a surface separating two topologically distinct magnetic regions; and a forced non-linear process in which the presence of a neutral line is assumed a priori, and one considers a stationary flow pattern in the surrounding region.

A. Tearing instability

Let us consider a magnetic field, $\tilde{B} = B(y)\hat{x}$, where B is some odd function of y that is roughly constant and equal to B_o for $|y| > a$ and is monotonically increasing between $y = -a$ and $y = +a$. There is a current, of density $j_z \sim cB_o/4\pi a$, confined to the layer $\{-a \leqslant y \leqslant +a\}$, in which an increase in plasma pressure balances the gradient of magnetic pressure (directed towards $y = 0$).

Suppose now that the lines of this configuration are pinched together by some velocity field, $v_y = u(y) \sin kx$, where u is an odd function of y. In the ideal MHD situation, the tension of the lines, which appears because of their bending by v_y, will exert a restoring force and nothing special happens. The situation is different if the resistivity, η, of the plasma is non-zero. The component of the field, B_y, which is created during the pinching with $B_y(y) = B_y(-y)$ and $B_y(0) = 0$, diffuses. This results in the appearance of a non-zero B_y at $y = 0$, and in a change in the topology of some of the field lines, which reconnect with lines on the other side of the boundary (Fig. 9). The important point is that, inside the "islands" thus created, there is a magnetic tension force which pulls the plasma towards the center, and so tends to

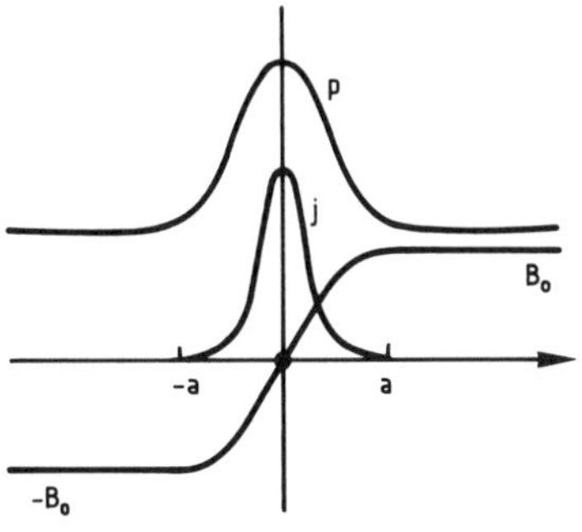

(9a)

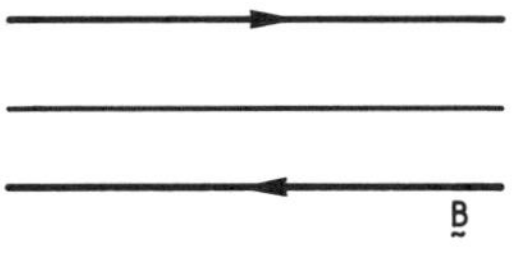

(9b)

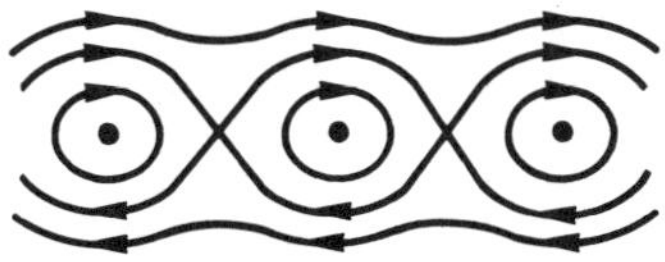

(9c)

Figure 9. Development of a tearing mode in a simple slab geometry.
9a: variations of the pressure, field, and current across a layer.
9b: initial configuration of the lines.
9c: appearance of islands when a perturbation is applied.

(10a) (10b) (10c)

Figure 10. Magnetic island coalescence.
Because of the attraction between currents of the same sign, configuration (a) may be unstable and evolve towards (b) (ideal MHD phase). The fields pressed together may reconnect, leading to (c). The process can start again, continuously diminishing the number of islands.

increase the perturbation. This force is responsible for the development of the so-called tearing mode.

The linear theory of the instability has been worked out by Furth et al.[69] Their analysis is based on the MHD equations in which the resistive term, $\eta \mathbf{j}$, is retained in Ohm's law, and incompressibility ($\nabla \cdot \mathbf{v} = 0$) is assumed (this may be checked a posteriori to be a valid approximation). These equations are applied to a generalization of the configuration described above to provide a heuristic description of the mechanism of a tearing mode. The assumption of a plane parallel current layer of thickness a is retained, but the field is allowed to have some shear: $\mathbf{B} = B_x(y)\hat{x} + B_z(y)\hat{z}$ (this field can reverse without going through zero in the layer). The perturbations, $f(\mathbf{r},t)$, are taken of the form $\tilde{f}(y)\exp\{i(\omega t+\mathbf{k}\cdot\mathbf{r}_\perp)\}$, where $\mathbf{r}_\perp = (x,z)$.

It turns out that the resistivity has important effects only in a boundary layer of thickness εa, with $\varepsilon \ll 1$, located on the plane where $\mathbf{k} \cdot \mathbf{B} = 0$. (When this condition is fulfilled, one of the terms in the equation, which in general prevents relative motion between field and plasma, becomes negligible.) The problem may be solved by classical singular perturbation techniques, which give, on matching the solutions inside and outside the boundary layer, a dispersion relation between ω and k. For long wavelengths ($ka < 1$), the relation is

$$\omega^{-1} \approx (9ka/2)^{2/5}\tau_A^{\,2/5}\tau_D^{\,3/5} , \tag{25}$$

where $\tau_D = 4\pi a^2/\eta c^2$ is roughly the time it would take for the field to diffuse through the layer if $\mathbf{v} = 0$, and $\tau_A = a/c_A \ll \tau_D$ is the transit time of an Alfvén wave across the layer (c_A is computed with the field outside the layer). Furth et al.[69] have shown the existence of other unstable modes (rippling and gravitational modes), but these will not be considered here.

In the linear theory, the width, δ, of a magnetic island increases exponentially with time. However, when δ becomes comparable to the width, εa, of the boundary layer [where $\varepsilon \approx (ka)^{-3/5}(\tau_A/\tau_D)^{2/5}$], non-linear effects begin to play a role and the growth rate of δ eventually saturates (see Bateman,[70] Pellat[71] and references therein).

The chain of islands thus produced may itself be unstable because of the attraction between the currents flowing in each island.[72] This is an ideal MHD instability, which leads to the aggregation of the islands on a timescale which is a decreasing function of δ.[73] Thus, this process should dominate when δ reaches some critical value, $\delta_c(k,a,B_o)$. In the presence of resistivity, two islands pressed together by this mechanism may reconnect; successive applications of this process lead to a single island (Fig. 10). This coalescence is clearly shown in numerical simulations,[74] and seems to lead to a high rate of reconnection.

The results we have presented about the tearing mode in a

plane shear layer have been extended to other situations. Different geometries have been considered: for example, much work has been done on tearing modes in cylindrical and toroidal regions, mainly for applications to fusion devices. In these geometries, the curvature of the lines is important and changes some of the characteristics of the instability. We refer the reader to more specialized introductions[70,71] to learn about these effects.

Actual astrophysical situations are much more complex than the simple models described so far. For example, velocity shears in the neighborhood of the reconnection surface are obviously present in most applications. Steps in that direction have recently been taken by Hofmann,[75] and by Dobrowolny et al.[76] In the latter paper, the effect of shearing motions parallel to the field is considered. It is shown that a larger growth rate may be obtained: $\omega^{-1} \alpha\ (\tau_A \tau_D)^{1/2}$ instead $\omega^{-1} \alpha\ \tau_A^{2/5}\ \tau_D^{3/5}$, given by Eq. (25).

B. <u>Forced stationary reconnection</u>

In the previous subsection, we were concerned with the time-development of spontaneous reconnection between the field lines of two regions that were initially magnetically unconnected. A different problem has also received much attention during the last twenty years, that of stationary reconnection. Typically, one considers configurations where two fields with different directions are pushed with some velocity v_R towards a neutral line, near which they reconnect. The plasma on the lines is ejected laterally after crossing the separatrix. Several models have now been constructed to describe this type of process. They are described in detail in Vasyliunas,[67] Parker,[61] and Sonnerup,[77] to whom we refer the reader for detailed references. The basic ideas are contained in the first models proposed,[78-80] and we will describe only these.

In the Sweet-Parker 2-D model, one considers two fields, of intensity B_o and opposite directions, which are pressed together over a length 2L (see Fig. 11), and annihilate resistively in a thin current layer of thickness 2ℓ. New field lines are continuously transported at velocity v_R from each side into the layer. The plasma, assumed incompressible, is accelerated in the layer and flows out along the lines from both ends of the annihilation region at the Alfvên velocity, $c_A = B_o\ /\ (4\pi\rho)^{1/2}$. This is understandable, since quasi-equilibrium across the lines requires that $p + B^2/8\pi = P_o$ = constant, while Bernouilli's law requires that $p + \rho v^2/2 = P_1$ along the lines; the value of P_1 depends on the field line in question. The lines on which matter emerges from near the center of current layer have B = 0 and v = 0; thus $P_1 \approx p(0) \approx P_o$ and $v^2 \approx B^2/4\pi\rho$.

From matter conservation, we have

$$\ell\ c_A = L\ v_R \quad . \tag{26}$$

Equality between the rate at which flux is brought into the layer and the rate at which it is destroyed by dissipation gives

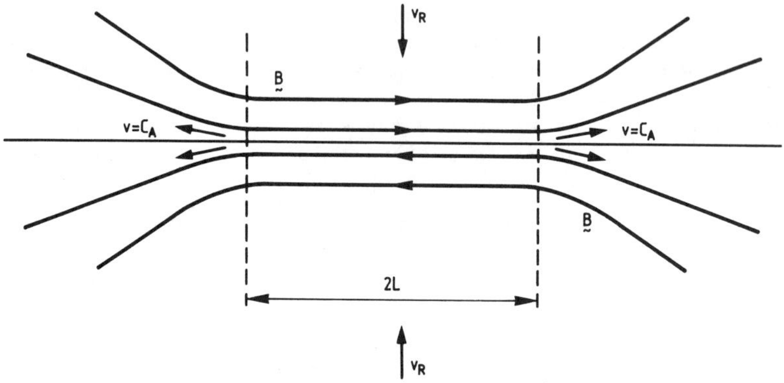

Figure 11. The Sweet-Parker reconnecting configuration.

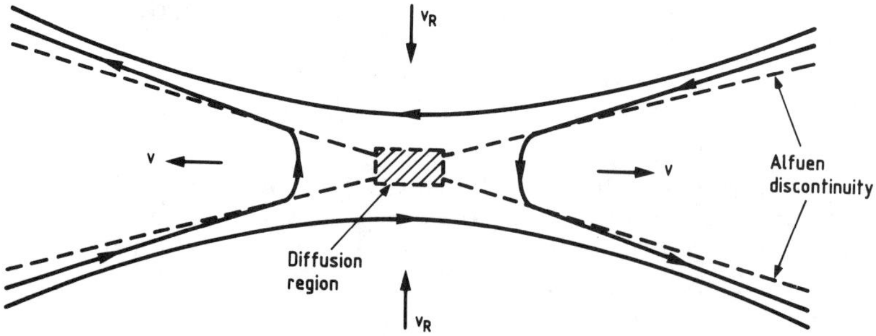

Figure 12. The Petschek reconnecting configuration.

$$v_R = D_B/\ell \quad , \tag{27}$$

where $D_B = \eta c^2/4\pi$ is the coefficient of diffusion of the magnetic field. Combining Eqs. (26) and (27), one immediately obtains

$$v_R = c_A R_m^{-1/2}, \qquad \ell = L R_m^{-1/2} \quad , \tag{28}$$

where the magnetic Reynolds number has been computed with L and c_A as the typical length and velocity (i.e. $R_m = L c_A/D_B$). For diffusion alone (with $\underset{\sim}{v} = 0$), one finds $v_R = 2\eta/L = 2c_A/R_m$; thus, dynamical reconnection has increased V_R by a factor of $R_m^{1/2}$.

Petschek[80] introduced new and important ideas to the theory of magnetic reconnection. He pointed out that the process can occur much more quickly if magnetic energy is annihilated not only by resistive dissipation but also by the propagation of an Alfvén wave. In this case there is no need for the width, λ, of the dissipation region to be of the same order as the typical length-scale, L, of the field. He proposed a 2-D model (Fig. 12) which has two waves originating from a small dissipation region around the neutral lines. In a stationary state these waves propagate away from the reversal region with velocity $B_n/(4\pi\rho)^{1/2}$ (B_n is the component of $\underset{\sim}{B}$ normal to the wave) relative to the matter. If α is the angle between the wave front and the general direction of the field, mass conservation demands that

$$M_A = v_R/c_A = \tan\alpha \quad . \tag{29}$$

This relation cannot be valid right up to the X-line, as that would imply a discontinuity of B_y at this point. Dissipative effects must overcome wave effects in a region of extent 2λ along Ox and 2ℓ along Oy. In this region, the relations (28) apply (with L replaced by λ). It is then possible to determine all the parameters of the system (α, λ, ℓ) if one fixes arbitrarily the value of M_A. However, arbitrarily large reconnection velocities are not possible, because there is an implicit assumption in our argument: we have supposed that the field in the wave region is the field at infinity, neglecting the effect of the current flowing in the dissipative layer. This current creates a field, B', which diminishes the original field, and appreciably slows the reconnection process when its value is of order $B_o/2$. This restriction yields the inequality

$$M_A \leq \frac{\pi}{4}\left[\mathrm{Log}\,\frac{2M_A^2}{R_m}\right]^{-1} \tag{30}$$

($R_m = Lc_A/D_B$, as above), providing an estimate for the maximum value of M_A which is much larger than that given by the Sweet-Parker model for $R_m \gg 1$.

It must be noted that neither model gives a definite value for the reconnection rate. Actually, this value is determined by the boundary conditions. There is, however, an upper bound to this rate when B_o is fixed. The question of what would happen if the boundary conditions imposed an inflow exceeding the upper bound then naturally arises. In that case, the field increases near the X-line, and this increases the value of the upper bound. It thus seems possible that a stationary state can always be achieved. This implies that there is no bound on the reconnection rate; however, if the field is too large near the X-line, the plasma may not reach it and there may still be limits on the rate of plasma transport into the reconnection region lines.[67]

For astrophysical applications, it is necessary to study models which are more general than those presented here. In particular, one often needs to evaluate the rate at which two fields with different intensities and directions, embedded in plasmas of different densities, reconnect. This situation is discussed in Sonnerup,[77] where it is suggested that two fields of intensity B_1 and B_2 ($B_1 < B_2$) cannot reconnect if the angle, θ, between their directions satisfies $\cos\theta > B_1/B_2$. If the reverse relation holds, reconnection may proceed, with a bound on the rate determined by c_A computed in the weaker field.

As shown by Vasyliunas,[67] the fast reconnection model described above may be extended to collisionless plasmas by keeping the inertial terms in Ohm's law (I-21) (the resistive term is zero in this case). This may be of interest for applications to the "force-free" part of the magnetosphere, where a low plasma density is expected (Section I-4-D).

C. Reconnection in an accreting magnetosphere

Reconnection may occur between the magnetospheric field and the field embedded in an adjoining plasma region. Such a region may be:

(i) the plasma layer above the magnetopause, in the case of spherical accretion,[19,21]
(ii) an accretion disk,[31]
(iii) plasmoids falling between the lines,[18]
(iv) a wind in direct contact with the magnetosphere,[81]
(v) the companion star, which has its own field.[47]

The rate at which the two fields reconnect is difficult to evaluate. However, there seems to be general agreement that the reconnection speed is some non-negligible fraction of the Alfvén speed: $v_R = \alpha_R c_A$, with $\alpha_R \approx 10^{-2} - 10^{-1}$, say ($c_A$ being computed in the weaker of the reconnecting fields; see above). Actually, v_R may depend sensitively on the conditions under which the reconnection occurs. When one starts from initial conditions in which no X-neutral lines exist, X-points may be created by the development of tearing modes. This mechanism, however, proceeds at a slow rate during the linear phase (from Eq. (25), we have $\tau_{TM}/\tau_A \sim R_m^{-3/5}$). During the non-linear phase, fast processes (e.g. island coalescence) may operate and greatly enhance the process; however, they are not yet well understood. Reconnection at a fraction of the Alfvén speed by Petschek's mechanism is possible, but, as emphasized by Vasyliunas,[67] its occurrence depends on the boundary conditions. It needs to be driven from the outside to be efficient. It is often argued that spontaneous reconnection may evolve into the Petschek regime, but the link between the two processes has yet to be firmly established. Petschek's mechanism could occur in case (iv), which is similar to the situation at the nose of the Earth's magnetosphere, where the two fields are pressed together by the solar wind. In the other cases, the presence of an external driving force is not evident. The generally adopted value of α_R needs much more work to be justified.

One of the properties of reconnection is that it approximately conserves magnetic flux. Because of the very localized character of the regions where effective dissipation takes place during reconnection, an element of matter will be threaded by about the same amount of flux after reconnection as before. Consider, for instance, an accretion disk which contains an embedded small-scale field frozen into the plasma. This field has, on average, zero flux through the midplane of the disk. Therefore, if it reconnects with the stellar field,[31] the amount of flux through the midplane should still be zero, unless some special mechanism operates.[10] It is interesting to note, however, that this argument, which would imply a negligible amount of stellar flux threading the disk and hence a weak disk/star interaction (in the absence of other threading processes), does not apply if the disk has an ordered magnetic field.

The possibility of having a large-scale field built up in a disk has been considered by Pudritz.[82] The mechanism involves an α-ω dynamo, which may work under conditions in which differential rotation and turbulence are present. The turbulence is made gyrotropic by the effects of differential rotation and vertical density gradients (the turbulence acquires a non-zero helicity, $\alpha = \langle \tilde{v} \cdot \nabla \times \tilde{v} \rangle \neq 0$), and this is what is required for a dynamo process to operate. An investigation of the possibility of having the dynamo-generated disk field reconnect with the stellar one would be very worthwhile.

Reconnection is likely to play an important role in limiting the growth of small-scale fields inside the disk.[83,84] Any field present in the disk is sheared by the differential rotation (this is a different mechanism from the dynamo discussed above) and any radial component will generate an azimuthal field, B_ϕ, of the same order in a Keplerian time, Ω_K^{-1}. In the absence of dissipation, B_ϕ could grow until its reaction on the motion stops the differential rotation. However, well before that happens, the field lines which are stretched by the motion are likely to develop X-points at which the field could reconnect and transfer part of its energy to the fluid. Another limiting mechanism which may compete with reconnection is magnetic buoyancy, which can lead to the expulsion of tubes of flux into the magnetosphere.[84-86]

LECTURE III
FORCE-FREE MAGNETOSPHERES AND QUASI-STATIC MOMENTUM-ENERGY TRANSFER BY FIELD-ALIGNED CURRENTS

In the first lecture, we saw that the magnetosphere of a compact object interacting with an accretion disk or a companion star may be regarded as being in a force-free state, except in those small regions where there is a strong accretion flow. In this third lecture, we will consider the structure of force-free configurations linking two dense regions of plasma. In addition, the exchange of momentum and energy which results from magnetospheric field-aligned currents flowing between these regions will be discussed.

In Section 1, we state our assumptions and establish the relevant equations and boundary conditions. In Section 2, we use a simple analytical model and the results of numerical calculations to prove that the amount of angular momentum that can be transferred between the two regions by field-aligned currents is bounded. In Section 3, we discuss the qualitative behavior of the magnetospheric force-free field when the feet of its lines (on the boundary of either region) are sheared indefinitely. In Section 4, we consider the problem of the interaction between a compact star and a dissipative accretion disk. Our main results are summarized in Section 5, where we also mention their relevance to the problem of disk corona activity and to some other problems.

1. The force-free approximation

A. Assumptions

Consider two conducting regions, V_1 and V_2, connected by magnetic field lines in a low-density plasma magnetosphere, V. We will assume that the field in V is always in a force-free configuration, i.e. one in which the currents flow along the field lines. This approximation is valid if:

(i) The energy density of the field in V is much larger than that of the plasma there, so

$$B^2/8\pi \gg \rho(v^2 + c_s^2) \quad . \tag{1}$$

(The density of plasma must, however, be large enough to support the electric currents computed in this approximation.)

(ii) In V_1 and V_2, where $B^2/8\pi \lesssim \rho(v^2 + c_s^2)$, the motion of the matter is sufficiently slow that the field in V can continuously adjust to the equilibrium state corresponding to the instantaneous value of the boundary conditions ("quasi-static" approximation). The structure of $\underset{\sim}{B}$ in V changes as a result of Alfvén waves propagating from the boundary, ∂V. Hence, the time taken for a wave to cross V must be much smaller than the characteristic timescale of the motions in V_1 and V_2.

The magnetospheric currents close through V_1 and V_2, where they are able to flow across the lines. The Lorentz forces thus exerted on V_1 and V_2 exchange momentum and energy between the two regions. This transfer mechanism should be contrasted with the one often considered in relation to the so-called "angular momentum problem" for a contracting protostellar cloud (see e.g. Mouschovias[87] and references therein). In this latter mechanism, angular momentum and energy are expelled to distant regions by torsional Alfvén waves propagating in the extra-cloud medium. In that case, the transfer is not quasi-static, but dynamic, since inertia plays an important role. Of course, in the situation considered here, this mechanism may be important in distributing the exchanged momentum and energy throughout V_1 and V_2.

B. Equations and boundary conditions for a force-free field

The equations which determine a force-free field in V are

$$\nabla \cdot \underset{\sim}{B} = 0 \tag{2}$$

and

$$(\nabla \times \underset{\sim}{B}) \times \underset{\sim}{B} = 0 \quad . \tag{3}$$

The second condition, which expresses the field-aligned character of the current, and thus the vanishing of the Lorentz force, may be transformed into

$$\nabla \times \underset{\sim}{B} = \alpha \underset{\sim}{B} \quad . \tag{4}$$

From Eq. (2), it follows that $\underset{\sim}{B} \cdot \nabla\alpha = 0$, so the function $\alpha(\underset{\sim}{r})$ is constant along each field line. Another convenient way of writing the force-free condition is

$$\nabla \cdot \left\{-\frac{B^2}{2} \underset{\approx}{I} + \underset{\sim}{B} \otimes \underset{\sim}{B}\right\} = 0 \tag{5}$$

[see Eq. (I-39)].

To compute $\underset{\sim}{B}$ in V, it is necessary to complete Eqs. (2) and (3) with some boundary conditions. From a purely mathematical point of view, the two following sets of conditions on ∂V define correct boundary value problems (BVP):

(i) BVP1: one gives the value of B_n, the component of $\underset{\sim}{B}$ normal to ∂V, and the value of α on that part of ∂V where $B_n > 0$.

(ii) BVP2: one gives the value of B_n and the "connectivity" of the field lines (all of which are assumed to intersect ∂V).

This means that one fixes a priori the point where each field line emerging from ∂V recrosses the boundary.

From a physical point of view, what can we regard as known? There are two possibilities, depending on whether the conditions in V_1 and V_2 are given, or have to be determined simultaneously with the magnetospheric structure. Consider, for instance, the interaction of a magnetic star, V_1, with its companion, V_2. In that case, the magnetic fields and the motions inside V_1 and V_2 may be assumed to be given, as they are produced by mechanisms independent of the force-free field in V, which can change conditions in V_1 and V_2 only on very long timescales. Thus we know the value of $B_n|_{\partial V}$ and the motion of the feet of the field lines on ∂V. The formulation BVP2 is directly relevant in this case. When the quasi-stationary state of $\underset{\sim}{B}$ in V has been computed, its long-term effect on the conditions on ∂V_1 and ∂V_2 may be determined.

The problem of the interaction of a magnetic star, V_1, with a dissipative accretion disk, V_2, is quite different. We can still assume that the value of B_n and the motion of the feet of the field lines are known on ∂V_1, but there is no way to prescribe a priori the values of these quantities on ∂V_2. These values are determined by matching between magnetospheric and disk phenomena on ∂V_2. In this case, BVP2 may still be a useful step in obtaining the configuration to which the disk and magnetosphere eventually evolve (Section 4).

BVP1 has no obvious physical meaning, because there is no way to impose the current, $j_n = c\alpha B_n/4\pi$, entering the magnetosphere. Nevertheless, because of its relative simplicity compared to BVP2, it is useful to consider it; it may often be used as an intermediate step in a calculation, and it provides many of the interesting structural properties of force-free magnetospheres.

C. A simple model

We will use the following simple model to gain insight into the structure of a force-free magnetosphere and the rate at which magnetic stresses exchange momentum and energy between two regions. Consider the two regions, V_1 and V_2, defined, in a cylindrical coordinate system $(\tilde{\omega}, \phi, z)$, by $V_1 = \{\tilde{\omega} \leqslant R; -h \leqslant z \leqslant h\}$ (we call it the "star"), and $V_2 \doteq \{\tilde{\omega} \geqslant R; \ -h \leqslant z \leqslant h\}$ (we call it the "disk"). The region, V, outside $V_1 \cup V_2$ (the magnetosphere) is assumed to be threaded by an axisymmetric, mirror symmetric $(z \leftrightarrow -z)$ force-free magnetic field.

As is well known, an axisymmetric field can be represented in terms of a flux function, $f(\tilde{\omega},z)$, and a function $B_\phi(\tilde{\omega},z)$ by

$$\underset{\sim}{B} = \underset{\sim}{B}_p + B_\phi \hat{\phi} = \frac{\nabla f \times \hat{\phi}}{\tilde{\omega}} + B_\phi \hat{\phi} . \tag{6}$$

The field lines, C, of $\underset{\sim}{B}$ lie on the surfaces, S(f), of constant $\underset{\sim}{f}$. Using Eq. (4) for a force-free $\underset{\sim}{B}$, one can easily show that (i) $\tilde{\omega} B_\phi$ is constant on any surface S(f), so that $\tilde{\omega} B_\phi = H(f)$,

(ii) $\alpha = H'(f)$ and (iii) f satisfies the Grad-Shafranov equation,

$$Lf \equiv -\left[\frac{\partial^2 f}{\partial \tilde{\omega}^2} + \frac{\partial^2 f}{\partial z^2} - \frac{1}{\tilde{\omega}}\frac{\partial f}{\partial \tilde{\omega}}\right] = H(f)\ H'(f) \quad . \tag{7}$$

To determine f in $V^+ = \{z > h\}$, we impose the value of $B_z(\tilde{\omega},h)$. The values of f in $\{z < -h\}$ will be given by the symmetry relation $f(z) = f(-z)$. According to the discussion in Section 1-B, B_z is indeed often known a priori. When it is not (for example, on the surface of a dissipative disk), consideration of the dependence on B_z of the results obtained may still yield some interesting properties (see Section 2 below). For simplicity, we require the function B_z to satisfy $B_z \geq 0$ for $\tilde{\omega} < R$, and $B_z \leq 0$ for $\tilde{\omega} > R$, so $0 \leq f(\tilde{\omega},h) \leq f_m$. We will also require that the field decrease to zero when $r \to \infty$ (it will be sufficient for the validity of our calculation to have $\lim_{r\to\infty} r^{3/2}B = 0$). Another condition--the value of $B_\phi(\tilde{\omega},h)$ (BVP1), or the position on ∂V^+ of the feet of the lines (BVP2)--will be imposed when needed.

2. Existence of an upper limit to the rate of momentum-energy transfer by magnetic stresses

A. Upper bound for the torque in the simple model

In our simple model, the magnetic torque imposed by the "disk" on the "star" is $\underset{\sim}{N}_1 = N_1\ \hat{z}$, where

$$N_1 = \fint_0^R B_\phi B_z\ \tilde{\omega}^2 d\tilde{\omega} = -\fint_R^\infty B_\phi B_z\ \tilde{\omega}^2 d\tilde{\omega} \tag{8}$$

(the bar on the integral sign just means that the value of the integrand is taken at the upper surface of the disk or star, $z = h$).

It is important to know if the torque could be arbitrarily large. The following simple argument shows that there is an upper bound, depending only on $B_z(\tilde{\omega},h)$, on the value of $|N_1|$. We apply Schwartz's inequality to the RHS of Eq. (8) to obtain

$$|N_1| \leq \left[\fint_0^R \tilde{\omega}^3\ B_z^{\ 2}\ d\tilde{\omega} \fint_0^R \tilde{\omega}\ B_\phi^{\ 2}\ d\tilde{\omega}\right]^{1/2} \quad . \tag{9}$$

Integrating the $\hat{z}$-component of Eq. (5) over $z > h$ and applying Gauss's formula, we find

$$\fint_0^\infty \tilde{\omega}\ (B_{\tilde{\omega}}^2 + B_\phi^2)\ d\tilde{\omega} = \int_0^\infty \tilde{\omega}\ B_z^{\ 2}\ d\tilde{\omega} \quad , \tag{10}$$

which combines with Eq. (9) to give

$$|N_1| \leq N_* = [\int_0^R \tilde{\omega}^3 B_z^2 d\tilde{\omega} \int_0^\infty \omega B_z^2 d\tilde{\omega}]^{1/2} \quad . \tag{11}$$

The interest of this inequality lies in the fact that no conditions have been imposed a priori on B_ϕ or on the shear of the configuration. Eq. (11) has to be satisfied by all possible force-free configurations corresponding to the given boundary function $B_z(\tilde{\omega},h)$.

Of course, there is no guarantee that the upper bound (11) will be reached, or even closely approached, by any actual configuration. The previous argument demonstrates the existence of a maximum value, N_c, for $|N_1|$ and provides an estimate, N_*, for it, but it says nothing about the usefulness of this estimate.

B. Discussion of the upper bound

To go further into this problem, we investigate the existence of solutions corresponding to a given value of $B_\phi(\tilde{\omega},h)$. A convenient way of doing this is to write the function $H(f) = \tilde{\omega} B_\phi$ in the form $H(f) = \lambda h(f)$, where $h(f)$ is a given function and $\lambda \geq 0$ is a parameter, and to follow a branch of solutions f_λ as λ increases from 0. Because of Eq. (10), we know a priori that the problem will have no solutions for

$$\lambda^2 \geq \lambda^2_* = [\int_0^\infty \tilde{\omega} \; B_z^2 d\tilde{\omega}] \; [\int_0^\infty h^2 \frac{d\tilde{\omega}}{\tilde{\omega}}]^{-1} \quad . \tag{12}$$

The existence problem may be treated to a certain extent by analytical methods, such as the technique used by Heyvaerts et al.[88] in their study of the corresponding problem in Cartesian geometry (x-invariant configurations). It can also be treated numerically by integrating Eq. (7) for a wide range of boundary conditions. This latter approach is followed by Aly, Cohn, Lamb and Zylstra,[89,90] who have recently undertaken a study of the magnetospheric structure around a disk accreting magnetic compact object, considering the force-free case as a first step (Fig. 13). The geometry they consider is more complex than the simple model adopted here. Their disk is infinitesimally thin and extends, like a real disk, from some $R_1 > R$ to some $R_2 \leq \infty$. They allow for the presence of holes in the disk. In these regions, the value of B_z at the midplane is not fixed, but $B_{\tilde{\omega}} = 0$, by symmetry. However, no fundamental differences between the properties of the two models have yet been found, and we present the findings of the analytical analysis of the simple model, M1, and the numerical analysis of the more complex one, M2, together.

The main finding is that the equation has at least one branch of solutions when λ is small, but no solutions are found for values of λ larger than some λ_c (for M1, Eq. (12) gives an estimate, λ_*, for λ_c). Roughly speaking, λ_c is reached when values of $|B_\phi/B_p|$ of order unity appear in some region of V^+ (here, B_p is the poloidal

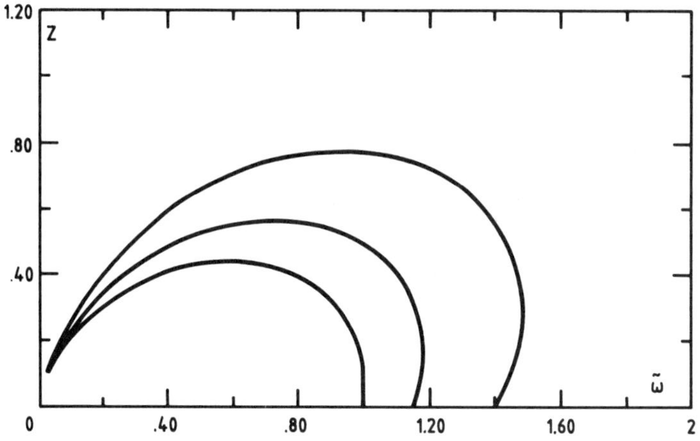

(13a)

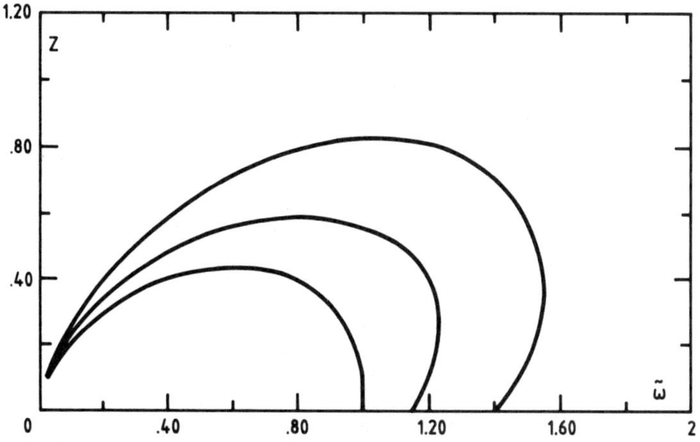

(13b)

<u>Figure 13. Force-free magnetospheric configurations around a disk-accreting magnetic compact object.</u>
The star (of radius 0.1) has a dipolar magnetic field and is located at the origin, while the disk lies in the plane $\{z=0\}$ and has $\tilde{\omega} \geq 1$. The value of B_z is imposed on the disk surface, as is the value of $\tilde{\omega}B_\phi$, which is taken to be of the form $\lambda h(f)$ ($\lambda \geq 0$ is a parameter). A solution for $\tilde{B}$ can be found when λ is less than some critical value, λ_c. The figures represent the intersection with a poloidal plane of three magnetic surfaces, S(f), corresponding to particular choices of $B_z^{(disk)}$ and h, with $\lambda=0$ (Fig. 13a) and $\lambda=\lambda_c$ (Fig. 13b). (From Aly, Cohn, Lamb and Zylstra[89]).

magnetic field). In addition, if $|B_\phi/B_z|$ is a decreasing function of radius on the disk surface, ∂V_2, and ℓ, the characteristic length scale for $B_z|_{\partial V_2}$, satisfies $\ell \gtrsim R\ (R_1)$ in model M1 (M2), it is found that $|B_\phi|$ and $|B_z|$ on ∂V_2 are of the same order when $\lambda=\lambda_c$. For model M1, it then follows that $N_c \lesssim N_*$. (If $|B_\phi/B_z|_{\partial V_2}$ first increases up to some radius R_1', the condition on ℓ becomes $\ell \gtrsim R_1'$.) If $\ell \ll R\ (R_1)$ for M1 (M2), $|B_\phi/B_z|_{\partial V_2}$ may be less than one even for $\lambda=\lambda_c$. This is easily understood: if the flux is concentrated near $\tilde{R}\ (R_1)$ in the disk, then a field line starting at $\tilde{\omega}_o(f)$ for z=0 is going to reach a maximum distance $\tilde{\omega}_m(f) \gg \tilde{\omega}_o(f)$, so $|B_\phi/B_p|_o \ll |B_\phi/B_p|_m$.

Thus, if we fix the amount of flux through the disk, the largest transfer of momentum seems to occur when the flux is distributed with a scale $R\ (R_1)$ for M1 (M2) or larger, and $|B_\phi/B_z|_{\partial V_2} \sim 1$. The existence of a maximum value for B_ϕ may seem quite strange. If, for example, one started rotating the disk at constant angular frequency, while the star is kept fixed, one might expect an indefinite increase in B_ϕ. What prevents this from happening? The calculations of Section 3 will give the answer to that question.

C. Questions of stability

For the maximum value of the torque predicted in Subsection A actually to be reached, the field configuration which realizes it must be stable in V. Force-free fields having all lines firmly tied to V_1 and V_2 are expected to be very stable. Their stability may be studied using the energy principle introduced in Section 1. The second variation of the energy is

$$\delta W = \int_V \{|\nabla \times \underset{\sim}{A}|^2 - \alpha \underset{\sim}{A} \cdot \nabla \times \underset{\sim}{A}\} \quad , \tag{13}$$

where $\underset{\sim}{A} = \underset{\sim}{\xi} \times \underset{\sim}{B}$ and the displacement, $\underset{\sim}{\xi}$, has been chosen such that $\underset{\sim}{\xi}|_{\partial V} = 0$ (line-tying). It can be shown[91] that, for any force-free field in $V^+ = \{z > h\}$, one has

$$\delta W \geqslant \{1 - \frac{2}{\sqrt{3}\,\pi^{2/3}} [(\int_{V+} \alpha^3 d\underset{\sim}{r})^{1/3} + (\int_{V+} [\int_C B ds \int_C |\nabla\alpha|^2 \frac{ds}{B}]^{3/2} d\underset{\sim}{r})^{1/3}]\} \int_{V+} |\nabla \times \underset{\sim}{A}|^2 d\underset{\sim}{r} \quad . \tag{14}$$

This immediately gives a sufficient condition for stability with respect to any small 3-D perturbations: $\alpha\ell \lesssim 1$, where ℓ is a characteristic linear dimension. Therefore, it is possible to approach the maximum torque closely in a stable way.

3. Shearing of a force-free magnetosphere when the "disk" is perfectly conducting

A. Statement of the problem and main results

Let us now turn to the other problem found to be important in our discussion of Section 1-B: the behavior of the force-free magnetosphere when shearing motions are applied to the feet of its lines on ∂V_1 and ∂V_2. To try to understand what happens in this situation, we consider again our simple model, so $B_z(\tilde{\omega},h)$ [or $f(\tilde{\omega},h)$] is fixed on ∂V^+ and $\lim_{r\to\infty} r^{3/2}B = 0$. We suppose that at time $t=0$ the field is in its potential state, f_o, i.e., it has no shear and no currents are flowing in V^+. If we assume that V_1 and V_2 rotate with angular frequencies $\Omega_1(\tilde{\omega})$ and $\Omega_2(\tilde{\omega})$, respectively, and that V_1, V_2 and V are perfectly conducting, then, in the quasi-static approximation of Section 1, any line, C, drawn on $S^+(f)$ will have the difference, $\delta\phi(f)$ (see Fig. 14), between the angular position of its feet on ∂V^+ increasing (in modulus) according to

$$\delta\phi(f) = H(f) \int_{C_p(f)} \frac{ds}{\tilde{\omega}^2 B_p} = (\Omega_2 - \Omega_1)(f)t = \Delta\Omega(f)t \quad , \tag{15}$$

where $C_p(f)$ is the intersection of $S^+(f)$ with a meridian plane and ds is the length element along C_p. The first equality in Eq. (15) just results from integrating along $C_p(f)$ the equation for a field line, $\tilde{\omega}d\phi/B_\phi = ds/B_p$. The goal now is to solve the Grad-Shafranov equation (7) along with Eq. (15), the boundary conditions on $f|_{\partial V^+}$ and $f|_\infty$ and the initial condition, $f_{t=0}(\tilde{\omega},z) = f_0(\tilde{\omega},z)$, to get a time-sequence of configurations $f_t(\tilde{\omega},z)$. Of course, the solution f_t must have the same topology as f_0 to be physically acceptable (because of the frozen-in law).

This problem was considered first by Sturrock and Barnes, who adopted a numerical approach and applied their results to solar flares[92] and active galactic nuclei.[93] It has been reconsidered recently by Aly,[94-96] who used "qualitative" analytical techniques. The main result of Aly's analysis may be summarized as follows (see Fig. 15). "If the problem stated above has a solution, $f_t(\tilde{\omega},z)$, for $t \in [0,\infty]$, then, as $t \to \infty$, f_t asymptotically approaches an open structure, with all the magnetospheric currents concentrated in an infinitesimally thin sheet. This means, in particular, that $\lim_{t\to\infty} B_\phi|_{\partial V^+} = 0$, while $\lim_{t\to\infty} V(f) = \infty$ for $f < f_1$, where $V(f)$ is the volume between $S(f)$ and ∂V^+, and $f_1 (\leq f_m)$ is the smallest value of f such that $\Delta\Omega(f') = 0$ for all f' in $[f_1,f_m]$."

It is worth noting that the proviso about the existence of f_t for all values of t, which has not yet been completely proved, is in contradiction to the idea favored by many solar physicists (see, for example, the review papers by Birn and Schindler[97] and Low[98]),

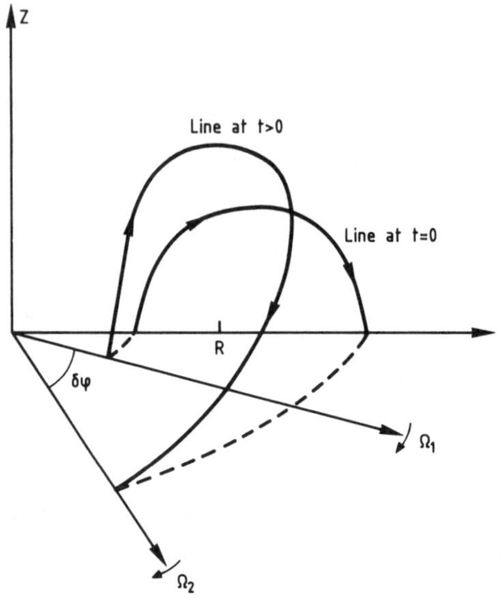

Figure 14. Shearing of a force-free field in the simple model.

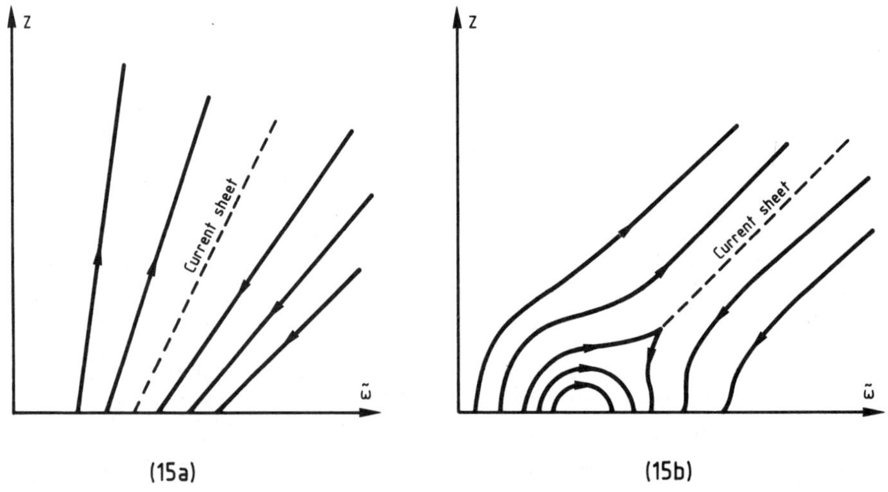

Figure 15. Possible asymptotic states approached by the field as $t \to \infty$.
The curves represent the intersection of magnetic surfaces, S(f), with a meridian plane. In 15a, $f_1 = f_m$, while, in 15b, $f_1 < f_m$.

according to whom there should exist a critical value, t_c, of t beyond which the problem no longer has any solution. (In the physics of the solar corona, t_c is identified by these authors with the time of onset of a flare.) However, there is no proof of this assertion, nor is there any known example exhibiting this kind of transition. In contrast, the view that f_t exists is supported by explicitly known solutions of similar problems.[99,100]

B. Schematic proof of the asymptotic property

Since the proof of the result quoted above is rather long, we give here only a brief outline showing the main steps.

(i) The energy of any force-free field in V^+ satisfies[101]

$$2W = \int_{V^+} \frac{B^2}{4\pi}\, d\underset{\sim}{r} = \oint \tilde{\omega}^2 B_\omega B_z d\tilde{\omega} \leqslant \left[\oint \tilde{\omega}^3 B_z^2\, d\tilde{\omega} \oint \tilde{\omega}\, B_z\, d\tilde{\omega}\right]^{1/2} , \quad (16)$$

which shows that it is not possible to store an arbitrarily large amount of energy in the field when $B_z|_{\partial V^+}$ is fixed. The equality in Eq. (16) is just the virial theorem, obtained by integrating $(\nabla \cdot \underset{\approx}{T}) \cdot \underset{\sim}{r} = 0$ over V^+. The inequality comes from the application of Schwartz's inequality and Eq. (10) to the RHS of the virial theorem.

(ii) Using Eq. (I-37), one finds that the energy of the magnetospheric field increases steadily during the shearing according to

$$dW/dt = 2W_\phi/t \quad , \qquad (17)$$

where

$$W_\phi = \int_{V^+} \frac{B_\phi^2}{8\pi}\, d\underset{\sim}{r} = t \int_0^{f_m} H(f)\Delta\Omega(f)df \quad . \qquad (18)$$

(iii) The value of W_ϕ must tend to zero as $t\to\infty$, because otherwise W would increase without bound, in contradiction to inequality (16). Because $(H\delta\phi)(f) \geqslant 0$, (see Eq. (15)), we must have $\lim_{t\to\infty} t\, B_\phi|_{\partial V^+} = 0$ for almost all values of $\tilde{\omega}$. (This argument assumesthe existence of limits for B_ϕ and tB_ϕ. A more sophisticated argument, which does not make these assumptions, may be developed. It leads to $\lim_{t\to\infty} B_\phi|_{\partial V^+} = 0$ for all values of $\tilde{\omega}$.)

(iv) From Eq. (15), one has

$$t\, |\Delta\Omega(f)| \leqslant |H(f)| \frac{1}{\inf_s \tilde{\omega}^2(f,s)} \int_{C_f(f)} \frac{ds}{B_p} = \frac{|H(f)|}{\inf_s \tilde{\omega}^2(f,s)}\left|\frac{dV}{df}\right| \quad . \qquad (19)$$

Because the LHS tends to infinity for all values of f for which $\Delta\Omega(f) \neq 0$, while $H(\underset{\sim}{f}) \to 0$ in the RHS, we must have either $|dV/df| \to \infty$ or $\inf_s \tilde{\omega}(f,s) \to 0$. This last condition, however, is impossible (because of flux conservation, it would mean an infinite increase of B_z and of the energy near the z-axis). Therefore, $|dV/df| \to \infty$ for all values of f for which $\Delta\Omega(f) \neq 0$, and $V(f) \to \infty$ for $f < f_1 \leq f_m$, where f_1 is the smallest value of f such that $\Delta\Omega(f') = 0$ for all f' in $[f_1, f_m]$. The magnetic structure then opens up as $t \to \infty$.

(v) In the last step, one shows that the ϕ-current tends to zero everywhere except in a region of arbitrary small extent near the surface $S(f_1)$. In the asymptotic state (which clearly must be singular), this current is concentrated in a sheet, which starts from $S(f_1)$ and goes to infinity. Along this sheet, one has, of course, $B_n = 0$, while $\underset{\sim}{B}_t = \underset{\sim}{B}_{pt}$ reverses (so $[B_t^2] = 0$). The sheet is in equilibrium, as it must be, since it is the limit of a sequence of equilibrium configurations.

C. Physical discussion

We can understand now what prevents B_ϕ from increasing indefinitely. The expansion of the field, which becomes quite rapid when the shear is large, causes B_ϕ to decrease at a faster rate than that at which it would increase from shearing alone. We see that this effect dominates when $|B_\phi/B_p|$ is of order unity. Before that, there is a phase during which $|B_\phi|$ increases steadily, with little change in the poloidal structure of the field.

The effect of shearing is to cut the magnetic connection between the disk and the star! This looks like an awful result, as we are interested in transferring momentum and energy between the two regions. However, it is clear from what we said about reconnection that this configuration is unstable if resistivity (which is never completely zero) is taken into account. As soon as the magnetic field is stretched out beyond some limit, reconnection starts and reduces the shear below that value; then the shear increases again and the process repeats, dissipating the energy of V_1 and V_2 in the magnetosphere at the average rate

$$\langle W_{m\ diss} \rangle = \int_0^{f_m} \Delta\Omega(f)\ \langle \tilde{\omega}\ B_\phi \rangle\ df \quad . \tag{20}$$

Thus, the situation is not so bad, because, on average, one can expect $\langle B_\phi \rangle|_{\partial V^+}$ not to be too far from the value giving the maximum possible torque; the time the magnetosphere spends in the quasi-open state is expected to be very short, due to the large value of the magnetospheric Alfvén speed.

4. Shearing of a force-free magnetosphere when the "disk" is dissipative

Let us now try to devise a method for solving the shearing problem when the disk is dissipative and the magnetospheric field may diffuse throughout. We first suppose that we know the motion of the disk plasma [velocity $(v_{\tilde{\omega}}, v_\phi = \tilde{\omega}\,\Omega)$] and that the time, τ, needed for the field to diffuse across the disk is smaller than the shearing time. Then, at each instant of time, we have, from Ohm's law and the continuity of the tangential electric field on ∂V_2,

$$\frac{B_\phi}{B_z} \approx (\Omega - \Omega_m)\tau \tag{21}$$

and

$$\frac{B_{\tilde{\omega}}}{B_z} \approx (v_{\tilde{\omega}m} - v_{\tilde{\omega}})\,\frac{\tau}{\tilde{\omega}} \quad , \tag{22}$$

where $(v_{\tilde{\omega}m}, \tilde{\omega}\,\Omega_m)$ is the velocity of the magnetospheric plasma on ∂V_2. On the other hand, we have, in the perfectly conducting magnetosphere,

$$\frac{d}{dt}\,\delta\phi = \frac{d}{dt}\left[(\tilde{\omega} B_\phi) \int_{C_f} \frac{ds}{\tilde{\omega}^2 B_p}\right] = (\Omega_m - \Omega_s) \tag{23}$$

$$\frac{\partial B_z}{\partial t} + \frac{1}{\tilde{\omega}}\frac{\partial}{\partial \tilde{\omega}}(\tilde{\omega}\, v_{\tilde{\omega}m}\, B_z) = 0 \quad . \tag{24}$$

If we assume that we know the values of $\delta\phi$ and B_z at some time t, we can compute $\underset{\sim}{B}$ in V by solving BVP2 and so obtain B_ϕ and $B_{\tilde{\omega}}$ on ∂V_2. Equations (21) and (22) then yield Ω_m and $v_{\tilde{\omega}m}$, and, by Eqs. (23) and (24), we find $d\delta\phi/dt$ and dB_z/dt. The system of equations is thus complete and may be used to determine $\underset{\sim}{B}$ at all times.

Now consider an actual disk, in which the motion of the plasma is not given, but must be determined self-consistently. We must add to our system the equations describing the dynamics of the disk. The hypothesis on τ may no longer be satisfied. However, if we are concerned only with the stationary state, we can still use the above scheme for computational purposes, although, of course, the computed intermediate states now have no physical meaning. In the final stationary state, we expect to have two distinct regions: an internal one (which may be nonexistent), in which amplification of the field by shearing is balanced by diffusion throughout the disk, the magnetosphere corotates with the star $(\Omega_m = \Omega_s)$ and $v_{\tilde{\omega}m} = 0$; and an external one, which has been blown open by the shearing, in which the magnetosphere corotates with the disk $(\Omega_m = \Omega;\ B_\phi = 0)$ and $v_{\tilde{\omega}m} = 0$. The external part cannot persist for long, because of its likely reconnection instabilities. One should take $B_\phi \approx \langle B_\phi \rangle$ there when computing the angular momentum transfer.

If the asymptotic state does not have open lines, it may also be reached by solving another scheme based on the BVP1 of Section 1-B. In that scheme, starting from a value of $B_z|_{\partial V_2}$, one (1) solves the disk equations to find $B_\phi|_{\partial V_2}$ and $B_{\varpi}|^{d}_{\partial V_2}$, (2) uses $B_z|_{\partial V_2}$ and $B_\phi|_{\partial V_2}$ to obtain $\underset{\sim}{B}$ in V (BVP1) and $B_{\varpi}|^{m}_{\partial V_2}$, (3) determines the $B_z|_{\partial V_2}$ which gives $B_{\varpi}|^{d}_{\partial V_2} = B_{\varpi}|^{m}_{\partial V_2}$, and so on. The main problem here is that BVP1 may not have solutions in some cases (Section 2).

5. Summary

In this lecture, we have proved the following results for a force-free magnetosphere:

(i) there is an upper bound on the rate at which momentum and energy may be exchanged between two regions by magnetospheric field-aligned currents. This upper bound may be estimated in a way which is independent of the details of the physical processes (which are, in general, not well known) at work in the system.

(ii) shearing of a force-free magnetosphere whose lines are anchored in two perfectly conducting regions leads to dissipation of energy in the magnetosphere, probably by a recurrent flaring process in which the field is stretched out and then reconnects.

(iii) shearing of a force-free magnetosphere when one of the regions is dissipative may lead to dissipation inside the disk (very dissipative disk), inside the magnetosphere, as in (ii) (less dissipative disk), or both.

The situation in (ii) is likely to describe what happens in the interaction of a magnetic star with a stellar companion, and (iii) the interaction between a magnetic star and an accretion disk (see F. K. Lamb's lectures). Another possible application is the problem of the coronal activity of an accretion disk. If field loops emerge from the surface of a disk (e.g. as a result of buoyancy instability) and connect externally (see, for example, Galeev et al.[85]), the shearing of these coronal loops by differential rotation in the disk will produce the kind of activity we have described above. Part of the dissipation which is necessary to make a disk accrete may well occur in the magnetosphere. There are many other astrophysical applications of the ideas developed here, since the situation in which a low-density magnetosphere is sheared by the motion of dense plasma regions is quite universal (e.g. coronae of the sun and the stars (flares), galactic coronae, etc.).

LECTURE IV
ACCRETION FLOWS

In this last lecture, we consider the structure of axisymmetric stationary accretion flows onto magnetic compact objects. In Section 1, we establish the equations governing this type of flow and, in Section 2, we consider their mathematical properties. Solutions valid near the compact star are presented in Section 3 and some aspects of the stability of accretion flows are studied in Section 4.

1. Basic equations

Consider an axisymmetric stationary accretion flow onto a magnetic compact object. According to the discussion presented in the first lecture, such a flow may be described, in most situations of interest here, by the ideal MHD equations. From Section I-3-A, we have, assuming homoentropicity for simplicity,

$$\nabla \cdot \rho \underset{\sim}{v} = 0 \quad , \tag{1}$$

$$\nabla(v^2/2 + \Phi + w) + (\nabla \times \underset{\sim}{v}) \times \underset{\sim}{v} = (\nabla \times \underset{\sim}{B}) \times \underset{\sim}{B}/4\pi \quad , \tag{2}$$

$$\nabla \times (\underset{\sim}{v} \times \underset{\sim}{B}) = 0 \quad , \tag{3}$$

$$\underset{\sim}{B} = \underset{\sim}{B}_p + B_\phi \hat{\phi} = \frac{\nabla f \times \hat{\phi}}{\tilde{\omega}} + B_\phi \hat{\phi} \quad , \tag{4}$$

where we have introduced the specific enthalpy, $w = w_o(\rho/\rho_o)^{\gamma-1}$. Equation (4), in which the flux function, $f(\tilde{\omega},z)$, appears, is equivalent to $\nabla \cdot \underset{\sim}{B} = 0$ (see Section III-1-C). The gravitational potential of the star, of mass M, is $\Phi = -GM/r$.

From Eq. (3) and the continuity of the tangential electric field at the surface of the conducting star, which is assumed to be rotating rigidly with angular frequency Ω_s, one immediately obtains

$$\underset{\sim}{v} = \kappa \underset{\sim}{B} + \tilde{\omega}\, \Omega_s \hat{\phi} \quad , \tag{5}$$

where $\kappa(\underset{\sim}{r})$ is some scalar function. In the absence of flow ($\kappa = 0$), Eq. (5) shows that the magnetosphere simply corotates with the star (Ferraro's law[102]). For $\kappa \neq 0$, the flow is exactly field-aligned in the corotating frame (and $\tilde{\underset{\sim}{E}} = -\underset{\sim}{v} \times \tilde{\underset{\sim}{B}}/c = 0$ there).

Combining Eq. (5) with the continuity equation (1), one finds that $\kappa\rho$ is constant on S(f) ($\underset{\sim}{B} \cdot \nabla(\kappa\rho) = 0$), so

$$\kappa\rho = F(f) \quad . \tag{6}$$

The quantity F(f) represents the accretion rate per unit of magnetic flux.

From the $\hat{\phi}$-component of the momentum equation (2) one obtains another first integral (we set $v_\phi = \tilde{\omega}\Omega$),

$$\tilde{\omega}^2\Omega - \frac{\tilde{\omega}B_\phi}{4\pi F(f)} = \ell(f) \quad , \tag{7}$$

which expresses the conservation of angular momentum. $\ell(f)$ is the angular momentum, per unit of accreted mass, transported along a line by the infalling matter (first term in the LHS) and the magnetic stresses (second term in the LHS). Equation (7) is a generalization of the relation $\tilde{\omega}B_\phi = H(f)$ for a force-free field (Section III-1).

The component of Eq. (2) along $\underset{\sim}{B}_p$ gives

$$\frac{1}{2}(v_p^2 + \tilde{\omega}^2\Omega^2) + w + \Phi - \Omega_s\Omega\,\tilde{\omega}^2 = E(f) \quad , \tag{8}$$

or

$$\frac{1}{2}(\kappa\underset{\sim}{B})^2 + w + (\Phi - \frac{1}{2}\tilde{\omega}^2\Omega_s^2) = E(f) \quad . \tag{9}$$

The second equality, Eq. (9), expresses the conservation of energy in the rotating frame (the term $-\tilde{\omega}^2\Omega_s^2/2$ is the centrifugal potential). In the inertial frame [Eq. (8)], the energy of the accreting matter is not conserved because of the work done by the Lorentz force ($\underset{\sim}{v} \cdot (\underset{\sim}{j} \times \underset{\sim}{B})/c \neq 0$ in general), which appears in Eq. (8) as the term $\tilde{\omega}^2\Omega\Omega_s$.

The algebraic equations (5)-(8) allow one to compute the quantities ρ, κ, $\underset{\sim}{v}$ and B_ϕ as functions of the arbitrary F(f), $\ell(f)$, E(f) (which are determined by the boundary conditions and so are assumed to be given) and the unknown f (or $\underset{\sim}{B}_p$). The remaining equation (2) determines f. The ∇f-component of Eq. (2) is

$$(1-4\pi\kappa F)\, L(f) + 4\pi F\nabla\kappa \cdot \nabla f - \tilde{\omega}^2 B_\phi^2\, F'/F$$
$$+ 4\pi F\ell'\tilde{\omega}\, B_\phi - 4\pi\rho\tilde{\omega}^2 E'(f) = 0 \quad , \tag{10}$$

where the operator L is as in Section III-1-C, and a prime

indicates differentiation with respect to f.

The above derivation is classical, and may be found in many papers concerned either with wind or accretion problems. Early references are Woltjer[103] (who seems to be the first to have derived the equations), and Mestel.[104,105] For the accretion problem, see Chia and Henriksen,[106] Inoue,[107] Ghosh, Lamb and Pethick,[108] Horiuchi,[45] Aly, Cohn, Lamb and Zylstra[89] and Zylstra.[90] Equations (1)-(10) have also been used in connection with laboratory plasma physics.[109]

2. Analysis of the equations. Critical points

A. The equation for f

As remarked in Section 1, the quantities κ, ρ, Ω and B_ϕ may be computed algebraically as functions of f, ∇f, and $\underset{\sim}{r}$. Their gradients may thus be computed as functions of f, ∇f, $\nabla\nabla f$, and $\underset{\sim}{r}$, and Eq. (10) is a second-order non-linear equation for f. It is interesting to present explicitly the form in which the second derivatives of f enter the equation. After a lengthy calculation, Eq. (10) may be written as

$$(1-4\pi\kappa F)\ \Big\{\Big(X + \frac{4\pi\kappa^3 F}{\tilde{\omega}^2}\, f'^{\,2}_{\tilde{\omega}}\Big) f''_{\tilde{\omega}^2} + \frac{8\pi\kappa^3 F}{\tilde{\omega}^2}\, f'_{\tilde{\omega}}\, f'_z\, f''_{\tilde{\omega} z}$$

$$+ \Big(X + \frac{4\pi\kappa^3 F}{\tilde{\omega}^2}\, f'^{\,2}_z\Big)\, f''_{z^2}\Big\} = G(f,\ \nabla f,\ \underset{\sim}{r}) \quad , \tag{11}$$

where

$$X = -\,v_p^4 + 2\,v_p^2\,(c_s^2 + c_A^2) - 4\,c_s^2\,c_{Ap}^2 \ . \tag{12}$$

The equation is linear with respect to the second derivatives (it is a so-called quasi-linear equation). This type of equation may be classified according to the sign of the discriminant, $\Delta = \beta^2 - \alpha\gamma$, where α, 2β, γ are the coefficients of $f''_{\tilde{\omega}^2}$, $f''_{\tilde{\omega} z}$ and f''_{z^2}, respectively. A simple calculation gives

$$\Delta = \Big(1 - \frac{v_p^{\,2}}{C_{Ap}^2}\Big)^2 \Big[(v_p^{\,2} - c_s^2)\ \Big(1 - \frac{v_p^{\,2}}{C_{Ap}^2}\Big) + v_p^{\,2}\,\frac{B_\phi^{\,2}}{B_p^{\,2}}\Big]$$

$$\times\ \Big[c_s^2\ \Big(1 - \frac{v_p^{\,2}}{C_{Ap}^2}\Big) - v_p^{\,2}\,\frac{B^2}{B_p^{\,2}}\Big] \quad , \tag{13}$$

where $c_{Ak}^2 = B_k^2/4\pi\rho$. The roots of Δ correspond to

$$\text{(i)}\quad v_p^2 = c_{Ap}^2 \tag{14}$$

$$\text{(ii)}\quad v_p^2 = \frac{1}{2}\left\{(c_s^2 + c_A^2) \pm [(c_s^2 + c_A^2)^2 - 4c_s^2\, c_{Ap}^2]^{1/2}\right\} = \begin{cases} (c_A^s)^2 & (-) \\ (c_A^f)^2 & (+) \end{cases} \tag{15}$$

$$\text{(iii)}\quad v_p^2 = \frac{c_s^2\, c_{Ap}^2}{c_s^2 + c_A^2} = \left(c_A^{(e)}\right)^2 \leqslant \left(c_A^s\right)^2 . \tag{16}$$

The sign of Δ, as well as the nature of equation (11), is shown in the following table.

Table II

Nature of the equation for f.

v_p^2	0		$\left(c_A^{(e)}\right)^2$		$\left(c_A^{(s)}\right)^2$		$\left(c_A^{(f)}\right)^2$	
Δ		$-$	0	$+$	0	$-$	0	$+$
type of equation		elliptic		hyperbolic		elliptic		hyperbolic

It is worth noting that the sign of Δ does not change at a point where $v_p^2 = c_{Ap}^2$. At such a point, one actually has $\alpha = \beta = \gamma = 0$ and the equation is singular.

B. The meaning of the critical points. Elliptic vs. hyperbolic equations

The existence of changes in the character of an equation describing a stationary flow (i.e. of transitions between elliptic and hyperbolic regions on critical lines where the flow velocity equals some characteristic speed of the medium) is a well-known phenomenon in pure hydrodynamics (in that case, the critical velocity is the sound speed, c_s). The physical difference between an elliptic region and a hyperbolic one may be understood by considering a small region of fluid surrounding a point, P, which is emitting sound pulses. Spherical wave fronts produced at P propagate through the fluid and, at the same time, are carried along by the flow. There are thus two possible situations. If the local speed, v, of the flow is less than the local sound speed, c_s, the wave fronts emitted at different times all surround P and do not intersect. Alternatively, if $v > c_s$, P lies outside the wave fronts, which intersect one another. The former case is characterized geometrically by the impossibility of drawing, at time δt, tangents from P to the wave front emitted at t=0, now centered on the point $P + \tilde{v}\,\delta t$. In that case, all surrounding points can be reached by signals emitted at P, which belongs to an elliptic region. In the latter case, where such tangents can be drawn (they are the so-called characteristic curves and form the "Mach cone"), there are points in any small neighborhood of P which cannot be reached by signals from P (one says that these points are outside the "zone of influence" of P), and P is in a hyperbolic region. Thus, elliptic and hyperbolic regions are characterized physically by the possibility or impossibility of linking arbitrary points by waves, and the nature of a region may be determined using an elementary geometrical construction.

Returning the axisymmetric MHD problem, we assume, for simplicity, that $B_\phi = 0$ and $v_\phi = 0$ (so the flow is field-aligned). Following the above discussion, we investigate the possibility of linking neighboring rings (with axes parallel to $\hat{z}$) by axisymmetric waves (which may be now of three types: Alfvén, slow and fast), and we consider the possible existence of characteristic curves. Wave fronts at time δt, emitted at t = 0 by a ring passing through a point P, are found by rotating around $\hat{z}$ one of the wave-front diagrams in Fig. 2, after scaling the diagram and placing its origin at distance $\tilde{v}\,\delta t = \kappa\tilde{B}\delta t$ from P. It is obvious from the figure that drawing tangents (i.e. characteristic curves) from P to the wave front is possible if $c_s^2c_A^2\,(c_s^2+c_A^2)^{-1} < v^2 < c_s^2$ or $c_A^2 < v^2$, where we have assumed for definiteness that $c_s^2 < c_A^2$. If one of these conditions is satisfied, P belongs to a hyperbolic region. If, on the other hand, $0 \leqslant v^2 < c_s^2c_A^2/(c_s^2+c_A^2)^{-1}$ or $c_s^2 < v^2 < c_A^2$, P is in an elliptic region. Thus we have found by a different method the critical velocities which emerged from the analysis of Subsection A above (relations (15) and (16), with $c_{Ap} = c_A$). Note that the critical "c_A" above should be interpreted not as the speed of the Alfvén mode, but as the speed of the fast one; the two speeds happen to be equal when $\tilde{v}\parallel\tilde{B}$. The special character of the Alfvénic point, which is not a transition point but a singular one, may also be understood by considering the diagrams in Fig. 2. The Alfvén axisymmetric wave front reduces to two rings located on the field surface passing through the emitting ring. There is always an Alfvén zone of dependence associated with the ring through P (and so no transition associated with this mode), but this zone is of zero volume: it coincides either with the whole magnetic surface passing through P if $v < c_A$, or with the part of that surface lying downstream of P if $v > c_A$. The singular condition $v = c_A$ corresponds to the permanent coincidence of one of the half-Alfvén fronts with the source.

When B_ϕ and v_ϕ are non-zero, the situation is more involved. As above, we consider the propagation through the moving fluid of the axisymmetric wave fronts emitted by a ring. However, the angle between the ring axis and the field is nonzero, and the critical speeds [Eqs. (15) and (16)] differ from those obtained above. Clearly, the value of v_ϕ is irrelevant to the existence of the characteristic curves, as this component is parallel to the wave fronts. This is why $\underset{\sim}{v}_p$, and not $\underset{\sim}{v}$, appears in Eqs. (15) and (16). The above remarks concerning the Alfvénic point (which now no longer coincides with the fast critical point) still apply.

Elliptic and hyperbolic equations have quite different mathematical properties. These are developed in detail in many textbooks (e.g. Courant and Hilbert[110]), to which we refer the reader. We simply quote here one of the most important differences for physical applications: while the solutions of elliptic equations (with analytic coefficients) are analytic, solutions of hyperbolic equations may admit discontinuities (i.e. shocks).

C. Accretion flows along a given magnetic field

For a given field, $\underset{\sim}{B}_p$, the algebraic equations (5)-(8) can be solved for the variables κ, ρ, Ω and B_ϕ. It is convenient to use the parameter

$$\chi = M_{Ap}^2 = \frac{4\pi\rho\, v_p^2}{B_p^2} , \tag{17}$$

in terms of which the expressions for κ, ρ, Ω and B_ϕ are

$$\kappa = \chi\ (4\pi F)^{-1} , \tag{18}$$

$$\rho = \chi^{-1}\ (4\pi F^2) , \tag{19}$$

$$\Omega = \frac{\Omega_s - \chi\ell/\tilde{\omega}^2}{1-\chi} , \tag{20}$$

and

$$B_\phi = \frac{4\pi F\tilde{\omega}}{1-\chi}\,(\Omega_s - \ell/\tilde{\omega}^2) . \tag{21}$$

Substituting these expressions into Eq. (8), one then obtains an equation for χ which is a fourth-degree algebraic equation for a cold flow, and a $(4q + p)$-degree equation for an adiabatic flow of index $\gamma = p/q + 1$.

The solutions for Ω and B_ϕ have a singularity if the flow passes through the critical point where $v_p^2 = c_{Ap}^2$ $(\chi = 1)$, unless the distance from this point to the $\hat{z}$-axis is given by

$$\tilde{\omega}^2 = \ell/\Omega_s \ . \tag{22}$$

This relation can be satisfied only if $\ell\,\Omega_s > 0$, i.e. if the angular momentum transfer on the line and the angular velocity of the star have the same sign.

If a field surface, S(f), is confined between radii $\tilde{\omega}_1(f)$ and $\tilde{\omega}_2(f)$, Eq. (22) can hold only if

$$\tilde{\omega}_1^2(f)\,\Omega_s \leq \ell(f) \leq \tilde{\omega}_2^2(f)\,\Omega_s \ , \tag{23}$$

where we have taken $\Omega_s > 0$ and $\ell > 0$. If this relation is not satisfied by the arbitrary function ℓ, then either the flow never passes through the Alfvênic point or the problem has no solution. As an example, let us consider an accretion flow between a disk and a star. In that case, $\tilde{\omega}_2(f)$ cannot be very much larger than the inner radius, r_i, of the disk, where matter starts flowing out along the field lines (since the pressure is very small, $c_s^2 \ll GM/r$, there is no outward driving force to make the flow extend to large radii). In this situation, a smooth flow either does not go through the Alfvênic point, or, if it does, the angular momentum transported by the lines must satisfy

$$\ell \lesssim r_i^2\Omega_s \ . \tag{24}$$

In the first case, $\rho\, v_p^2/2 < B_p^2/8\pi$ throughout the magnetosphere, so the currents required to balance the forces due to accretion have only a small effect on the structure of the magnetosphere. The currents which generate B_ϕ, however, may be strong, as in the force-free situation described in Lecture III.

The other critical points do not appear explicitly in the equations (they are not singular points like the Alfvênic point), but they appear in the differential equations for κ, ρ, Ω and B_ϕ along the lines of $\underset{\sim}{B}_p$.

3. Accretion flow near a compact object

In the region near the compact star, the energy of the magnetic field dominates the energy density of the plasma and the field is nearly force-free. Therefore, the field does not depart much from the vacuum field generated by the currents flowing inside the star. We may thus assume that we know f and $\underset{\sim}{B}_p$, and we will take for $\underset{\sim}{B}_p$ the field of a dipole of magnetic moment $\underset{\sim}{\mu} = \mu\hat{z}$. The solution to the equations for this flow has been computed by Ghosh, Lamb and Pethick,[108] and we will follow their arguments.

As shown above, all the flow variables may be computed if we know f and we have chosen the arbitrary functions F(f), ℓ(f), and

E(f). From Eq. (8), we have

$$\frac{1}{2}(v_p^2 + \tilde{\omega}^2\Omega^2) + w - \frac{GM}{r} - \Omega_s\Omega\,\tilde{\omega}^2$$

$$\approx \frac{1}{2}\tilde{\omega}_o^2\,\Omega_o^2 + w_o - \frac{GM}{r_o} - \Omega_s\,\Omega_o\,\tilde{\omega}_o^2 \quad , \tag{25}$$

where the index 0 refers to the point at which the flow is matched onto the outer accretion flow. In that boundary region, all the terms on the RHS of Eq. (25) are smaller than GM/r_o or of the same order. For $r \ll r_o$, the RHS of Eq. (25) is small compared to GM/r and may be taken equal to zero. Inside the magnetosphere, the terms $\tilde{\omega}^2\Omega^2/2$ and $\Omega_s\Omega\tilde{\omega}^2$ in the LHS of Eq. (25) may also be neglected. This follows because Ω_s is smaller than $(GM/r_o^3)^{1/2}$ (a necessary condition for accretion) and $|\tilde{\omega}(\Omega - \Omega_s)| = |v_p(B_\phi/B_p)| \ll |v_p|$ ($B_\phi \ll B_p$, since B_ϕ is created by the motion of the matter, whose energy is smaller than that of the poloidal field, by assumption). Moreover, the pressure term is also negligible (cooling is fast enough to prevent mirror forces from becoming large; see Ghosh, Lamb and Pethick[108]). Not too surprisingly, the poloidal velocity of the matter in the inner part of the magnetosphere is then given by the free-fall value,

$$v_p^2 = \frac{2GM}{r} \quad . \tag{26}$$

On a dipolar line, with equation $r = r_o \sin^2\theta$, we have

$$M_A = \frac{|v_p|}{c_{Ap}} = M_o(f)\,\frac{\sin^5\theta}{[4 - 3\sin^2\theta]^{1/2}}$$

$$= M_o(f)\left(\frac{r}{r_o}\right)^{5/2}\frac{1}{[4 - 3(r/r_o)]^{1/2}} \quad , \tag{27}$$

where $M_o(f)$ is a constant on any field line. This equation, along with Eqs. (20) and (21), gives the values of Ω and B_ϕ.

In the situations usually considered, Ω_s and ℓ have the same sign. If the star is slowly rotating ($|\Omega_s| < |\ell/r_A^2|$, where r_A is the radius at which M_A, given by Eq. (27), is equal to one), it follows that the sign of Ω is opposite to that of Ω_s throughout that part of the magnetosphere in which the above approximations are valid. Physically, this is due to the fact that, when $\ell\,\Omega_s > 0$, the lines must have a forward pitch to accelerate the rotation of the star (momentum transfer in the inner magnetosphere is essentially magnetic). Matter falling along the lines therefore rotates backward in the corotating frame; it also rotates backward in the fixed frame if Ω_s is not too large. In the opposite situation, where $\ell\,\Omega_s < 0$, the matter always rotates in the same sense as the neutron star, and in the opposite sense to the angular momentum transfer.

4. Stability of accretion flows

The stability of accretion flows has not yet been investigated in much detail. However, one important point has been made about the transition from a super-Alfvénic field-aligned flow to a sub-Alfvénic one at the Alfvén surface (where $v^2 = c_A^2$). It has been found that such a transition is always unstable.[111] No other conclusions have yet been established. However, it may be that some of the numerous results obtained in studies of sheared flows in various other contexts are relevant to our problem.

A. Stability/instability of trans-Alfvénic flows

Consider a field-aligned flow which passes from a region where it is super-Alfvénic ($v^2 > c_A^2$) to one where it is sub-Alfvénic ($v^2 < c_A^2$). As indicated above, the flow is unstable in the transition region. The physical explanation of this fact is quite simple. Suppose that, because of perturbations in the sub-Alfvénic domain, Alfvén waves are created and escape in both directions along the field lines. Consider a wave which propagates upstream (outwards) at the speed $c_A - |v|$ with respect to the fixed frame. As the Alfvén surface, S_A, is approached, the propagation speed decreases, reaching zero on S_A (actually, the wave never reaches S_A; see below). Energy thus accumulates continuously in the neighborhood of S_A, and the situation is unstable.

The reverse transition, which occurs, for example, in a stellar wind, is stable. In that case, perturbations in the inner region are transported downstream at the speed $c_A + v$ and pass the transition zone without difficulty. On the other hand, perturbations created upstream with respect to the Alfvén surface cannot propagate downstream ($v > c_A$), so they can never perturb the neighborhood of S_A. It is worth noting that a similar phenomenon is well known in gas dynamics. A smooth transition from a subsonic region to a supersonic one is possible (and observed), while the reverse transition necessarily implies the presence of shocks. This is well explained by the above argument, with sound waves substituted for Alfvén waves.[112]

The mathematical proof of the instability in the MHD case has been given by Williams,[111] using a very simple spherical model in which

(i) the field is radial, $\underset{\sim}{B} \propto r^{-2}\,\hat{r}$, and

(ii) the unperturbed flow is field-aligned and has velocity $\underset{\sim}{v} = -\,Kr^{-s}\,\hat{r}$, where $K > 0$ (non-rotating accretion flow).

The magnetic field has no effect on the basic flow, which is just the classical Bondi flow describing spherical accretion.[113] The Alfvén Mach number is given by

$$M_A^2 = \frac{4\pi\rho v_r^2}{B^2} \propto r^{2-s} \quad , \tag{28}$$

where use has been made of the constancy of $\rho v_r r^2$ for a stationary spherically symmetric flow (conservation of matter). If $s < 2$, (e.g. for a free-fall flow, with $s = 1/2$), M_A is zero at infinity, and tends to infinity at the origin; there is thus a point where $M_A = 1$.

If the perturbations are not purely radial, their propagation is affected by the field (Section I-3-C). These perturbations, if small, are solutions to the linearized ideal MHD equations, which take a simple form if $\partial/\partial\phi = 0$. In that case, the equations for δv_ϕ and δB_ϕ decouple from the others. For this system, one can show the existence of a "potential" function (from which one may determine δv_ϕ and δB_ϕ) which obeys a second-order partial differential equation. By computing its characteristic curves, one finds that an outward-moving disturbance takes an infinite time to reach the Alfvén surface. The equation can be transformed into characteristic form and solved by a WKB method, yielding the values of δv_ϕ and δB_ϕ and the energy associated with these quantities,

$$\delta W = \int \{ \frac{1}{2} \rho(\delta v_\phi) + \frac{(\delta B_\phi)^2}{8\pi} \} \, d\underset{\sim}{r} \quad , \tag{29}$$

which increases exponentially in time. This proves the unstable character of the flow.

B. <u>Discussion of the instability</u>

Following Williams, we now discuss two important points: (i) does the result depend on the strong assumptions of the model, or is it quite general? and (ii) what actually happens if a flow has to pass through an Alfvén critical point?

(i) The above simple model assumes that the curvature of the lines is zero and that the flow is field-aligned. The introduction of curved lines does not change the situation, because it is always possible to consider regions near S_A which are sufficiently small that the 1-D model applies. The flow is thus "locally" unstable--the global geometry of the lines has no effect. If the flow is not field-aligned, but is axisymmetric, one can transform to the rotating frame where $\underset{\sim}{v} \parallel \underset{\sim}{B}$. In this way, Williams deduces that the transition is still unstable (however, it could be argued that, in the rotating frame, the equations of the simple model are no longer valid because of the presence of the Coriolis force, which couples δv_ϕ and δv_θ and changes the propagation of the waves). On the other hand, the model considers only the Alfvén waves, and it is interesting to address the possibility of a "piling-up" instability corresponding to the other modes (slow and fast). The point is that these waves are not constrained to propagate along the field lines and in a geometry more complex than the spherical one they may be able to escape "across the lines." This case is much more complicated, and new computations are needed.

(ii) What happens if a flow is forced through the super- to sub-Alfvênic transition? One possibility is that a shock develops, as in the corresponding situation in gas dynamics. However, a smooth trans-Alfvênic shock is unstable.[114] The most likely possibility is that the transition would take place through an extended turbulent region. Williams gives the following picture of the flow: (a) an outer region, in which the field has small effects; (b) a shock, followed by a turbulent region in which the flow becomes sub-Alfvênic; (c) a stable sub-Alfvênic flow in the magnetosphere.

C. Other possible sources of instability

The instability discussed above appeals to one particular mechanism. There may be many other mechanisms also leading to instability. Accretion is an example of sheared flow, a class of flows known to be unstable in many cases (e.g. the KHI discussed in Section II-2, which may occur when all the shear is concentrated in a thin sheet). As an example of work on the stability of sheared flows which may be relevant to accretion flows, we quote the paper by Lau and Liu,[115] in which a sufficient criterion for stability is proved. A field-aligned flow, $\underset{\sim}{v} = v_o(y)\,\hat{z}$, is found to be stable if the maximum of $|v_o|$ is less than twice the Alfvén speed, irrespective of the shape of the profile, v_o. (In their model, the plasma is incompressible and the magnetic field is uniform.) A similar result is found by Adam.[116]

On the other hand, when plasma flows along curved lines, a centrifugal force appears which tends to accentuate the concavity of the flow lines. The net gravitational and centrifugal force may act to destabilize the plasma, roughly as described in Section II-1. This may, however, be balanced by other effects of the flow, and more work is needed before a firm conclusion can be reached.

CONCLUSION

In this series of lectures we have studied some MHD problems which are important in constructing self-consistent models of systems containing an accreting magnetic compact object.

One of the first problems to emerge is the nature of the mechanisms which permit the infalling matter to cross the stellar magnetic field lines, enter the magnetosphere and mix with the field. Numerous works have suggested that large-scale MHD instabilities (in particular RTI and KHI) play an important role in this process. Until now, most studies have concentrated on the linear theory of these instabilities. This is important in proving their relevance, but is not sufficient if one wants to follow the evolution of the system, which necessarily involves the development of large amplitudes and the appearance of non-linear effects. These effects are difficult to handle analytically and, to go beyond the current simple qualitative descriptions of the non-linear stages, it is necessary to appeal to extensive numerical simulations of the kind undertaken by Wang et al.[54,55]

Another important process is the reconnection of magnetic field lines. This has been abundantly studied for two decades in relation to laboratory plasmas (in fusion devices, etc.) and astrophysical ones (in particular, solar flares and the Earth's magnetosphere). Although a great deal is now understood about this process, most of the analytical and numerical work has considered special configurations and may be difficult to apply in the context of accreting magnetic compact objects. In these systems, reconnection is believed to occur between field lines belonging to regions where conditions are quite dissimilar and which have large relative motions. More specific studies, taking these effects into account, are needed.

In an accretion disk, all the ingredients required for the development of a dynamo process (differential rotation, turbulence and seed magnetic fields) are present. Much should be learned from the developments which may be expected in the next few years in the active field of dynamo theories. The possibility of generating a large-scale disk magnetic field,[82] which could then reconnect with the stellar field, is an appealing one, and is likely to be important in understanding accretion onto a magnetic compact object. In addition, the effects of internal turbulence on the threading of a disk (or a star) by an external field require more study. This problem has only been considered on the basis of intuitive arguments. The use of existing computer codes to determine what actually happens would be welcome.

In Lecture III, we considered the structure of a force-free magnetosphere. When this is posed as a boundary-value problem, with very little of the physics retained, one has a well-defined mathematical problem to solve. Analytic studies of the type we have reported may give, if not a closed-form solution to any particular BVP, at least the main qualitative features of the solutions (existence or non-existence, stability, characteristics of the evolution of a field when parameters are changed, etc.). However, these studies are quite difficult, as they deal with a domain of mathematics (the theory of non-linear partial differential equations) in which many basic problems remain to be solved. The development of numerical codes is another important factor, as they are the only way to obtain explicit solutions to the concrete problems of interest. A 2-D code now exists for computing the structure of a disk-accreting magnetosphere,[89,90] and should be used next to solve the self-consistent problem in which the structure of the disk is computed simultaneously with that of the magnetosphere. A 3-D code, which would allow the correct treatment of the interaction of a magnetic star with its companion, or the interaction of an oblique rotator with a disk, is still a dream.

Another basic problem in the physics of accreting magnetic compact objects is the determination of the structure of the plasma flow between the outer flow region and the star. For most of the accreting matter to fall on the polar caps, as required by the observed pulsed radiation, the plasma must be threaded by the field lines and must flow along magnetic surfaces. In Lecture IV, we considered this type of flow by using a simple model which assumed axisymmetry and stationarity. Even in this case, little has been done beyond a simple analysis of the equations of the problem: the problem has been reduced to a set of algebraic equations and a quasi-linear second-order partial differential equation of mixed type (i.e., there are some regions where the equation is elliptic and some where it is hyperbolic). To emphasize the difficulties involved, it may be enough to look at the number of papers which are still published every year about transsonic flows in hydrodynamics, a problem which has been studied for decades. Numerical codes of the type currently being developed by Aly, Cohn, Lamb, and Zylstra[89,90] are necessary to go further into the MHD problem.

Although our knowledge of the basic mechanisms at work in accreting magnetic compact objects has increased dramatically during the last ten years, we are still far from having self-consistent models, and much work remains to be done.

ACKNOWLEDGEMENTS

I would like to thank the organizers of the Taos Workshop, R. Epstein, D. Evans, W. Feldman and F. K. Lamb, for enabling me to present these lectures. I am very grateful to F. K. Lamb for the numerous valuable discussions about accretion theories I had with him during several stays at the University of Illinois at Urbana-Champaign, when I had the opportunity to collaborate with him and several members of his research group: H. Cohn, M. Cook, P. Ghosh and G. Zylstra, whom I would also like to thank. Many of the results discussed here were obtained during this collaboration. The typing of the manuscript by J. Christensen is gratefully acknowledged.

The preparation of these notes was supported by CNRS, France, Los Alamos National Laboratory, and NSF grant PHY 80-25605 and NASA grant NSG 7653 at Illinois.

REFERENCES

1. Krall, N. A., and Trivelpiece, A. W., "Principles of Plasma Physics" (New York: McGraw-Hill Book Company, 1973).
2. Roberts, P. H., "An Introduction to Magnetohydrodynamics", (New York: Elsevier, 1967).
3. Rossi, B., and Olbert, S., "Introduction to the Physics of Space", (New York: McGraw Hill, 1970).
4. Thompson, W. B., "An Introduction to Plasma Physics", (Reading, MA: Addison-Wesley, 1962).
5. Pomraning, G. C., "The Equations of Radiation Hydrodynamics", (Oxford: Pergamon Press, 1973).
6. Jeffrey, A., and Taniuti, T., "Non-Linear Wave Propagation" (New York: Academic Press, 1964).
7. Bernstein, I. B., Frieman, E. A., Kruskal, M. D., and Kulsrud, R. M., Proc. Roy. Soc. (London), A244, 17 (1958).
8. Lamb, F. K., in "Magnetospheric Boundary Layers", Proceedings of the Sydney Chapman Conference, Alpbach, Austria, ed. B. Barrick (ESA SP Series, 1979).
9. Lamb, F. K., in "High Energy Transients in Astrophysics" (Santa Cruz, CA, 1983), ed. S. E. Woosley, AIP Conference Proceedings no. 115 (1984).
10. Vasyliunas, V. M., Space Sci. Rev., 24, 609 (1979).
11. Fortner, B., Lamb, F. K., and Zylstra, G., in Numerical Astrophysics, eds. J. Centrella, J. LeBlanc, M. LeBlanc, and R. I. Bowers (in press) (1983).
12. Arons, J., Burnard, D. J., Klein, R. I., Mckee, C., and Pudritz, R., in "High Energy Transients in Astrophysics", (Santa Cruz, CA., 1983), ed. S. E. Woosley, AIP Conference Proceedings no. 115 (1984).
13. Pringle, J. E., and Rees, M. J., Astron. Astrophys., 21, 1 (1972).
14. Lamb, F. K., Pethick, C. J., and Pines, D., Astrophys. J., 184, 271 (1973).
15. Davidson, K., and Ostriker, J. P., Astrophys. J., 179, 585 (1973).
16. Arons, J., and Lea, S. M., Astrophys. J., 207, 914 (1976).
17. Arons, J., and Lea, S. M., Astrophys. J., 210, 792 (1976).
18. Arons, J., and Lea, S. M., Astrophys. J., 235, 1016 (1980).
19. Elsner, R. F., Ph.D. Thesis, University of Illinois at Urbana-Champaign (1976).
20. Elsner, R. F., and Lamb, F. K., Astrophys. J., 215, 897 (1977).
21. Elsner, R. F., and Lamb, F. K., Astrophys. J., 278, 326 (1984).
22. Basko, M. M., Sov. Astron., 21, 595 (1977).
23. Baan, W. A., Astrophys. J., 214, 245 (1977).
24. Baan, W. A., Astrophys. J., 227, 987 (1979).
25. Michel, F. C., Astrophys. J., 213, 836 (1977).
26. Michel, F. C., Astrophys. J., 214, 261 (1977).

27. Michel, F. C., Astrophys. J., 216, 838 (1977).
28. Burnard, D. J., Lea, S. M., and Arons, J., Astrophys. J., 266, 175 (1983).
29. Ichimaru, S., Astrophys. J., 208, 701 (1976).
30. Scharlemann, E. T., Astrophys. J., 219, 617 (1978).
31. Ghosh, P., and Lamb, F. K., Astrophys. J., 232, 259 (1979).
32. Ghosh, P., and Lamb, F. K., Astrophys. J., 234, 296 (1979).
33. Anzer, U., and Börner, G., Astron. Astrophys., 83, 133 (1980).
34. Anzer, U., and Börner, G., Astron. Astrophys., 122, 73 (1983).
35. Aly, J. J., Astron. Astrophys., 86, 192 (1980).
36. Kundt, W., and Robnik, M., Astron. Astrophys., 91, 305 (1980).
37. Riffert, H., Astrophys. Space Sci., 71, 195 (1980).
38. Lipounov, V. M., Sov. Astron., 24, 722 (1980).
39. Lipounov, V. M., Astrophys. Space Sci., 82, 343 (1982).
40. Lipounov, V. M., and Shakura, N. I., Sov. Astron. Lett., 6, 14 (1980).
41. Lipounov, V. M., and Shakura, N. I., Sov. Astron. Lett., 7, 72 (1980).
42. Lipounov, V. M., Semënov, E. S., and Shakura, N. I., Sov. Astron., 25, 439 (1981).
43. Horiuchi, R., and Tomimatsu, A., Prog. Theor. Phys., 62, 400 (1979).
44. Horiuchi, R., Kadonaga, T., and Tomimatsu, A., Prog. Theor. Phys., 66, 172 (1981).
45. Horiuchi, R., Prog. Theor. Phys., 68, 541 (1982).
46. Joss, P. C., Katz, J. I., and Rappaport, S. A., Astrophys. J., 230, 176 (1979).
47. Lamb, F. K., Aly, J. J., Cook, M. Lamb, D. Q., Astrophys. J. Lett., 274, 271 (1983).
48. Campbell, C. G., Mon. Not. R. Astron. Soc., 205, 1031 (1983).
49. Chanmugam, G., and Dulk, G. A., in "Cataclysmic Variables and Related Objects" (IAU Colloquium no. 72, Haifa, Israel, 1982), ed. M. Livio and G. Shaviv, p. 223 (1983).
50. Goldreich, P. and Julian, W. M., Astrophys. J., 157, 869 (1969).
51. Michel, F. C., and Dessler, A. J., Astrophys. J., 251, 654 (1981).
52. Lipounov, V. M., Sov. Astron., 22, 702 (1978).
53. Chandrasekhar, S., "Hydrodynamic and Hydromagnetic Stability", (Oxford, Clarendon, 1961).
54. Wang, Y. M., and Nepveu, M., Astron. Astrophys., 118, 267 (1983).
55. Wang, Y. M., Nepveu, M., and Robertson, J. A., Astron. Astrophys., 135, 66 (1984).
56. Gerwin, R. A., Rev. Mod. Phys., 40, 652 (1968).
57. Todd, L., Phys. Fluids, 9, 814 (1966).
58. Plesset, M. S., and Hsieh, D. Y., Phys. Fluids, 7, 1099 (1964).

59. Southwood, D. J., in "Magnetospheric Boundary Layers", Proceedings of the Sydney Chapman Conference, Alpbach, Austria, 1979, ed. B. Battrick (ESA SP Series, 1979).
60. Aly, J. J., Ghosh, P., and Lamb, F. K., unpublished (1980).
61. Parker, E. N., "Cosmical Magnetic Field", (Oxford: Clarendon Press, 1979).
62. Wang, Y. M., and Welter, G. L., Astron. Astrophys., 113, 113 (1982).
63. Northrop, T., Phys. Rev., 103, 1150 (1956).
64. Phillips, O. M., "The Dynamics of the Upper Ocean", (Cambridge: Cambridge University Press, 1977).
65. Wolff, R. S., Goldstein, B. E., Yeates, C. M., J. Geophys. Res., 85, 7697 (1980).
66. Wang, Y. M., Astron. Astrophys., 74, 253 (1979).
67. Vasyliunas, V. M., Rev. Geophys. Space Phys., 13, 303 (1975).
68. Pellat, R., Space Science Rev., 23, 359 (1979).
69. Furth, H. P., Killeen, J., and Rosenbluth, M. N., Phys. Fluids, 6, 459 (1983).
70. Bateman, G., "MHD Instabilities" (Cambridge: MIT Press, 1978).
71. Pellat, R., in "Solar Phenomena in Stars and Stellar Systems", ed. R. M. Bonnet and A. K. Dupree (Dordrecht: Reidel Publishing Company, 1981).
72. Finn, J. M., and Kaw, P. K., Phys. Fluids, 20, 72 (1977).
73. Pellat, R., Sov. J. Plasma Phys., 9, 124 (1983).
74. Bhattacharjee, A., Brunel, F., and Tajima, T., Phys. Fluids, 26, 3332 (1983).
75. Hofmann, I., Plasma Phys., 17, 143 (1975).
76. Dobrowolny, M., Veltri, P. and Mangeney, A., J. Plasma Phys., 29, 393 (1983).
77. Sonnerup, B. V. Ö, , in "Solar System Plasma Physics", Vol. III, ed. L. T. Lanzeratti, C. F. Kennel, and E. N. Parker (Amsterdam: North-Holland Publishing Company, 1979).
78. Sweet, P. A., Nuovo Cimento Suppl., 8, 188 (1958).
79. Parker, E. N., J. Geophys. Res., 62, 509 (1957).
80. Petschek, H. E., in "The Physics of the Solar Flares", AAS - NASA Symposium (Greenbelt, MD, 1963), NASA SP. 50, ed. W. N. Hess, p. 425 (1964).
81. Neugebauer, M., and Tsurutari, B. T., Astrophys. J., 226, 494 (1979).
82. Pudritz, R. E., Mon. Not. R. Astron. Soc., 195, 881 and 897 (1981).
83. Eardley, D. M., and Lightman, A. P., Astrophys. J., 200, 187 (1975).
84. Coroniti, F. V., Astrophys. J., 244, 587 (1981).
85. Galeev, A. A., Rosner, R., and Vaiana, G. S., Astrophys. J., 229, 318 (1979).
86. Stella, L., and Rosner, R., Astrophys. J., 277, 312 (1984).
87. Mouschovias, T. C., in "Protostars and Planets", ed. T. Gehrels (Tucson: University of Arizona Press), p. 209 (1978).

88. Heyvaerts, J., Lasry, J. M. Schatzman, M., and Witonsky, P., Astron. Astrophys., 111, 104 (1982).
89. Aly, J. J., Cohn, H., Lamb, F. K., and Zylstra, G., in preparation (1984).
90. Zylstra, G., Ph.D. Thesis, University of Illinois at Urbana-Champaign (1985).
91. Aly, J. J., preprint (1984).
92. Barnes, C. W., and Sturrock, P. A., 1972, Astrophys. J., 174, 659 (1972).
93. Sturrock, P. A., and Barnes, C. W., Astrophys. J. 176, 31 (1972).
94. Aly, J. J., in "Unstable Current Systems and Plasma Instabilities in Astrophysics," M. R. Kunda and G. D. Holman, Ed., (Dortrecht: Reidel Publishing Company), p. 221 (1985).
95. Aly, J. J., Astron. Astrophys., 143, 19 (1985).
96. Aly, J. J., in preparation.
97. Birn, J., and Schindler, K., in "Solar Flares Mangetohydrodynamics", ed. E. R. Priest (New York: Gordon and Breach), p. 337 (1981).
98. Low, B. C., Rev. Geophys. Space Phys., 20, 145 (1982).
99. Birn, J., Goldstein, H., and Schindler, K., Solar Phys., 57, 81 (1978).
100. Priest, E. R., and Milne, A. M., Solar Phys., 65, 315 (1980).
101. Aly, J. J., Astrophys. J., 283, 349 (1984).
102. Ferraro, V. C. A., Mon. Not. R. Astron. Soc., 97, 458 (1937).
103. Woltjer, L., Astrophys. J., 130, 405 (1959).
104. Mestel, L., Mon. Not. R. Astron. Soc., 122, 473 (1961).
105. Mestel, L., Mon. Not. R. Astron. Soc., 138, 359 (1968).
106. Chia, T. T., and Henriksen, R. N., Astrophys. J., 177, 699 (1972).
107. Inoue, H., Publ. Astron. Soc. Japan, 28, 293 (1976).
108. Ghosh, P., Lamb, F. K., and Pethick, C. J., Astrophys. J., 217, 578 (1977).
109. Zehrfeld, H. P., and Green, B. J., Nucl. Fusion, 12, 569 (1972).
110. Courant, R., and Hilbert, D., "Methods of Mathematical Physics", Vol. 2, (New York: Interscience Publishers, 1962).
111. Williams, D. J., Mon. Not. R. Astron. Soc., 171, 537 (1975).
112. Meyer, R. E., Quart. J. Mech. Appl. Math., 5, 257 (1952).
113. Bondi, H., Mon. Not. R. Astron. Soc., 112, 195 (1952).
114. Shercliff, J. A., "A Textbook of Magnetohydrodynamics" (Oxford: Pergamon Press, 1965).
115. Lau, Y. Y., and Liu, C. S., Phys. Fluids, 23, 939 (1980).
116. Adam, J. A., Quart. J. Mech. Appl. Math. (1978).

GLOBAL ASPECTS OF STREAM EVOLUTION IN THE SOLAR WIND

J. T. Gosling
University of California, Los Alamos National Laboratory
Los Alamos, NM 87545

ABSTRACT

A spatially variable coronal expansion, when coupled with solar rotation, leads to the formation of high speed solar wind streams which evolve considerably with increasing heliocentric distance. Initially the streams steepen for simple kinematic reasons, but this steepening is resisted by pressure forces, leading eventually to the formation of forward-reverse shock pairs in the distant heliosphere. The basic physical processes responsible for stream steepening and evolution are explored and model calculations are compared with actual spacecraft observations of the process. Solar wind stream evolution is relatively well understood both observationally and theoretically. Tools developed in achieving this understanding should be applicable to other astrophysical systems where a spatially or temporally variable outflow is associated with a rotating object.

INTRODUCTION

Eclipse photographs reveal that the solar corona is highly non-uniform, being structured by the complex solar magnetic field into a series of arcades, rays, holes, and streamers. Thus it is not surprising that the solar wind at 1 AU is highly variable since solar rotation insures a progression of different coronal structures at the central meridian of the sun. Spacecraft observations reveal that the major excursions of solar wind speed, density, and pressure conform to a characteristic pattern of variability,[1] which has come to be known as solar wind stream structure.

Figure 1, which shows the result of superposing data from 25 streams[2], illustrates the typical appearance of a solar wind stream at Earth's orbit. At the leading edge there is a rapid rise in flow speed, which is followed by a much slower decline back to low values. As the speed rises, the density quickly attains a maximum several times greater than its average value. As the speed falls, the particle density falls to abnormally low values. The gas pressure also maximizes as the speed rises. On the leading edge of the stream, the flow is deflected first to the east (i.e. flow from the east of the sun) and then to the west.

The above pattern of variability is the inevitable consequence of the evolution of a stream as it progresses outward from the sun. Such evolution continues as a stream moves beyond Earth's orbit, producing dramatic changes in stream structure in the distant

0094-243X/86/1440124-21$3.00

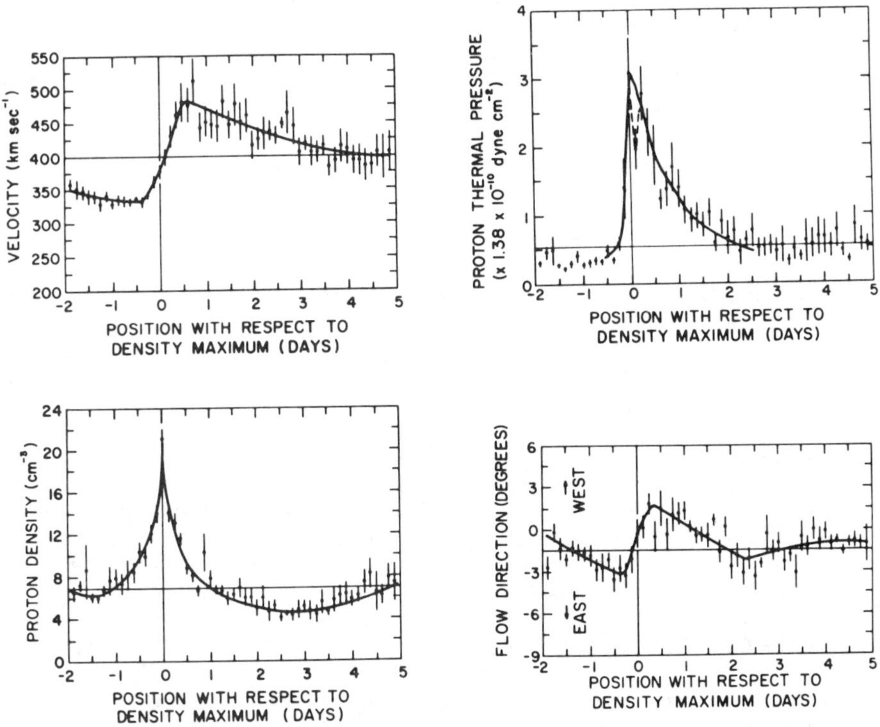

Fig. 1. Characteristic temporal variations in flow speed, density, pressure, and flow direction associated with a solar wind stream at Earth's orbit. These are average profiles of 25 streams which contained relatively large density compressions.[2]

heliosphere. Gas dynamic[3,4] and magnetohydrodynamic[5,6] codes of varying degrees of sophistication have successfully modeled the stream evolution process. Indeed, the agreement between model predictions and spacecraft observations over a wide range of heliocentric distances is one of the major triumphs of solar wind physics. In this paper we shall be concerned with providing a simple physical description of the problem and in exploring consequences of stream evolution in the distant heliosphere. The paper is intended to be a mini-tutorial rather than a review, and no attempt has been made to provide comprehensive referencing. More detailed treatments of the subject can be found elsewhere.[7,8]

SOLAR ORIGINS OF STREAM STRUCTURE AND KINEMATIC STREAM SPEEPENING

As previously mentioned, the solar wind expansion is spatially variable, with high speeds generally originating in magnetic unipolar regions of low coronal density known as coronal holes[9,10] and low speeds originating in regions of high density near field polarity reversals such as found in coronal streamers.[11,12,13] Alternating slow and fast plasma is directed Earthward as the sun rotates. A snapshot of the speed of Earthward-bound plasma as a function of heliocentric distance inside of 1 AU at some initial time, t_o, might appear as in the upper panel of Figure 2. The dots in the snapshot identify discrete parcels of plasma moving at different speeds; these parcels originated from different positions on the sun at different times. (A similar radial variation in speed would result if the coronal expansion were spatially uniform but temporally variable on a time scale of several days.) The faster-moving plasma at the crest of the stream overtakes and collides with the slower plasma ahead while simultaneously running away from slower-moving plasma behind it as the stream moves out from the sun. As a result, the stream evolves towards a sawtooth form with increasing heliocentric distance as illustrated in the middle and bottom panels of Figure 2. In addition, material within the stream is rearranged as the stream steepens; parcels of plasma near the leading edge are compressed and heated while parcels on the trailing edge are rarified and cooled. Some of the slow moving plasma ahead is swept up by the stream. The temporal variations of solar wind speed, density, and pressure shown in Figure 1 are readily identified with such evolved streams.

DYNAMICAL RESPONSE OF THE SOLAR WIND TO KINEMATIC STREAM STEEPENING

To this point, the evolution of a solar wind stream has been discussed purely as the kinematical result of a nonconstant radial velocity profile, which in turn is a consequence of nonuniform coronal expansion and solar rotation. In actual fact, there is a dynamical response to the steepening process which both limits the steepening and leads to new features at very large distances from the sun. Consider, for example, a fluid parcel approximately halfway up the leading edge of the stream. If the stream amplitude is v_o, as in Figure 3, then this parcel sees plasma streaming towards it from both the sunward and the anti-sunward directions at speeds ranging up to $v_o/2$ (middle panel). Because the plasma is constrained from interpenetrating by virtue of the weak magnetic field imbedded in it, the pressure (lower panel) builds up to resist this inflow of material from both sides. The onrushing plasma from both sides is decelerated, compressed, and heated as it encounters the pressure gradient. In the reference frame of the Earth or the sun these decelerations appear as an acceleration of the low speed plasma near the trough of the stream and a deceleration of plasma

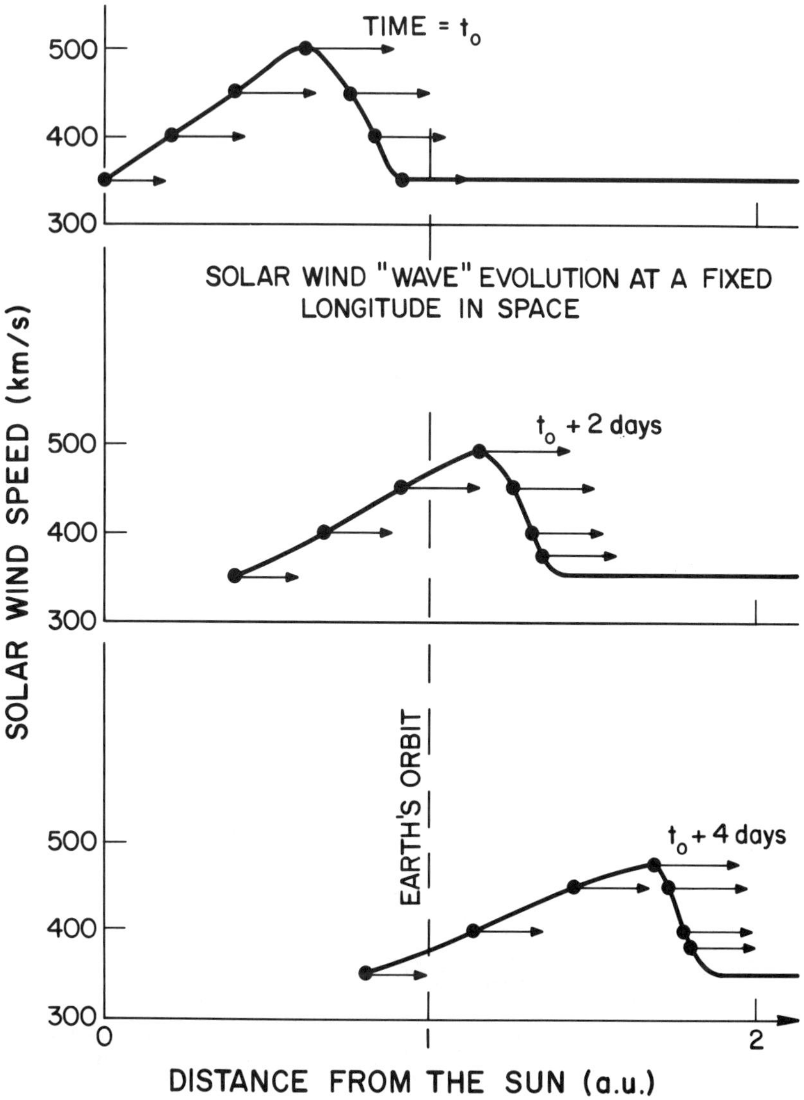

Fig. 2. A solar wind stream steepens kinematically. Plasma at the crest of the stream is moving faster than plasma in the trough and thus gradually catches up with it. As the stream steepens, parcels of plasma (indicated by dots) on the leading edge of the stream are compressed together, while those on the rear are spread out into a rarefaction.

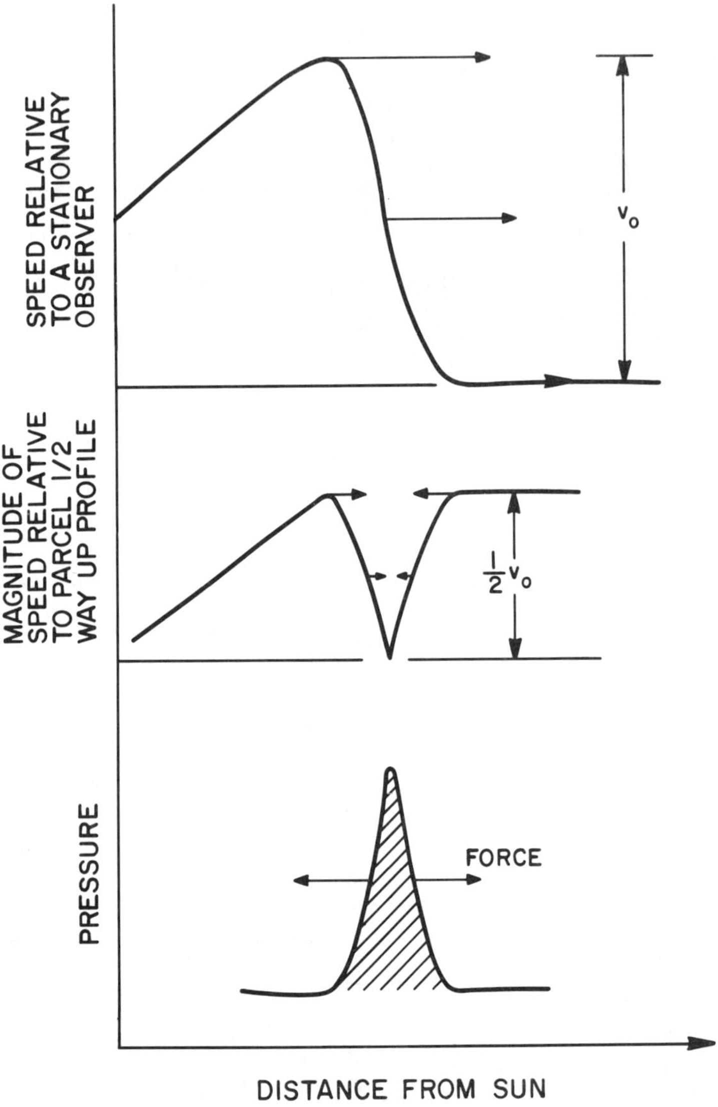

Fig. 3. Kinematic stream steepening leads to an enhancement in pressure on the leading edge of a stream (lower panel). This pressure increase serves to limit the steepening of the stream. The forces associated with the pressure gradients transfer momentum from the fast plasma at the crest to slow plasma in the trough. When the stream amplitude, v_o, is large relative to the fast mode speed, a pair of shocks forms on either side of the pressure enhancement. Typically this shock formation occurs well beyond the orbit of Earth.

near the crest. The result is a transfer of momentum and energy from the fast-moving plasma to the slow-moving plasma.

SHOCK FORMATION

The pressure ridge induced by stream steepening initially propagates outward into the onrushing plasma at the fast mode speed, c_f, which at 1 AU is ~70 km s^{-1}. So long as $c_f - v_o/2 > 0$, the stream will gradually damp out as described above. However, for large amplitude streams where $c_f - v_o/2 < 0$ the pressure ridge can not expand outward rapidly enough at the fast mode speed to warn the onrushing plasma of the impending collision. When this occurs the pressure grows nonlinearly as the stream steepens and a pair of shocks form on either side of the high pressure region. One of these shocks propagates backward towards the crest of the stream, while the other propagates forward towards the trough. However, both shocks are convected away from the sun by the very high bulk flow of the wind. Observations indicate that few streams steepen sufficiently inside 1 AU to cause shock formation by the time streams cross the Earth's orbit.[2] Nevertheless, because c_f decreases with increasing heliocentric distance, all solar wind streams of large amplitude should eventually contain a double shock structure at very large distances from the sun.

A SIMPLE MODEL CALCULATION OF STREAM EVOLUTION BEYOND 1 AU

Using "reasonable" boundary conditions near the sun, numerical models incorporating the physics discussed above can reproduce the coupled variations in speed, density, and pressure which are observed in the solar wind at 1 AU. Further, the codes can be used in conjunction with the 1 AU observations to predict stream evolution beyond Earth's orbit. Figure 4 shows the result of such a calculation of stream evolution[14] using a simple 1-dimensional gas dynamics code.[3] Three successive repetitions of the "average" stream structure shown in Figure 1 with peak velocities separated by eight days were used as the input signal at 1 AU to calculate the stream structure at larger distances. The figure displays solar wind speed and density as functions of heliocentric distance at a time when the three streams introduced at 1 AU are spread out between 1 and 6 AU. Clearly visible in this display are the formation of the forward-reverse shock pair bounding the compressed shell of plasma on the rising portion of each stream. These shocks give the velocity profile the appearance of a double sawtooth. The stream amplitude decreases and the compression region expands with increasing heliocentric distance as the reverse shock propagates back into the crest of the stream and the forward shock propagates into the lower speed plasma ahead. Thus, damping of the stream occurs by removal of the fastest and slowest material. By 5 AU a large fraction of the mass in the solar wind flow should be found within compression regions at the leading edges of streams.

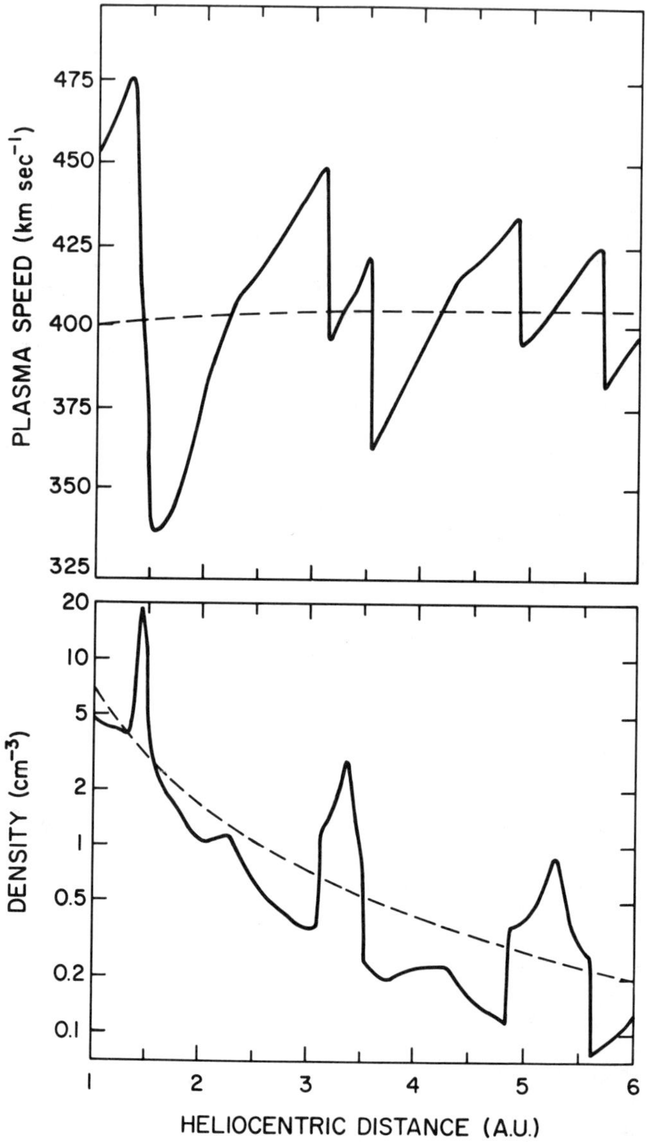

Fig. 4. Solar wind speed and density versus heliocentric distance predicted by a 1-dimensional gas dynamics model for three identical solar wind streams introduced at 1 AU. The light dashed lines indicate the variation with heliocentric distance expected for a steady, structureless flow.[14]

OBSERVATIONS OF SOLAR WIND STREAMS IN THE DISTANT HELIOSPHERE

Observations of solar wind streams in the distant heliosphere by instruments aboard Pioneers 10 and 11 and Voyagers 1 and 2 are in reasonably good agreement with the predictions of this simple 1-dimensional model. Figure 5 shows two successive 25-day sequences of hourly solar wind speed values obtained between 4.03 and 4.23 AU.[14] It is quite clear that each stream during this interval was sawtooth-like in appearance, there being one or more abrupt rises in speed on the rising portion of each stream. More complete observations, including measurements of density, temperature, and field strength establish that the abrupt jumps in speed are, in fact, the shocks predicted by the model.[15,16]

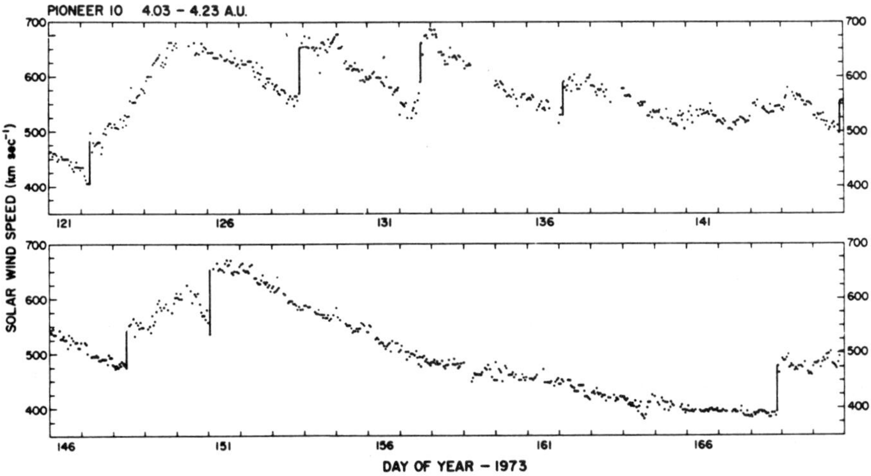

Fig. 5. One-hour values of the solar wind speed observed by Pioneer 10 during a 50-day period when the spacecraft moved between 4.03 and 4.23 AU. The vertical lines emphasize abrupt speed jumps which are shocks.[14]

EXPERIMENTAL TEST OF A STREAM EVOLUTION MODEL

This comparison has been made more explicit by taking advantage of intervals when one of the distant Pioneers or Voyagers was nearly radially aligned with the sun and Earth. In such cases measurements made near Earth can be used to predict stream structure further out, and this predicted structure can be compared in detail with the actual observations. Figure 6 illustrates the result of such a comparison.[17] The upper panel contains the solar wind speed profile

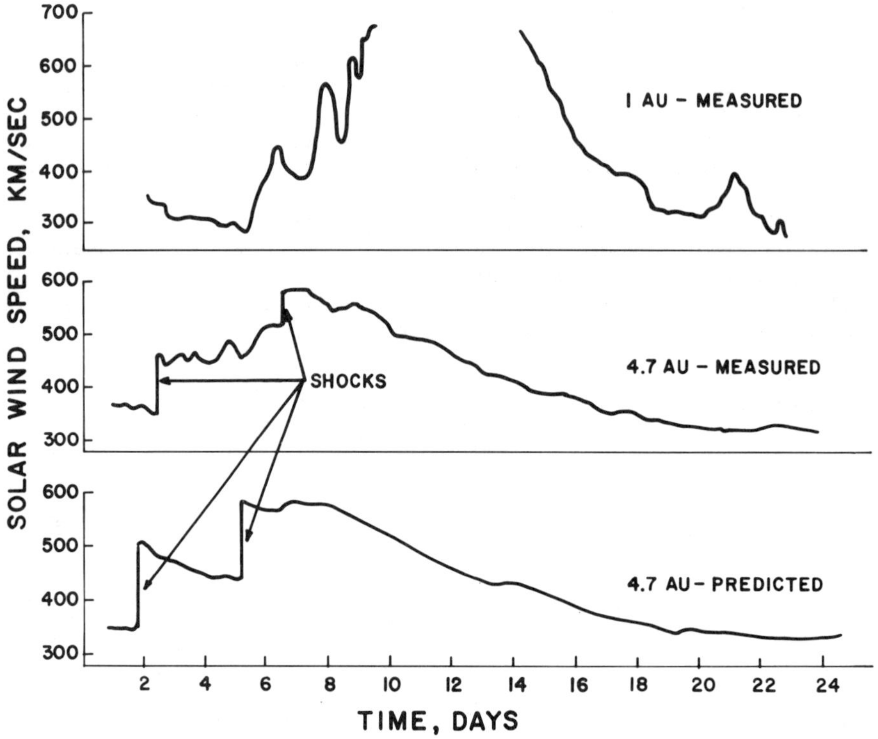

Fig. 6. A comparison of the speed profiles of a solar wind stream observed first at 1 AU by IMP 7 and later at 4.65 AU by Pioneer 10 together with the stream profile at 4.65 AU predicted by a 1-D gas dynamical model. The 1 AU observations were used as the boundary conditions for the model calculations.[17]

measured near Earth by IMP 7. (A gap in the data is caused by passage of IMP 7 into the Earth's magnetosphere where the solar wind can not be directly sampled.) The middle panel shows the speed profile of this same stream measured approximately 16 days later at 4.65 AU by Pioneer 10. In order to compare the two profiles, the Pioneer 10 data have been shifted forward in time in such a manner that the leading edge of the stream would be coincident at both spacecraft had the plasma on the leading edge of the stream propagated at constant speed from the Earth out to Pioneer 10. In fact, the stream arrived two days "early" at Pioneer 10, a consequence of the advance of the leading edge of the stream into the slower plasma ahead. In addition, the Pioneer 10 profile is more sawtooth-like and contains two shocks on the leading edge which were not present at 1 AU. Finally, the high frequency structure present at 1 AU between days 6-10 is considerably damped in the Pioneer 10 data.

The bottom panel of Figure 6 shows the predicted speed profile at Pioneer 10 using the IMP 7 measurements of speed, density, and pressure as inputs to the aforementioned 1-dimensional gas dynamics code. (A linear interpolation was performed across the data gap.) Agreement between the predicted and the observed speed profiles is quite good, although the model overestimates the amplitude of the shock transitions and underestimates the width of the shell of compressed gas between the shocks. (More sophisticated codes, for example, 2 or 3-dimensional MHD, improve the agreement between model predictions and observations.[18]) Note that the model predicts that the short wavelength structure present on the leading edge of the stream as it passed 1 AU should be damped out by the time the stream reaches 4.65 AU, as observed. Local velocity fluctuations are quickly wiped out as each sub-stream exchanges momentum with its immediate surroundings and is swept up by the longer wavelength stream. Thus the stream evolution process makes the solar wind act like a "low-pass filter" in the sense that only the longest wavelength velocity structures survive at large heliocentric distances.

STREAM STRUCTURE IN TWO DIMENSIONS

Thus far we have concentrated on stream evolution at a nominally fixed longitude in space. An appreciation of the global extent of this evolution can be gained from Figure 7, which shows a model calculation of the equatorial density structure of the stream in Figure 6 at a particular instant in time assuming that the coronal expansion is time stationary (but spatially variable) in a frame of reference corotating with the sun.[17] Curved arrows in the figure are streamlines, which trace out the locus of all parcels of plasma originating at common positions on the sun, but at different times. The various shadings represent density levels differing by factors of four and normalized to that for a steady structureless flow (r^{-2} decline). Darker shadings correspond to the higher densities. This entire pattern corotates with the sun; however, it is worth emphasizing that only the pattern rotates--each parcel of solar wind plasma moves outward nearly radially.

In Figure 7 plasma flows into the compressed shell from both sides. In this particular example, maximum compression is attained at about 3 AU. Beyond 3 AU the region of greatest compression expands slowly. The total region of compression expands more rapidly as the strong forward-reverse shock pair propagates away from the center of compression. By 5 AU streamlines originally separated by 180° in solar longitude are contained within the compression region bounded by the shocks.

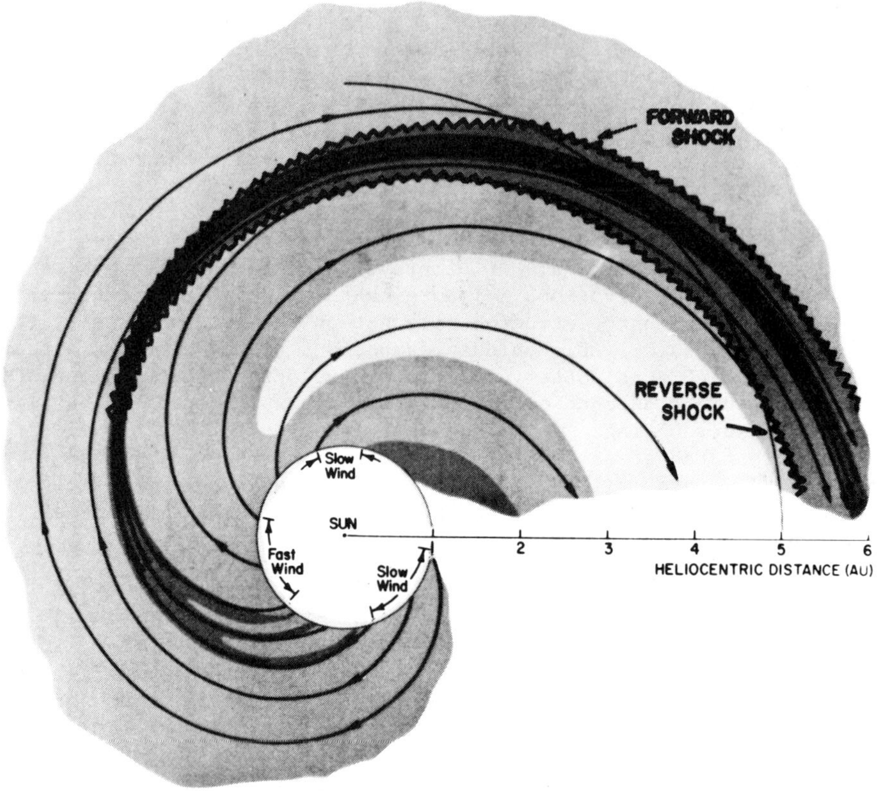

Fig. 7. Calculated equatorial density structure of the high speed solar wind stream of Fig. 6. The various shadings represent density levels differing by factors of four and normalized to the r^{-2} variation expected for structureless flow. The curved arrows are stream lines and the heavy, jagged lines are the shocks formed by stream steepening.[17]

In the absence of any other streams in the equatorial plane of the sun, the forward shock eventually overtakes the reverse shock from the same stream on the previous rotation,[20] as illustrated in Figure 8. Thus when the coronal expansion is time-stationary, all streamlines eventually lie within the compression region. In the example shown in Figure 8, the shocks intersect beyond 20 AU from the sun. When two or more streams are present in the equatorial plane, the forward and reverse shocks from the various streams intersect before the above closure can occur. For example, Figure 9 illustrates that for two streams identical to the one shown in

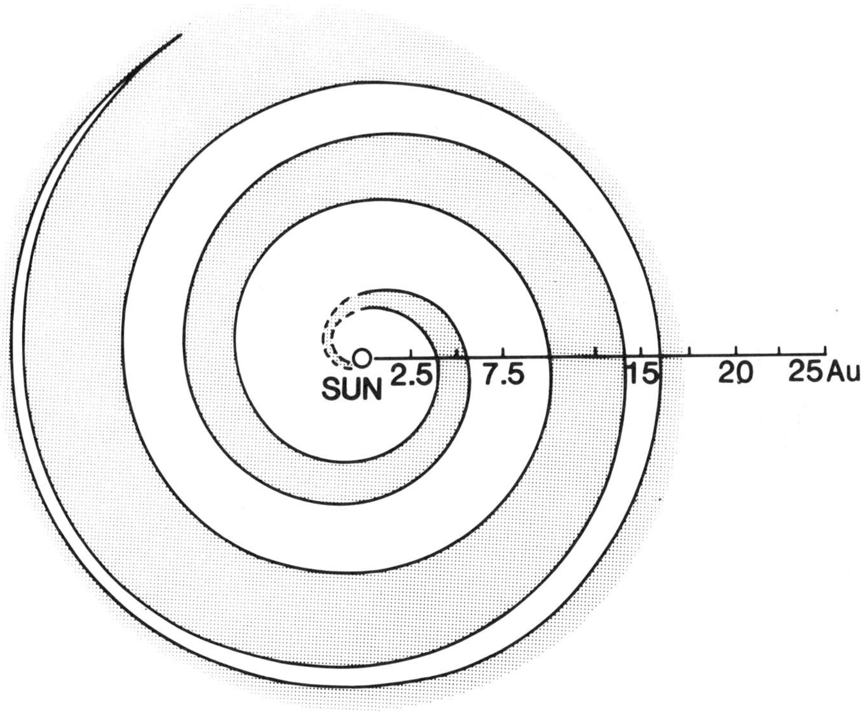

Fig. 8. Stream compression region in the solar equatorial plane for a stream similar to that of Figs. 6 and 7 assuming steady state for four solar rotations. The compression region beyond ~3 AU is bounded by a forward-reverse shock pair (solid lines), and hence broadens with increasing heliocentric distance. Beyond ~20 AU the forward and reverse shocks intersect one another in this example.

Figure 8 all streamlines contain compressed plasma beyond about 10 AU from the sun.

The basic structure of the solar wind in the distant heliosphere thus differs considerably from that observed at 1 AU. Stream amplitudes are severely reduced and high frequency velocity structures are damped out. What is left is expanding regions of high gas and field pressure[21] fronted by shocks which intersect and interact with one another.[22]

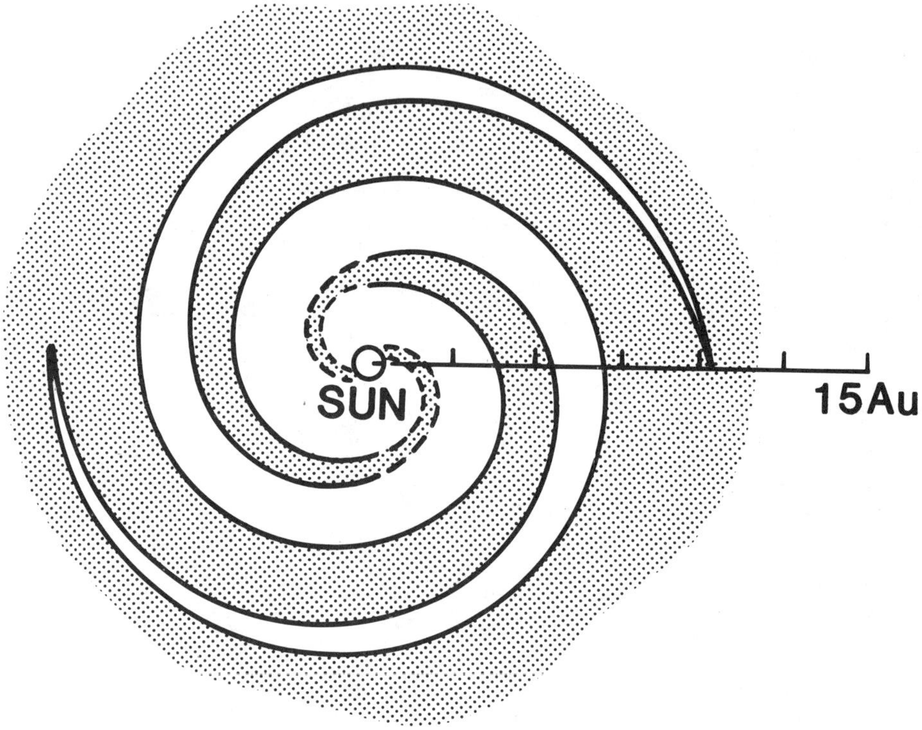

Fig. 9. Stream compression regions in the solar equatorial plane for two identical, stationary streams. The compression regions for the two streams begin to overlap at ~10 AU in this example; beyond that point all of the plasma in the equatorial plane has been compressed in interplanetary space at least once.

STREAM EVOLUTION IN THREE DIMENSIONS

For time-stationary solar wind streams, it is solar rotation that ultimately drives the evolution of stream structure with increasing heliocentric distance. Thus we might expect stream evolution to proceed at different rates at different solar latitudes; streams should evolve most rapidly in the solar equatorial plane, and more slowly at higher solar latitudes.[23,24]

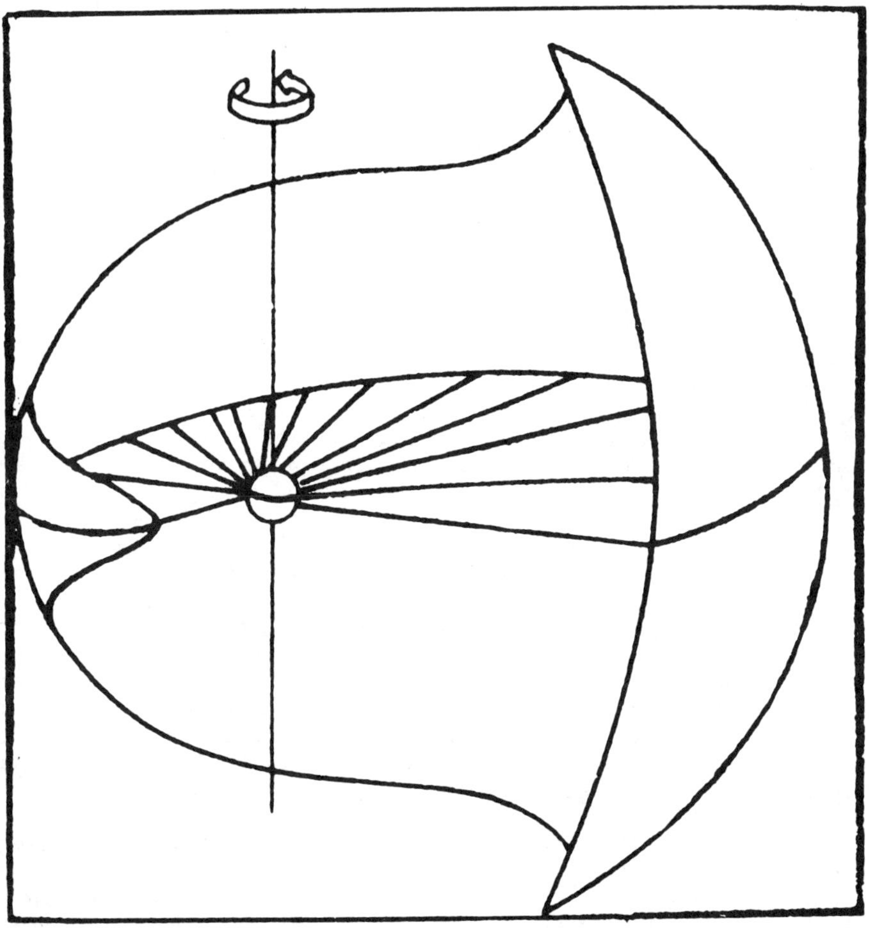

Fig. 10. A models's prediction of a single corotating shock surface in three dimensions. The location of the equatorial plane is indicated by lines radiating from the sun.[24]

Figure 10 shows a model calculation of the possible 3-dimensional configuration of a shock surface resulting from stream evolution.[24] In this particular example a stream uniform in latitude but having a square-wave variation of speed with longitude was assumed near the sun. For such a stream the rate of streamline convergence induced by solar rotation is a maximum in the equatorial plane and is zero over the poles. As a result, the shocks associated with the evolution of the stream form closest to the sun in the equatorial plane, and the shock surface takes the form of a Chinese pennant.

Other shock surface geometries are obtained for different initial stream configurations. Nevertheless, the result is general: Streams should evolve at a slower rate at high solar latitudes. As spacecraft observations to date have been limited to near the ecliptic plane, this prediction has not yet been verified observationally. However, tests should be possible with data obtained from the International Solar Polar Mission, now scheduled for launch in 1986.

NUMERICAL CODE COMPARISON: 1-D, 2-D, 3-D; MHD, HD

From the foregoing it is apparent that stream evolution is inherently 3-dimensional in character. Further, a magnetic field is imbedded within the solar wind, suggesting that magnetic forces play a role in stream dynamics. Nevertheless, as we have seen, 1-dimensional (1-D) gas dynamics codes are useful for obtaining an understanding of much of the basic physics involved and provide predictions in reasonable agreement with observations. The 1-D codes do, however, always predict too strong an interaction at the leading edge of a stream. In multi-dimensional codes (and in interplanetary space) the plasma can partially relieve the stresses induced by kinematical stream steepening by simply slipping aside. Such slippage is responsible, for example, for the east-west deflection in the flow on the leading edge of a stream[19] (see Figure 1). Further, MHD codes enhance this slippage since the magnetic field provides an additional pressure force and also increases the characteristic speed (fast-mode) with which the pressure signal can expand into the surrounding volume. In this section we wish to examine briefly the relative advantages of multi-dimensional codes with and without inclusion of the magnetic field.

Before proceeding it is useful to remember that in the present context it is solar rotation which drives stream evolution. The interaction front at the leading edge of a time-stationary stream is aligned nearly along a spiral (see Figures 7, 8, 9) and the largest pressure gradients all are perpendicular to this front. Thus the pressure forces associated with stream steepening produce primarily azimuthal and radial deflections of the flow. (Azimuthal deflections are more important inside of 1 AU where the spiral is generally inclined less than 45° to the radial direction and radial deflections are more important outside of ~1 AU where the spiral becomes inclined more nearly perpendicular to the radial direction.) Meridional deflections, on the other hand, are a second order effect.

Figure 11 shows hypothetical variations in solar wind flow speed, density, and temperature at 0.3 AU.[25] The pattern is periodic in longitude and the magnetic field is of uniform strength (4.5×10^{-4} Gauss). Figure 12 shows the result of projecting this stream out to 1 AU in the ecliptic plane using 1-D, 2-D, and 3-D MHD

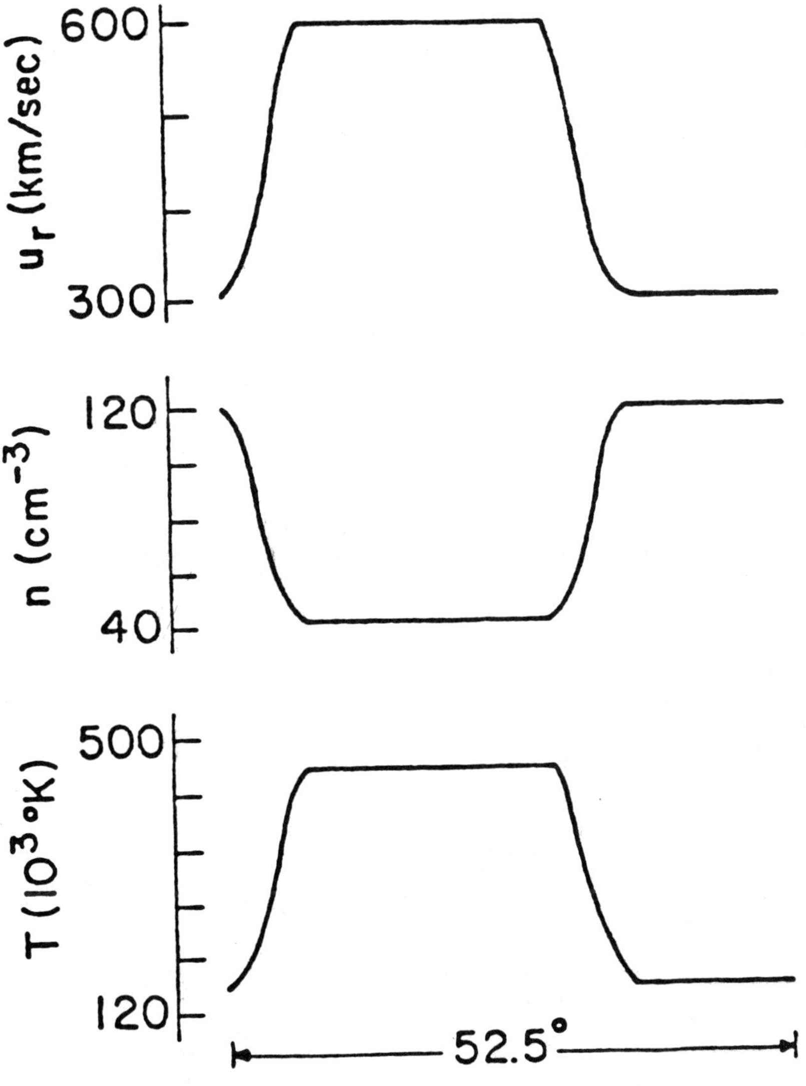

Fig. 11. Hypothetical stream structure at 0.3 AU as a function of heliographic longitude. These initial conditions constitute the input for the model calculations shown in Figs. 12 and 13.[25]

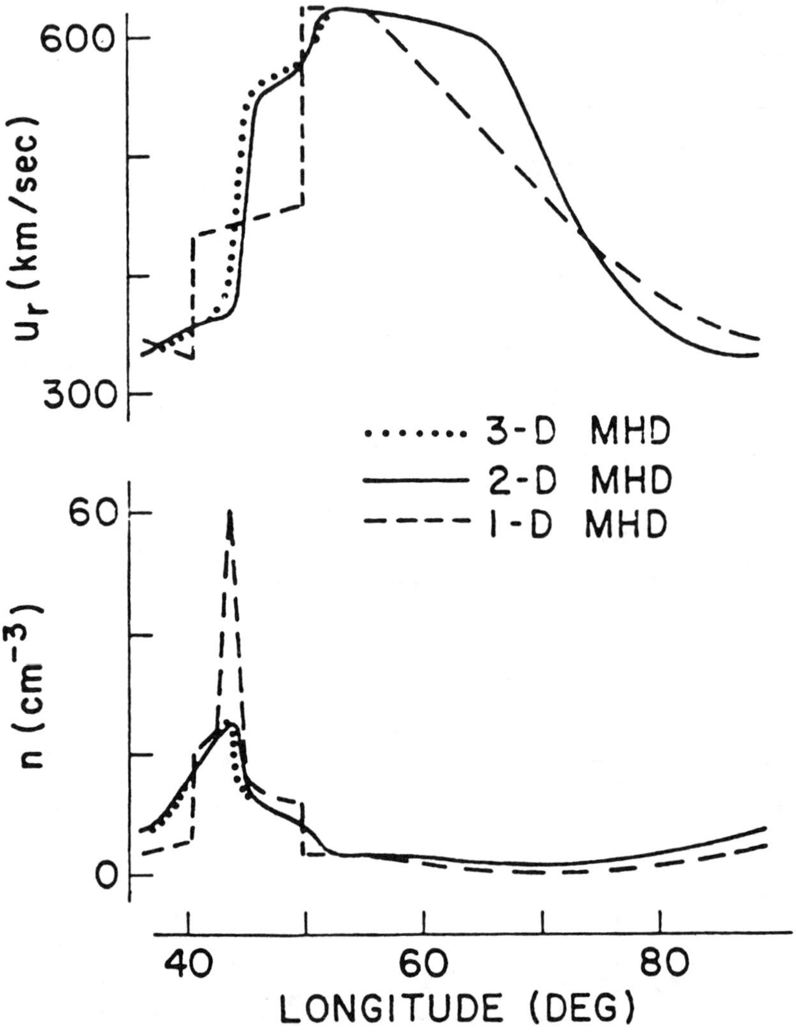

Fig. 12. The effect of geometry on stream evolution and shock formation. Three models differing only in their dimensionality are used to project the stream structure of Fig. 11 out to 1 AU in the solar equatorial plane. Speed is at top; number density is below.[25]

models.[25] (The 1-D code relates the 0.3 AU longitudinal signal of Fig. 11 with a temporal signal at that same distance by equating ϕ and Ωt, where ϕ is solar longitude, Ω is the angular rotation rate of the sun, and t is the time.) In going from 1-D to 2-D, relatively large changes in the predicted stream profile occur. The 1-D calculation (dashed line) overestimates the degree of compression and for this stream produces the familiar forward-reverse shock pair well inside of 1 AU. On the other hand, as expected from the above discussion, differences between the predictions of the 2-D and the 3-D calculations are minor. Both calculations indicate that azimuthal shears can delay substantially formation of the forward-reverse shock pair. However, since the spiral interaction front becomes more nearly perpendicular to the radial direction at larger heliocentric distances, both the 2-D and 3-D models eventually develop the shocks. The relatively minor differences between the 2-D and 3-D calculations clearly indicate that the computationally simpler 2-D codes are adequate for predictions in the solar equatorial plane where most measurements are made.

Figure 13 provides a similar comparison of the effect of including and neglecting the magnetic field.[25] In both examples illustrated, the stream of Fig. 11 was projected out to 1 AU in the equatorial plane using a 2-D model. In the HD (hydrodynamic, i.e. gas dynamic) calculation the forward-reverse shock pair again develops inside of 1 AU and the plasma is more highly compressed than in the MHD calculation. The plasma is better able to resist the kinematic steepening of the stream in the presence of a magnetic field because the magnetic field both provides an additional pressure force and increases the characteristic speed with which the pressure wave expands into the surrounding plasma. As the boundary conditions shown in Fig. 11 appear reasonable based upon direct solar wind measurements in the inner heliosphere,[26] and because forward-reverse shock pairs associated with time-stationary streams are relatively rare at 1 AU,[2] it is clearly important to include the magnetic field in model calculations of solar wind stream evolution.

TEMPORAL EFFECTS

Throughout this paper we have tacitly assumed that the coronal expansion, which produces the solar wind, is time-stationary. This assumption is most valid during the approach to and at the minimum of the solar 11-year activity cycle.[27] At other phases of the activity cycle the coronal expansion usually is variable on a time scale shorter than one solar rotation (~27 days as viewed from Earth), and transient solar wind disturbances are more common. Under these conditions the structure of interplanetary space is more complex than that illustrated in Figs. 7-10. Nevertheless, the basic physical processes outlined in this paper continue to operate at all phases of the solar cycle. We have emphasized that kinematical stream steepening is an inevitable consequence of

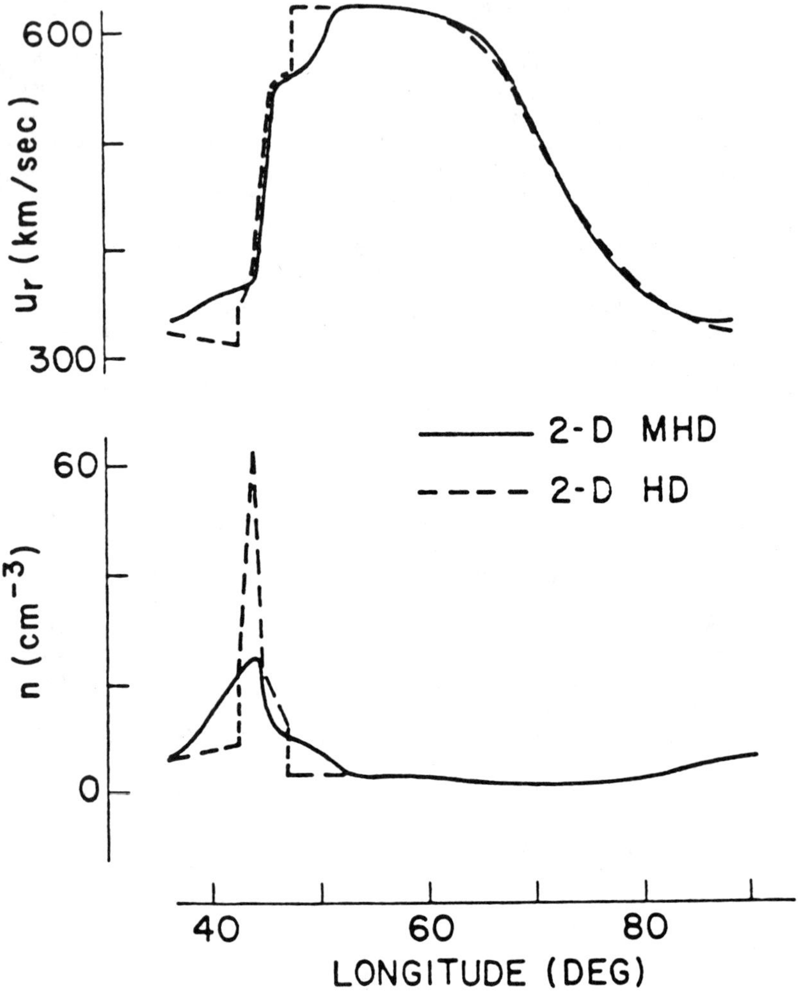

Fig. 13. The effect of magnetic field on stream evolution and shock formation. Here, two-2-D models, one with and one without the magnetic field, map the flow to 1 AU.[25]

variable flow speed along a radial line, and such steepening should occur regardless of what is responsible for establishing the radial velocity profile in the first place. For example, we mentioned near the beginning of the paper that the radial velocity profile of Fig. 2 could be the result of a temporally variable coronal expansion rather than a spatially variable one. In such a situation the compression region need not be aligned nearly along a spiral. In fact, for a spatially uniform but temporally variable coronal expansion the pressure ridge is an outward-moving spherical shell centered on the sun. When the temporal variation is rapid, shock formation occurs well inside of 1 AU because azimuthal slippage is inhibited in this geometry. More typically, high speed solar wind streams result from a combination of spatial and temporal variability, and the global structure of interplanetary space reflects such origins.

ACKNOWLEDGMENTS

I thank S. Fuselier, M. Thomsen, and R. Zwickl for their comments on a draft of this manuscript. This work was done under the auspices of the U.S. Department of Energy with NASA support under S-04039-D.

REFERENCES

1. C. W. Snyder, M. Neugebauer, and U. R. Rao, J. Geophys. Res., 68, 6361 (1963).
2. J. T. Gosling, A. J. Hundhausen, V. Pizzo, and J. R. Asbridge, J. Geophys. Res., 77, 5442 (1972).
3. A. J. Hundhausen, J. Geophys. Res., 78, 1528 (1973).
4. V. Pizzo, J. Geophys. Res., 85, 727 (1980).
5. R. S. Steinolfson, M. Dryer, and Y. Nakagawa, J. Geophys. Res., 80, 1223 (1975).
6. V. Pizzo, J. Geophys. Res., 87, 4374 (1982).
7. A. J. Hundhausen, Coronal Expansion and Solar Wind, Springer-Verlag, New York (1972).
8. V. Pizzo, in Collisionless Shock Waves in the Heliosphere, ed. R. Stone and B. Tsurutani, AGU publication (1985).
9. A. S. Krieger, A. F. Timothy, and E. C. Roelof, Solar Phys. 29, 505 (1973).
10. N. R. Sheeley, J. W. Harvey and W. C. Feldman, Solar Phys., 49, 271, 1976.
11. G. Borrini, J. T. Gosling, S. J. Bame, W. C. Feldman, and J. M. Wilcox, J. Geophys. Res., 86, 4565 (1981).
12. W. C. Feldman, J. R. Asbridge, S. J. Bame, E. E. Fenimore, and J. T. Gosling, J. Geophys. Res., 86, 5408 (1981).
13. J. T. Gosling, G. Borrini, J. R. Asbridge, S.J. Bame, W. C. Feldman, and R. T. Hansen, J. Geophys. Res., 86, 5438 (1981).
14. A. J. Hundhausen and J. T. Gosling, J. Geophys. Res., 81, 1436 (1976).
15. E. J. Smith and J. H. Wolfe, Geophys. Res. Lett., 3, 137 (1976).
16. E. J. Smith and J. H. Wolfe, Space Sci. Rev., 23, 217 (1979).

17. J. T. Gosling, A. J. Hundhausen, and S. J. Bame, J. Geophys. Res., 81, 2111 (1976).
18. M. Dryer, Z. K. Smith, E. J. Smith, J. D. Mihalov, J. H. Wolfe, R. S. Steinolfson, and S. T. Wu, J. Geophys. Res., 83, 4347 (1978).
19. G. L. Siscoe, J. Geophys. Res., 77, 27 (1972).
20. J. T. Gosling and A. J. Hundhausen, Sci. Am., 236, 36 (1977).
21. L. F. Burlaga, J. Geophys. Res., 88, 6085 (1983).
22. M. Dryer and R. S. Steinolfson, J. Geophys. Res., 81, 5413 (1976).
23. S. T. Suess, A. J. Hundhausen, and V. Pizzo, J. Geophys.Res., 80, 2023 (1975).
24. G. L. Siscoe, J. Geophys. Res., 81, 6235 (1976).
25. V. Pizzo, in Solar Wind Four, ed. H. Rosenbauer, MPAE-W-100-81-31, p.153 (1981).
26. H. Rosenbauer, R. Schwenn, E. Marsch, B. Meyer, H. Miggenrieder, M. D. Montgomery, K.H. Muhlhauser, W. Pilipp, W. Voges, and S. M. Zink, J. Geophys., 42, 561 (1977).
27. J. T. Gosling, J. R. Asbridge, S. J. Bame, and W. C. Feldman, J. Geophys. Res., 81, 5061 (1976).

MAGNETIC RECONNECTION IN THE TERRESTRIAL MAGNETOSPHERE

William C. Feldman
Los Alamos National Laboratory
Los Alamos, NM 87545

ABSTRACT

An overview is given of quantitative comparisons between measured phenomena in the terrestrial magnetosphere thought to be associated with magnetic reconnection, and related theoretical predictions based on Petschek's simple model.[1] Although such a comparison cannot be comprehensive because of the extended nature of the process and the relatively few in situ multipoint measurements made to date, the agreement is impressive where comparisons have been possible. This result leaves little doubt that magnetic reconnection does indeed occur in the terrestrial magnetosphere. The maximum reconnection rate, expressed in terms of the inflow Mach number, M_A, is measured to be $M_A = 0.2 \pm 0.1$.

INTRODUCTION

Magnetic reconnection in a magnetoplasma is a process in which plasma flows across a boundary that separates topologically distinct magnetic cells. This process appears in a variety of settings and proceeds at rates which depend sensitively on the boundary conditions as well as on the ambient plasma state.

There are many equivalent ways of characterizing magnetic reconnection. The definition adopted in Space Plasma Physics requires a nonzero $E_{\parallel}$, the electric field parallel to the magnetic separator. Here the magnetic separator, often called the neutral line, is the line of intersection of two separatrix surfaces. These surfaces define the boundary between distinct magnetic cells.[2,3,4,5] Alternatively, stressed plasmas containing a nonpotential magnetic field can relax to a more potential configuration while generating hot plasma jets.[6] When such a condition exists, convected magnetic energy density in the form of a Poynting flux is converted to convected plasma energy density in the form of enthalpy flux and bulk convection flux.

The sudden heating of solar plasma provides the gross observational feature of solar flares. It seemed natural, therefore, to seek an explanation of solar flares in terms of magnetic reconnection.[7] The earliest quantitative theories[8,9,10,11] considered a single-step process in which magnetic energy was converted to plasma energy through the resistive diffusion of plasmas carrying antiparallel magnetic fields. These fields are in turn forced towards a central current sheet by external boundary conditions. However, reconnection rates predicted by these theories, expressed equivalently either in terms of $E_{\parallel}$ or the upstream inflow Alfvénic Mach number, M_A, were too slow to account for the rates of energy release observed in solar flares. Here, energy was thought to be converted throughout a thin but long layer

0094-243X/86/1440145-12$3.00

requiring a conversion time proportional to the square root of the length.

A solution to this dilemma was provided by Petschek.[1] In his model, the energy conversion process was split into two parts. In the first part, field lines were torn and reconnected in a relatively small diffusion region containing the separator or neutral line. This tearing is accompanied by relatively little energy conversion because of the small volume of the diffusion region. The second part consisted of magnetic field annihilation across the fronts of obliquely propagating intermediate- or slow-mode waves. These waves are expected to develop steep fronts to form rotational discontinuities or slow-mode shocks, respectively. Practically all of the magnetic energy that was converted to plasma energy through reconnection occurred across these fronts because of their much larger areas. The resultant large increase in reconnection rate afforded by this model reflects the large difference between wave speeds and diffusion speeds characteristic of most cosmical plasmas. Reconnection rates allowed by this model were now sufficiently large to account for solar flares.

Because of the difficulty of measuring microscopic details of the electric, magnetic and plasma flow fields within reconnection sites embedded in solar active regions, progress in reconnection theories of solar flares has been relatively slow. Such progress has therefore shifted to theories of the terrestrial magnetosphere where such difficulties are not as severe. Here, as within solar active regions, magnetic reconnection has been suggested as a primary energy-conversion process. Indeed, an early attempt to account for measured convection patterns in the ionosphere found a ready explanation in terms of magnetic reconnection leading to an open magnetosphere.[12] Although the concept of an open magnetosphere has yet a few remaining doubters, an overwhelming mass of observations exists that is well organized by theoretical predictions based on an open magnetosphere. Furthermore, these observations are difficult to understand as a whole if the terrestrial magnetosphere is closed.[4,13,14,15,16,17,18,19]

Two sites for magnetic reconnections are likely; 1) the blunt-nosed surface separating the terrestrial plasma from the solar wind, the subsolar magnetopause, and 2) the current sheet separating north and south lobes of the geomagnetic tail embedded within the plasma sheet. A schematic picture of the terrestrial magnetosphere showing these sites is reproduced from Sonnerup[3] in Figure 1. An overview of our current quantitative understanding of magnetic reconnection at the subsolar magnetopause and at the plasma sheet is given in the next two sections, respectively. This overview will not be comprehensive but instead be tailored to the purpose of this workshop; to identify processes which may operate in both solar-terrestrial and astrophysical plasmas. We may then be able to scale relevant solar-terrestrial phenomena to various astrophysical systems, thereby deepening our understanding of both systems. Attention will therefore be confined to a few quantitative results which, in the authors opinion, may prove useful in this regard.

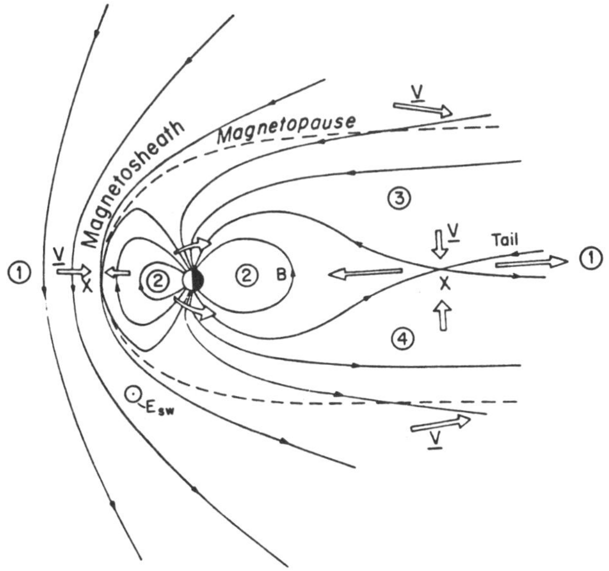

Fig. 1. A schematic representation of the interaction between the solar wind with the terrestrial magnetosphere in the noon-midnight meridian when the magnetic field in the magnetosheath is directed due south. Two sites for magnetic reconnection are identified by X points in the figure, 1) the subsolar region of the magnetopause and 2) the central plane of the magnetotail. This figure is reproduced from Sonnerup.[3]

RECONNECTION AT THE MAGNETOPAUSE

The first application of Petschek's model to the magnetopause was given by Levy et al.[20] Their schematic representation is shown in Figure 2. In order to match the shocked solar wind flow (which carries a generally turbulent magnetic field having variable orientation) to the terrestrial magnetosphere (a low β plasma having generally northward pointing magnetic field firmly anchored in the earth) the boundary must consist of a pair of wave fronts. Each of these fronts consists of a standing Alfvén wave overlying an inner slow-mode expansion wave. Here β is the ratio of plasma pressure to magnetic pressure. Most of the deflection and acceleration of the incident solar wind flow is effected by the $\underset{\sim}{J} \times \underset{\sim}{B}$ force within the front of the Alfvén wave. The Alfvén waves are expected to be sharply crested, having the appearance of rotational discontinuities. They connect at a relatively small, central volume, the diffusion region, where incident magnetosheath and magnetospheric field lines tear and reconnect to one another.

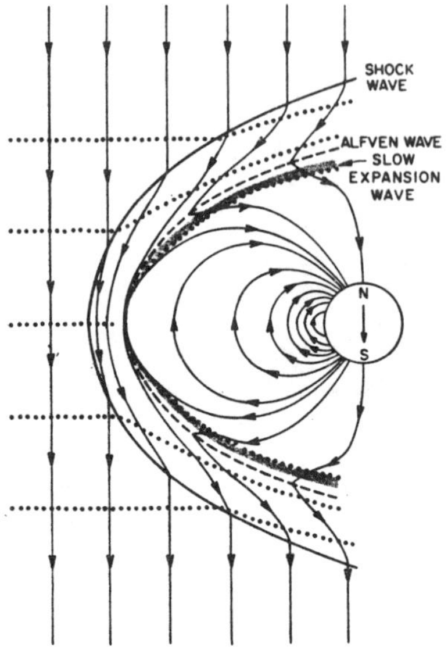

Fig. 2. A schematic representation of reconnection at the subsolar magnetopause identifying the various MHD structures expected. The forward edge of the interaction between the solar wind and terrestrial magnetosphere is a standing fast-mode shock. This is followed by a magnetopause consisting of a leading Alfvén wave, which in actuality is a rotational discontinuity, and a trailing slow-mode expansion wave. This expansion wave has been identified observationally as a solar wind-magnetosphere boundary layer. This figure is reproduced from Levy et al.[20]

A continuous transition between plasma and field conditions within the diffusion region and the far wave fields, can only be achieved if the plasma inflow rate V_n, is less than about 0.1 to 0.2 of the Alfvén speed, V_A. Both speeds are referenced to conditions just upstream of the magnetopause in the magnetosheath.[20] However, because the magnetosheath and all other cosmical plasmas are compressible, this result does not require a maximum merging rate with respect to conditions imposed at the outer boundaries of two adjacent magnetic cells. Indeed, external pressures sufficiently large can cause a compression of the plasma near the interface thereby raising the local upstream Alfvén speed at the magnetopause to the point where the inflow Mach number at the interface, $M_A = (V_n/V_A) \lesssim 0.2$[21]

Much of the original analysis of magnetopause structure near the subsolar region of the terrestrial magnetosphere has since been experimentally verified.[4,17,18] In particular, detailed analyses of magnetopause structure showed that (1) the upstream boundary of the sunward-facing magnetosphere was a rotational discontinuity at least some of the time,[22,23] and that (2) it is often followed by a boundary layer (a) having general characteristics intermediate between those of the magnetosheath and the magnetosphere[24,25,26] but, in a few special cases, (b) composed of plasma jets having transverse momentum sufficient to balance the transverse magnetic stress at the discontinuity.[27,28,29]

Although clear examples have been found of an open magnetopause having characteristics consistent with the Levy et al. model,[20] many examples have also been found of a closed boundary having the form of a tangential discontinuity.[30] This result is also consistent with Petschek's model and is in accord with many indirect solar wind-magnetosphere parameter correlation. These correlations indicate that the magnetopause acts as a half-wave rectifier of the interplanetary magnetic field.[16,31] When the interplanetary magnetic field has a southward component, reconnection at the magnetopause occurs, but when it has a northward component, reconnection ceases. However, an additional feature of the observations not foreseen in early theoretical analyses was the patchy nature of the magnetic interconnection process[32,33,34] Reconnection at the magnetopause is often evident as discrete events having a characteristic physical scale size of about one earth radius, R_E. It is estimated that roughly half the magnetic flux transferred from the solar wind to the magnetotail occurs through such discrete flux transfer events.[18]

Two approaches have been used to assess the efficiency with which energy is fed into the magnetosphere through magnetic reconnection at the magnetopause. The first is a lower limit which compares the total energy dissipated in the near-earth magnetotail with the total solar wind energy incident on the entire sunward face of the magnetopause.[35,36,37,38] The lower-limit aspect of this comparison stems from two disparate effects. A combination of the curvature of the magnetopause, the dipole-like character of the terrestrial magnetic field, and the relatively small scale, large-amplitude nature of the magnetosheath turbulence, all together prevent a uniformly maximal efficiency for magnetic reconnection over the entire sunward face of the magnetopause. The second effect is that a fraction of the total energy input to the magnetotail through the magnetopause was not counted in these estimates. In particular, a substantial fraction of the stored magnetotail energy density is released back to the solar wind in the form of plasmoids[39,40,41] and jetting wake plasma.[42] Nevertheless, the resultant estimate can provide an order of magnitude value which may prove useful in applications to specific astrophysical systems. On the average, the maximum solar wind-magnetosphere energy-coupling efficiency through the reconnection process, can range from less than about 1% to about 5%.[36,37,38] This efficiency is not readily compared with the maximum magnetic reconnection rate deduced by Levy et al.[20,21]

A closer estimate of this maximal rate was given by Holzer and Slavin[43,44] who compared the rate of net magnetic flux transfer from the forward hemisphere of the magnetosphere to the geomagnetic tail, with the rate of magnetic flux transported by the solar wind to the sunward face of the magnetopause. The result of their analysis yielded a ratio of inflow to transported rates of about 0.2. This ratio is very close to the maximum inflow Mach number predicted by Levy et al.[20] Although the foregoing observed and predicted rates are not directly comparable because the observations are referenced to the far upstream solar wind rate and the theoretical predictions are referenced to the inflow Mach number evaluated just upstream of the magnetopause, the very close agreement is encouraging.

RECONNECTION IN THE GEOMAGNETIC TAIL

Petschek's symmetric model of magnetic reconnection[1] is expected to find its closest realization in the geomagnetic tail.[3] Here the interaction between the solar wind and the earth's magnetic dipole fashions two nearly symmetric, low-β lobes of plasma, having oppositely-directed magnetic fields. These fields are pressed together by the solar wind along the center plane of the tail. A schematic view of this topology, in the form of a meridional cut through the tail, is shown in Figure 3.

Tearing of north and south lobe fields occurs at a neutral line in the center of the figure. The field lines just tearing are the separatrices which form the boundaries of neighboring, topologically distinct magnetic cells. According to the Petschek model, the locus of lines parallel to the neutral line along which the fields downstream of the separatrices bend sharply, are four slow-mode shocks which intersect at the neutral line. Relatively slow moving plasma flows southward from the north and northward from the south into the shocks. Within the shock fronts they are redirected, heated, and accelerated to form plasma jets travelling toward (to the left of the neutral line in Figure 3) and away (to the right of the neutral line in Figure 3) from the Earth, respectively. The magnitude of the magnetic field also decreases abruptly within the shocks. If in addition, a neutral line is formed within the near-Earth plasma sheet, (region 2 of Figure 1) then the tailward component of initially reconnected plasma sheet field lines forms nested closed loops, represented in Figure 3 as a tailward-jetting plasmoid.

Many observations of plasma phenomena in the near-Earth magnetic tail are well organized by the foregoing reconnection model of the plasma sheet and surrounding lobes.[13,14,15,45,46,47] Most notable is the strong correlation between 1) earthward jetting plasmas and energetic particles coupled with northward magnetic field, and 2) tailward jetting plasmas and energetic particles coupled with southward magnetic field. This occurs near the center plane of the tail as pictured in Figure 3.

Recent measurements of the deep geomagnetic tail have confirmed this picture. They have also allowed a quantitative comparison with the Petschek model. In particular, dynamical changes in the deep geomagnetic tail have been observed to correlate closely with those occurring in the near tail as

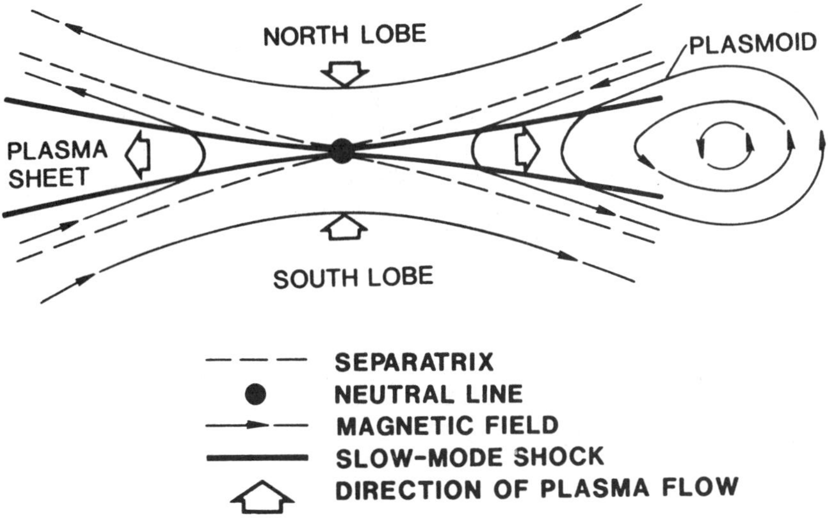

Fig. 3. A schematic representation of reconnection in the geomagnetic tail identifying the boundaries between the various flow regimes within the four separate magnetic cells. A description is given in the text.

predicted by magnetic reconnection.[48] This near-Earth activity was also shown to correlate closely with the detection in the distant magnetotail of 1) tailward jetting plasmoids,[39,40] 2) tailward propagating energetic particles,[41,49,50,51] and 3) standing slow-mode shocks forming a semi-permanent boundary between plasma sheet and lobe plasma populations.[52,53]

The semi-permanent feature of slow-mode shocks observed in the distant geomagnetic tail has allowed the best quantitative comparison with Petschek's simple model of symmetric reconnection since the theory assumes time stationarity. Most important in this regard is the inflow Alfvén Mach number discussed previously. The comparison is excellent. The average inflow Alfvén Mach number observed just upstream of 26 of the least ambiguous examples of slow-mode shocks observed by ISEE 3 near 200 R_E downstream of the earth is $M_A = 0.19 \pm 0.08$.[42]

SUMMARY AND CONCLUSIONS

A host of in situ observations lend support to the Petschek[1] model of magnetic reconnection in the terrestrial magnetosphere.[15,47] Because of the time- and spatially-varying orientation of the solar wind magnetic field and the generally large-amplitude turbulence of the magnetosheath flow, magnetic reconnection at the sunward face of the magnetopause proceeds by way of a standing rotational discontinuity followed by a slow-mode expansion fan.[17] However, a significant fraction of magnetic reconnection at the magnetopause proceeds in discrete events, called flux transfer events,[18] having a distinct scale size of about 1 R_E.

Magnetic reconnection in the geomagnetic tail is, in two major respects, simpler to understand than that which proceeds at the sunward magnetopause. This simplicity stems from the closely equal magnitude yet antiparallel orientations of magnetic fields in the north and south lobes of the magnetotail. This topology is fixed by the interaction between the supersonically streaming solar wind and the terrestrial magnetic dipole. In consequence, there is no need for a rotational discontinuity to match boundary conditions. All that remains is a slow-mode compression fan which steepens to form four slow-mode shocks intersecting along a magnetic neutral line. Most of the transformation of magnetic energy in the form of a Poynting flux, to plasma convection and enthalpy flux, occurs in a quasi-steady state fashion at these shocks.[42] The rest occurs in a time-dependent fashion at the onset of reconnection to form earthward jetting plasmas and energetic particles[13,15,46,47] leading to a host of related near-earth phenomena,[36] as well as tailward jetting plasmoids[39,40] embedded within a sheet of magnetically open, jetting wake plasma.[42]

Although quantitative analyses of reconnection at the sunward magnetopause[35,36,37,38,43,44] yield rates consistent with the Levy et al.[20] analysis of the Petschek[1] mechanism, they are not precisely comparable. The strong time and spatial variations of conditions at the magnetopause resulting from inherent variations in the solar wind flow are difficult to account for in detail. This situation is not the case in the deep geomagnetic tail. Here the terrestrial dipole imposes an order to the lobe plasma topology not present at the magnetopause. In consequence, magnetic reconnection in the deep geomagnetic tail occurs in a more geometrically simple fashion. In addition, because of the distance from the neutral line, it occurs in a more time stationary fashion. A quantitative comparison shows that the measured reconnection rates[42] are in excellent agreement with theoretical predictions,[20] both giving an Alfvénic inflow Mach number, $M_A = 0.2 \pm 0.1$.

ACKNOWLEDGMENTS

This work was performed under the auspices of the U.S. Department of Energy with partial NASA support under S-54496A.

References

1. Petschek, H. E., AAS NASA Symposium on the Physics of Solar Flares, NASA Spec. Publ., SP-50, (NASA, 1964), p.425.

2. Vasyliunas, V. M., Rev. Geophys. Space Phys., 13, (1974), p.303.

3. Sonnerup B. U. Ö., Solar System Plasma Physics, Vol. III, ed. by C. F. Kennel, L. J. Lanzerotti, and E. N. Parker, (North-Holland Amsterdam 1979), p. 45.

4. Sonnerup, B. U. Ö., Magnetic Reconnection in Space and Laboratory Plasmas, ed. by E. W. Hones, Jr., Geophysical Monograph 30(AGU Press, 1984), p. 92.

5. Axford, W. I., Magnetic Reconnection in Space and Laboratory Plasmas, ed. by E. W. Hones, Jr., Geophysical Monograph 30 (AGU Press, 1984), p. 1.

6. Parker, E. N., Magnetic Reconnection in Space and Laboratory Plasmas, ed. by E. W. Hones, Jr., Geophysical Monograph 30 (AGU Press, 1984), p. 32.

7. Giovanelli, R. G., Mon. Not. Roy. Astron. Soc., 107, 338 (1947).

8. Sweet, P. A., Electromagnetic Phenomena in Cosmical Physics, ed. by B. Lehnert (Cambridge University Press, London, 1958).

9. Sweet, P. A., AAS-NASA Symposium on the Physics of Solar Flares, Spec. Publ. SP-50, (NASA, 1964), p. 409.

10. Parker, E. N., J. Geophys. Res., 62, 509 (1957).

11. Parker, E. N., Astrophys. J. Suppl., Ser. 8, 177 (1963).

12. Dungey, J. W., Phys. Rev. Lett., 6, 47 (1961).

13. Russell, C. T. and R. L. McPherron, Space Sci. Rev., 15, 205 (1973).

14. Nishida, A., Space Sci. Rev., 34, 185 (1983).

15. Nishida, A., Magnetic Reconnection in Space and Laboratory Plasmas, ed. by E. W. Hones, Jr., Geophysical Monograph 30 (AGU Press, 1984), p. 159.

16. Reiff, P. H., Magnetic Reconnection in Space and Laboratory Plasmas, ed. by E. W. Hones, Jr., Geophysical Monograph 30(AGU Press, 1984), p. 104.

17. Paschmann, G., Magnetic Reconnection in Space and Laboratory Plasmas, ed. by E. W. Hones, Jr., Geophysical Monograph 30(AGU Press, 1984), p. 114.

18. Russell, C. T., Magnetic Reconnection in Space and Laboratory Plasmas, ed. by E. W. Hones, Jr., Geophysical Monograph 30(AGU Press, 1984), p. 124.

19. Baker, D. N., S.-I. Akasofu, W. Baumjohann, J. W. Bieber, D. H. Fairfield, E. W. Hones, Jr., B. Mauk, and R. L. McPherron, Solar Terrestrial Physics - Present and Future, NASA Ref. Publ. 1120, ed. by D. M. Butler and K. Papadopolous, (NASA, 1984), p. 8-1.

20. Levy, R. H., H. E. Petschek, and G. L. Siscoe, AIAA Journal 2, 2065, 1964.

21. Sonnerup, B. U. Ö., J. Geophys. Res., 79, 1546 (1974).

22. Sonnerup, B. U. Ö., and L. J. Cahill, Jr., J. Geophys. Res., 72, 171 (1967).

23. Sonnerup, B. U. Ö., J. Geophys. Res., 76, 6717 (1971).

24. Hones, E. W., Jr., J. R. Asbridge, S. J. Bame, M. D. Montgomery, S. Singer, and S.-I. Akasofu, J. Geophys. Res., 77, 5503 (1972).

25. Akasofu, S.-I., E. W. Hones, Jr., S. J. Bame, J. R. Asbridge, and A. T. Y. Lui, J. Geophys. Res., 78, 7257 (1973).

26. Rosenbauer, H., H. Grünwaldt, M. D. Montgomery, G. Paschmann, and N. Sckopke, J. Geophys. Res., 80, 2723 (1975).

27. Paschmann, G., B. U. Ö. Sonnerup, I. Papamastorakis, N. Sckopke, G. Haerendel, S. J. Bame, J. R. Asbridge, J. T. Gosling, C. T. Russell, and R. C. Elphic, Nature, 282, 243 (1979).

28. Sonnerup, B. U. Ö., G. Paschmann, I. Papamastorakis, N. Sckopke, G. Haerendel, S. J. Bame, J. R. Asbridge, J. T. Gosling, and C. T. Russell, J. Geophys. Res., 86, 10049 (1981).

29. Gosling, J. T., J. R. Asbridge, S. J. Bame, W. C. Feldman, G. Paschmann, N. Sckopke, and C. T. Russell, J. Geophys. Res., 87, 2147 (1982).

30. Papamastorakis I., G. Paschmann, N. Sckopke, S. J. Bame, and J. Berchem, J. Geophys. Res., 89, 127 (1984).

31. Burton, R. K., R. L. McPherron, and C. T. Russell, Science, 189, 717 (1975).

32. Haerendel, G., G. Paschmann, N. Sckopke, H. Rosenbauer, and P. C. Hedgecock, J. Geophys. Res., 83, 3195 (1978).

33. Russell, C. T. and R. C. Elphic, Space Sci. Rev., 22, 681 (1978).

34. Russell, C. T. and R. C. Elphic, Geophys. Res. Lett., 6, 33 (1979).

35. Akasofu, S.-I., Physics of Magnetospheric Substorms, (Reidel, Hingham, MA, 1977), p. 274.

36. Akasofu, S.-I., Space Sci. Rev., 28, 121 (1981).

37. Baker, D. N., E. W. Hones, Jr., P. R. Higbie, R. D. Belian, and P. Stauning, J. Geophys. Res., 86, 8941 (1981).

38. Bargatze, L. F., R. L. McPherron, and D. N. Baker, J. Geophys. Res., in press, 1985.

39. Hones, E. W., Jr., D. N. Baker, S. J. Bame, W. C. Feldman, J. T. Gosling, D. J. McComas, R. D. Zwickl, J. Slavin, E. J. Smith, and B. T. Tsurutani, Geophys. Res. Lett., 11, 5 (1984).

40. Hones, E. W., Jr., Magnetic Reconnection in Space and Laboratory Plasma, ed. by E. W. Hones, Jr., Geophysical Monograph 30 (AGU Press, 1984). p. 178.

41. Scholer, M., G. Gloeckler, D. Hovestadt, B. Klecker, and F. M. Ipavich, J. Geophys. Res., 89, 8872 (1984).

42. Feldman, W. C., D. N. Baker, S. J. Bame, J. Birn, E. W. Hones, Jr., S. J. Schwartz, and R. L. Tokar, Geophys. Res. Lett., 11, 1058 (1984).

43. Holzer, R. E., and J. A. Slavin, J. Geophys. Res., 83, 3831 (1978).

44. Holzer, R. E., and J. A. Slavin, J. Geophys. Res., 84, 2573 (1979).

45. Hones, E. W., Jr., Physics of Solar Planetary Environments, ed. by D. J. Williams, (AGU Press, 1976), p. 558.

46. Hones, E. W., Jr., Dynamics of the Magnetosphere, ed. by S.-I. Akasofu, D. Reidel Publ. Co. (Dordrecht, Holland, 1979). p. 545.

47. Baker, D. N., Magnetic Reconnection in Space and Laboratory Plasmas, ed. by E. W. Hones, Jr., Geophysical Monograph 30 (AGU Press, 1984), p. 193.

48. Baker, D. N., S. J. Bame, R. D. Belian, W. C. Feldman, J. T. Gosling, P. R. Higbie, E. W. Hones, Jr., D. J. McComas, and R. D. Zwickl, J. Geophys. Res., 89, 3855 (1984).

49. Scholer, M., G. Gloeckler, D. Hovestadt, F. M. Ipavich, B. Klecker, and C. Y. Fan, Geophys. Res. Lett., 10, 1203 (1983).

50. Scholer, M., G. Gloeckler, B. Klecker, F. M. Ipavich, D. Hovestadt, and E. J. Smith, J. Geophys. Res., 89, 6717 (1984).

51. Cowley, S. W. H., R. J. Hynds, I. G. Richardson, P. W. Daly, T. R. Sanderson, K. P. Wenzel, J. A. Slavin, and B. T. Tsurutani, Geophys. Res. Lett., 11, 275 (1984).

52. Feldman, W. C., S. J. Schwartz, S. J. Bame, D. N. Baker, J. Birn, J. T. Gosling, E. W. Hones, Jr., D. J. McComas, J. A. Slavin, E. J. Smith, and R. D. Zwickl, Geophys. Res.. Lett., 11, 599 (1984).

53. Feldman, W. C., D. N. Baker, S. J. Bame, J. Birn, J. T. Gosling, E. W. Hones, Jr., S. J. Schwartz, J. A. Slavin, and R. D. Zwickl, J. Geophys. Res., 90, 233 (1985).

HEATING AND GENERATION OF SUPRATHERMAL PARTICLES AT COLLISIONLESS SHOCKS

M. F. Thomsen
Los Alamos National Laboratory
Los Alamos, NM 87545

ABSTRACT

Collisionless plasma shocks are different from ordinary collisional fluid shocks in several important respects. They do not in general heat the electrons and ions equally, nor do they produce Maxwellian velocity distributions downstream. Furthermore, they commonly generate suprathermal particles which propagate into the upstream region, giving advance warning of the presence of the shock and providing a "seed" population for further acceleration to high energies. Recent space observations and theory have revealed a great deal about the heating mechanisms which occur in collisionless shocks and about the origin of the various suprathermal particle populations which are found in association with them. An overview of the present understanding of these subjects is presented herein.

INTRODUCTION

Shock waves are formed by the nonlinear steepening of compressive wavemodes in a fluid. In contrast to nonlinear waves such as solitons, shocks provide a transition from one set of plasma parameters to another, the transition being irreversible by virtue of the action of dissipative processes. In general, shocks are formed when a disturbance (for example, an impenetrable object) travels through a fluid at a speed which exceeds the speed with which small amplitude waves can communicate the presence of the disturbance to the upstream medium. The role of the shock is to convert the upstream super-"sonic" flow to a sub-"sonic" flow which can then adjust itself to accommodate the disturbance.

In a collisionless plasma, the two compressive wavemodes which can form shocks are the fast and slow magnetosonic modes. Although slow mode shock waves have been reported in the geomagnetic tail and the solar wind, fast mode shocks in space plasmas have been much more extensively studied and at this point are considerably better understood. This review will therefore concentrate on fast mode shocks. Such shocks are ubiquitous in the solar-terrestrial environment. Collisionless bow shock waves stand upstream of the planets in the supersonic flow of the solar wind. High speed solar mass ejections drive interplanetary shocks ahead of them as they move outward in the interplanetary medium. And high speed solar wind streams steepen into "corotating" shocks at the interface with adjacent lower-speed streams (see Gosling this volume).

The job of converting supersonic flow to subsonic flow has two aspects: The reduction of the flow speed normal to the shock and the increase of the characteristic wave speed in the plasma (the fast magnetosonic speed). The two combine to reduce the Mach number of the flow to less than one as the plasma crosses the shock. The reduction of the flow speed is accomplished by a combination of a macroscopic electrostatic field within the shock transition, an outward pressure gradient, and magnetic forces. The increase in the magnetosonic speed is accomplished by a combination of compression (which increases the Alfvén speed, v_A) and heating (which increases the sound speed, c_s).

In space plasmas all of these changes are brought about through collisionless processes, and this very important fact leads to significant differences between these shocks and ordinary collisional fluid shocks. Two of the most important differences are 1) that the collisionless heating mechanisms do not in general produce Maxwellian velocity distributions in the downstream medium, nor do they necessarily heat the ions and electrons equally, and 2) that in addition to heating the plasma, collisionless shocks generate suprathermal particles which propagate back into the upstream medium. These suprathermal particles communicate an early warning of the presence of the shock to the incident plasma and thus broaden the transition region between the true upstream and downstream states. They also modify the upstream plasma conditions, which determine shock structure. Finally, they serve as a "seed" population subject to further acceleration to much higher energies.

In this paper we review what space observations and theory have revealed about the heating mechanisms which occur in collisionless shocks and about the origin of the various suprathermal particle populations which are found in association with such shocks. Recent and more detailed reviews of these subjects have been written by W. C. Feldman (electron velocity distributions near collisionless shocks), A. J. Klimas (the electron foreshock), J. T. Gosling and A. E. Robson (the evolution of ion distributions across collisionless shocks), M. F. Thomsen (suprathermal upstream ions), and M. Scholer (energetic particles at the bow shock and at interplanetary shocks). These and related reviews may be found in the Proceedings of the AGU Chapman Conference on Collisionless Shock Waves in the Heliosphere, Napa, CA (Feb. 1984). The contributed papers from the Napa Conference have appeared in the January 1985 issue of J. Geophys. Res.

As will become apparent below, the nature of the heating and suprathermal particle production varies with the upstream plasma parameters. Particularly important in this regard are the fast magnetosonic Mach number of the shock (M_f), the angle between the shock normal and the upstream magnetic field (θ_{Bn}), the ratio of the upstream thermal energy density to the magnetic field energy density ($\beta \equiv 8\pi nkT/B^2$, with $T = T_e + T_i$), and the upstream ratio of the electron temperature to the ion temperature (T_e/T_i). It is important to bear in mind that the shocks which nature supplies for *in situ* examination do not cover the full range of parameters which may be of astrophysical interest. Those parameters for which the coverage is most complete are θ_{Bn}, which varies in space shocks from near 90° to near 0°, and β, which is generally $\lesssim 1$ but sometimes as low as 0.01 and occasionally greater than 10. The temperature ratio is fairly well covered, being typically $\gtrsim 1$, but ranging from ~0.5 to ~10. The parameter for which the coverage is the most restricted is the Mach number. This varies from near 1 at some interplanetary shocks to somewhat greater than 10 on occasion at the Earth's bow shock. At the bow shocks of the outer planets the Mach numbers should be somewhat higher. The highest observed Alfvén Mach number has been $M_A \sim 20$ at Saturn. However, so far the outer planets missions have been too few and not ideally instrumented to shed much light on collisionless shock physics at $M \gtrsim 10$.

SHOCK HEATING

The shock jump equations (or Rankine-Hugoniot relations) consist of Maxwell's equations plus equations for the conservation of mass, momentum, and energy under certain approximations. If the upstream conditions are known, these relations can be used to estimate the total temperature change across a shock. However, they say nothing about the partition of the heating between the electron and ion species. This partition is determined by the details of the shock structure, i.e., by the macroscopic electric and magnetic fields within the shock and by the microscopic turbulence there. Laboratory and space observations indicate that this partition varies with the upstream parameters. Moreover, the heating of either species does not in general occur simply as an increase in the thermal spread of a Maxwellian velocity-space distribution. In fact, collisionless shocks in space are generally observed to produce distinctly non-Maxwellian distributions, and the production and evolution of such distributions depends on the upstream parameters.

A. Electrons

Laboratory studies of perpendicular ($\theta_{Bn} = 90^{o}$), nearly collisionless shocks have found very considerable electron heating, with upstream to downstream temperature ratios being a factor of 40 or more (e.g., Paul[1]). In contrast, at space shocks the electron temperature jump averages around 3 and rarely exceeds ~6.[2,3,4] A full parametric study of electron heating at space shocks has yet to be done, but it has been established that one of the primary determinants of electron heating is the strength of the shock, as measured for example by the downstream to upstream density ratio.[5] The temperature jump is generally larger for the stronger shocks, but there does not appear to be any one value of the "polytropic index," γ, which relates the temperature and density jumps. The difference between the large electron temperature changes observed in the laboratory and the relatively small changes observed in space is possibly due to a difference in the plasma instabilities operating in the two regimes and also probably to the strong heat flux observed at space shocks (see below).

Not only the amount of electron heating, but also the nature of the heating is dependent on the shock strength.[6] At the weakest shocks the heating appears almost exclusively in the direction perpendicular to the magnetic field, and the evolution of the distribution is basically consistent with conservation of the magnetic moment in the increased magnetic field downstream. At progressively stronger shocks, there appears increased heating in the direction parallel to the magnetic field. At times the heating in the parallel direction may even exceed that in the perpendicular direction. Once the parallel heating becomes significant, the downstream electron distributions have a distinctly non-Maxwellian nature. The most striking characteristic of the distributions downstream of these stronger shocks is a plateau at the lowest energies.

These features are illustrated in Figure 1.[6] Figure 1a shows cuts through the electron velocity distributions observed upstream and downstream of three relatively weak interplanetary shocks. For each shock, cuts are displayed for the directions perpendicular and parallel to the magnetic field. Figure 1a shows that for these shocks the up- and downstream distributions in the parallel direction are virtually indistinguishable; the only apparent heating is in the perpendicular direction. Figure 1b shows similar cuts through the velocity distributions observed up- and downstream of a slightly stronger interplanetary shock. The heating is still primarily in the perpendicular direction and is generally consistent with simple conservation of the magnetic moment, as indicated by the filled circles, but there is now also some apparent heating in the parallel direction. Finally, Figure 1c shows similar cuts observed downstream of two relatively strong interplanetary shocks. For these shocks the parallel heating clearly exceeds the perpendicular heating. Furthermore, the plateau or flat-top at low energies is very evident. This flat-top is a characteristic feature of the magnetosheath electrons downstream of the Earth's bow shock, which is typically quite strong ($M_A \gtrsim 4$).[2,3,7] The higher energy portion of the downstream electron distribution is typically a power-law in speed, $f(v) \propto v^{-q}$, where values of q have been found in the range $6 \lesssim q < 10$.[7] A more complete survey of the variation of q with the various shock parameters has not yet been done.

Because the Earth's bow shock is nearly at rest with respect to an Earth-orbiting satellite, it sometimes happens that a satellite crosses the bow shock so slowly that the detailed evolution of the particle distribution functions within the shock ramp itself can be resolved. During such slow traversals, a feature is often observed in the electron distribution which offers a clue to the mechanism by which the electrons are actually heated by the shock. Figure 2 shows an example of the distributions

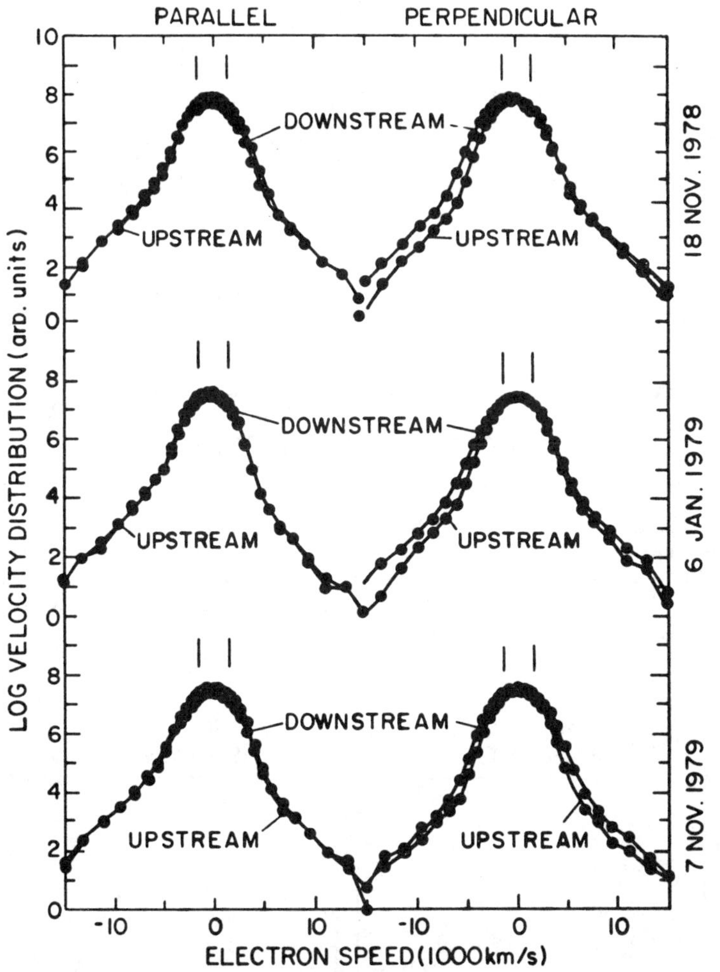

Fig. 1a. Cuts through the electron velocity distributions observed near several interplanetary shocks (from Feldman et al.[6]). "Parallel" and "perpendicular" refer to cuts through the distributions in the directions parallel and perpendicular to the ambient magnetic field. Comparison of cuts taken up- and downstream of three relatively weak interplanetary shocks, showing that the heating is in the perpendicular direction.

observed during one such slow crossing (from Feldman et al.[8]). The figure shows cuts through the observed electron velocity distribution, taken in the direction parallel to the local magnetic

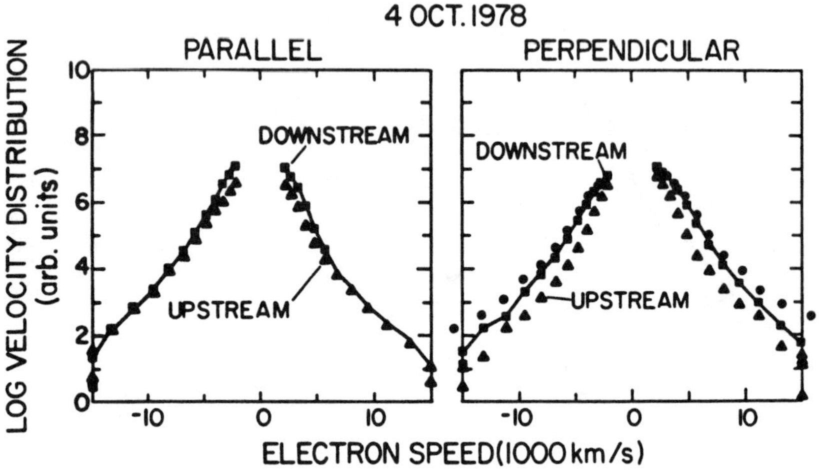

Fig. 1b. Comparison of cuts taken up- and downstream of a slightly stronger interplanetary shock, showing some slight heating in the parallel direction. The heating in the perpendicular direction is consistent with expectations based on conservation of magnetic moment (circles).

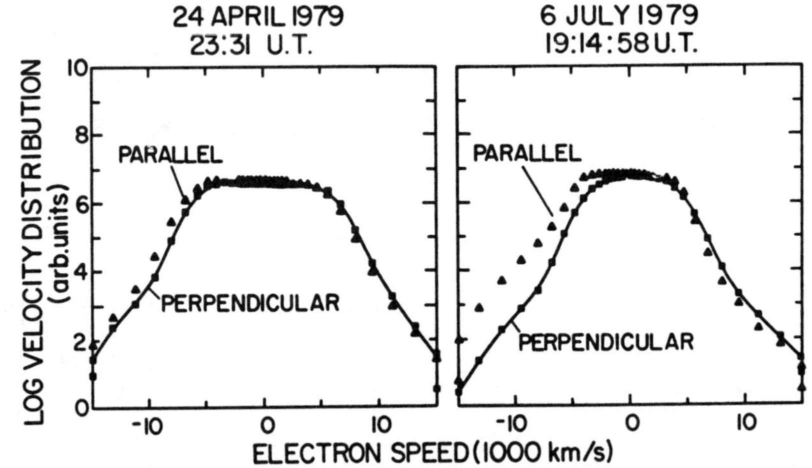

Fig. 1c. Comparison of cuts in the parallel and perpendicular directions downstream of two moderately strong interplanetary shocks. Note the flat top at low energies and the strong heating in the parallel direction.

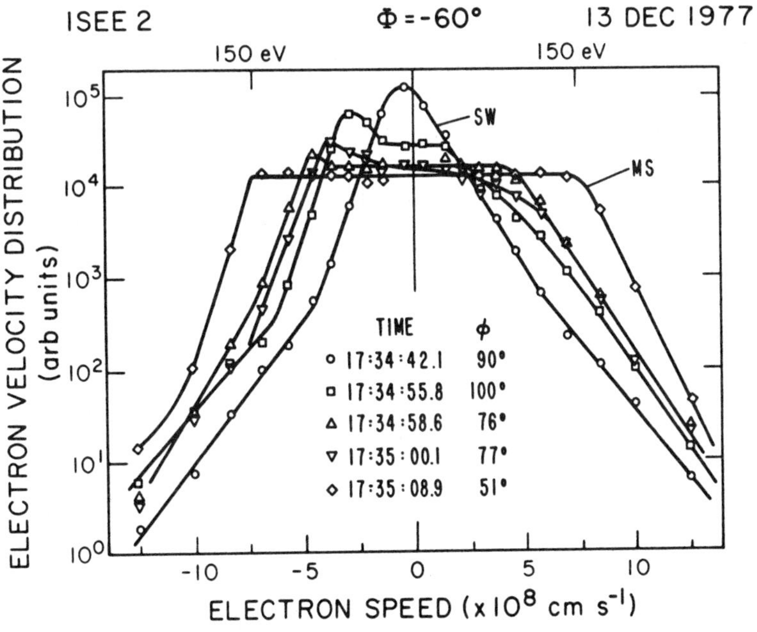

Fig. 2. Cuts taken through the electron velocity distribution, in the direction parallel to the magnetic field, during a fairly slow crossing of the Earth's bow shock (from Feldman et al.[8]). The spectrum at 17:34:42 (SW) was taken in the upstream solar wind. The spectrum at 17:35:09 (MS) was taken in the downstream magnetosheath and exhibits the characteristic flat top at low energies. The spectra in between were obtained within the magnetic ramp and show the presence of a field-aligned peak in the distribution at the ingoing edge of the developing flat-top.

field. The sequence starts at 17:34:42.1 in the upstream solar wind (labelled SW in the figure) and ends at 17:35:08.9 in the magnetosheath downstream of the shock (labelled MS). The other three spectra were obtained at various points within the shock ramp and exhibit the general evolution from the upstream Maxwellian to the downstream flat-topped distribution. In addition, there appears within the ramp a peak in the distribution, located at the edge of the developing flat-top and directed into the downstream region. As the spacecraft approaches the downstream medium, the peak moves to progressively larger speeds and simultaneously decreases in height.

This peak has been interpreted as the signature of the acceleration of solar wind electrons by the magnetic field-aligned component of the shock electrostatic field.[8] This electrostatic field is caused by charge separation and plays an important role in slowing and deflecting the incident flow. The electron acceleration process is illustrated schematically in Figure 3. As the incoming solar wind electrons pass through the potential jump of the shock, they are accelerated inward, forming an offset peak in the inward-moving direction (solid curves at positions 2 and 3). The flat-topped region at lower energies represents the population of electrons which is trapped behind the shock, unable to get out over the potential barrier. The solid curves show that in the absence of scattering, Liouville's theorem would predict that the height of the inward-accelerated peak remain constant. In reality (dashed curves), the peak diminishes, the signature that wave-particle scattering is occurring, which in turn feeds both the

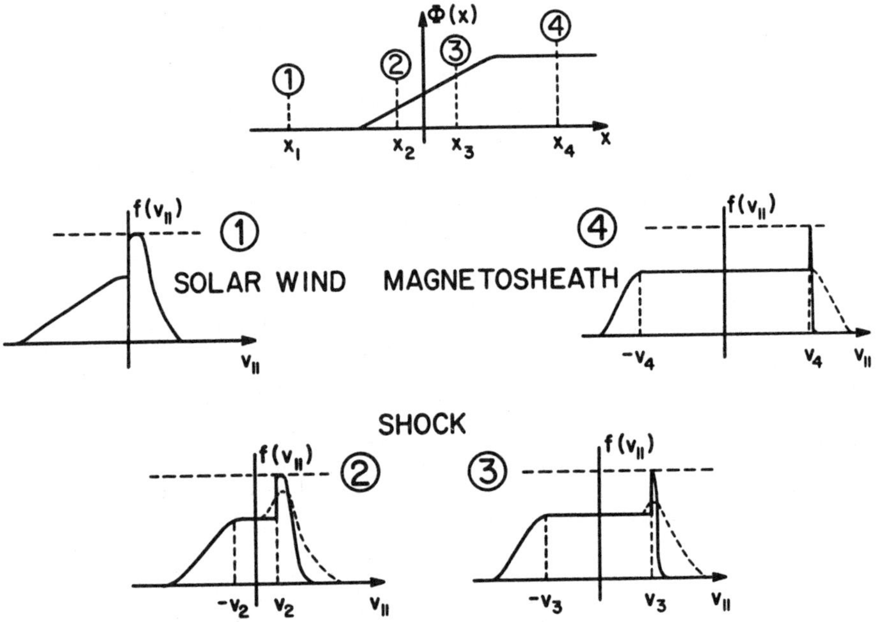

Fig. 3. Schematic illustration of the development of the parallel electron distribution function through the shock (after Feldman[83]). Incident electrons are accelerated into the shock by the jump in the electrostatic potential (top), forming an offset peak in the inward-moving direction. Wave-particle scattering erodes the peak, feeding both the higher energy population and the lower-energy population, which is trapped by the potential barrier and forms the flat top of the downstream distribution.

trapped population and the population of suprathermal electrons at energies above the break-point of the flat-top. (In this view, the break-point energy is equal to the electrostatic potential jump experienced by the electrons. There is an interesting reference-frame dependence to this relationship, which has been discussed by Goodrich and Scudder[9].) The suprathermal tail in the outgoing direction is able to surmount the potential barrier and escape into the upstream region, forming the heat flux away from the bow shock which is characteristically seen whenever an observing spacecraft is magnetically connected to the shock.[3,10,11,12]

B. Ions

Although no systematic study has yet been made of the variation in the ion heating at collisionless shocks as a function of the shock parameters, it is known that the amount and nature of the heating are certainly dependent on the Mach number of the shock. Early laboratory work, primarily with perpendicular shocks ($\theta_{Bn} = 90^o$), indicated that below some critical Mach number (typically $M_c \sim 3$), the ions heat very little.[1,13,14,15,16] Above M_c, the ions were inferred to heat quite strongly, an increase which was attributed to the onset of ion reflection at the shock, leading to dispersal of the ions in velocity space (e.g., Phillips and Robson[17]). Such a qualitative transition was also expected on theoretical grounds. Above the critical Mach number resistive dissipation due to microinstabilities is not adequate to satisfy the conservation requirements, and some sort of ion "viscosity" is needed in addition This viscosity is provided by ion reflection,[18,19,20] which sets in when the downstream flow speed is exceeded by the downstream sound speed (e.g., Edmiston and Kennel[21]).

Observations of collisionless shocks in space verified the previously inferred general Mach number dependence of the ion heating, with the exception that space shocks showed significant heating even at the lower Mach numbers where ion reflection is not yet important.[5,22,23] These lower Mach number shocks typically produce ion heating which is correlated with but exceeds the electron heating.[5] The downstream-to-upstream ion temperature jump can be as large as a factor of 20, and the increase occurs primarily in the bulk of the ion distribution, in the direction perpendicular to the magnetic field.[23] The difference between the low Mach number heating patterns (for both ions and electrons) observed in the laboratory and in space has been attributed to the fact that the laboratory shocks were usually nearly perpendicular while the space shocks are usually oblique, with a corresponding difference in the nature of the dominant plasma instabilities which provide the dissipation within the shock.[23]

As indicated above, the ion heating increases considerably as the Mach number rises above the critical Mach number for the initiation of ion reflection. (The critical Mach number can be computed from fluid theoretical arguments and depends on the upstream parameters, especially θ_{Bn} and β; see, e.g., Edmiston and Kennel[21]). At the Earth's bow shock, which is usually supercritical, typical downstream-to-upstream ion temperature ratios are ~20-100. The increase in ion temperature is always much greater than the associated increase in electron temperature. And as described below, the ion reflection process, which is responsible for most of the ion heating, results in downstream distributions which are distinctly non-Maxwellian.

As illuminated by laboratory studies,[17] numerical simulations,[24,25] and space observations,[26,27] at a supercritical shock a portion of the incident ions are reflected from the shock in a nearly specular manner ("specular" meaning that the component of motion parallel to the shock normal is reversed while the component parallel to the shock surface is unchanged). The orbit of a specularly reflected ion is illustrated schematically in Figure 4 for a perpendicular shock whose normal ($\hat{n}$) lies within the plane of the figure. Figure 4a shows the shock in physical space as the heavy solid line. The solar wind is incident from the left. When it encounters the shock, most of the incident ions are directly transmitted (and slowed and deflected as well), while some fraction of them are nearly specularly reflected back into the upstream region. These reflected ions then gyrate about the upstream magnetic field with large gyrovelocities ($\approx 2\ V_{in}$, where V_{in} is the velocity of the upstream flow parallel to the shock normal). They continue to convect with the solar wind and hence are swept back into the downstream region, as shown in the figure.

Figure 4b shows the two-dimensional velocity space signature one would expect to see at a position immediately upstream of the shock as indicated by the dashed line in Figure 4a. In addition to the incident solar wind, one would expect to see two "bunches" of reflected ions: Those newly reflected ions which are at point (1) in their trajectory, and ions reflected from another point on the shock which are about to reencounter the shock at point (2) in their trajectory (Fig. 4a). Velocity space bunches of gyrating ions such as these are commonly observed immediately upstream of supercritical, quasi-perpendicular ($\theta_{Bn} > 45^o$) shocks.[26,27] They are also seen in the ramp and immediately downstream. (Note that bunching in velocity space can occur without bunching in configuration space.)

From Figure 4b it is clear that the reflection process introduces a gross dispersion into the ion velocity distribution. This dispersion is ultimately responsible for the large ion heating at a supercritical shock. Downstream of the shock this dispersion shows up initially as a second peak in the ion energy distribution.[2] Gyrophase mixing and various plasma instabilities then combine to "thermalize" the distribution by sharing the large kinetic energy contained in the gyromotion of the reflected ions with the directly transmitted core. However, this thermalization process does not produce a Maxwellian distribution. Rather, what

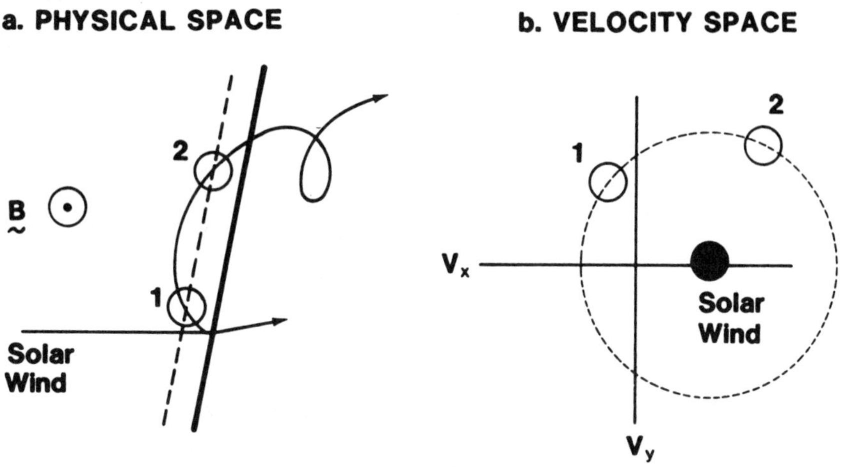

Fig. 4. Schematic illustration of specular ion reflection at a supercritical, perpendicular shock (after Sckopke et al.[27]). a) Diagram showing physical space trajectories of incident, transmitted, and reflected ions. The reflection introduces a net gyromotion relative to the convection. b) Two-dimensional velocity space signature one would expect to observe within the upstream "foot" of the shock at the position indicated by the dashed line in a). The magnetic field is parallel to v_z. In addition to the incident solar wind, one expects two velocity-space "bunches" of reflected ions which are at points (1) and (2) of their trajectory [see (a)].

is left is a persistent, non-Maxwellian shoulder to the ion distribution, as shown in Figure 5.[27] The figure presents a time sequence of energy distributions observed downstream of a shock crossing in which reflected, gyrating ions were quite prominent. Just behind the shock, at 22:52, there is a clear indication of the second peak attributable to reflected ions. Within only a few spectra downstream of the shock, the valley between the peaks is filled in. The resultant shoulder at ~1.0 keV persists well into the downstream medium, the last spectrum in Figure 5 having been obtained more than an hour after the shock crossing.

To lowest order, the above discussion of how ion reflection leads to downstream heating only applies for shocks with $\theta_{Bn} > 45^{\circ}$. This is because only for these quasi-perpendicular shocks is the net guiding center motion of specularly reflected ions towards the downstream region.[28] For $\theta_{Bn} < 45^{\circ}$, the net guiding center motion is back upstream away from the shock. Such backstreaming, specularly reflected ions have indeed been observed upstream of a

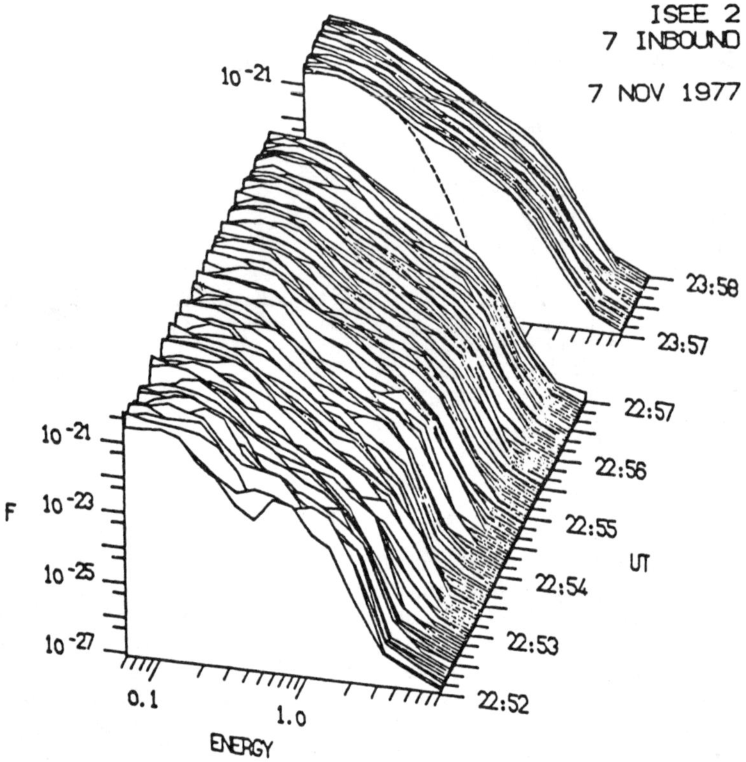

Fig. 5. Time sequence of ion energy spectra obtained immediately downstream of a crossing of the Earth's bow shock, showing the "thermalization" of the reflected ions. The resultant distribution is non-Maxwellian, with a persistent shoulder at ~1.0 keV.[27]

quasi-parallel bow shock crossing.[28] Nonetheless, there is evidence that even in the quasi-parallel regime ion reflection contributes importantly to ion heating downstream of supercritical shocks. It has been suggested[29,30,31,32] that this is a result of a self-regulating feedback system in which specularly reflected ions travelling back upstream away from the quasi-parallel shock generate large-amplitude MHD-like waves which convect toward the shock with the incident plasma. These large-amplitude waves periodically alter the local value of θ_{Bn} at the shock so that the incident plasma encounters the shock sometimes in a quasi-perpendicular orientation, allowing downstream transmission of reflected ions, and sometimes in a quasi-parallel orientation, allowing upstream escape and continued generation of the large-amplitude waves.

SUPRATHERMAL PARTICLE PRODUCTION

As mentioned above, one of the important differences between collisionless and collisional shocks is that collisionless shocks generate suprathermal particles which propagate back into the upstream region, giving advance notice of the presence of the shock and providing a "seed" population for further acceleration to high energies. In this section we categorize the various populations of suprathermal particles which have been observed upstream of collisionless space shocks and give an indication of the best present understanding of the origins of these populations. Most of the information regarding shock-generated suprathermals has been obtained in the region upstream of the Earth's bow shock, known as the "foreshock." In this review we concentrate primarily on foreshock findings, but it is important to bear in mind that the bow shock is highly curved compared to other space shocks (for example, travelling interplanetary shocks). The strong curvature has several consequences, which will be pointed out below.

The nominal foreshock geometry is illustrated in Figure 6, which shows the bow shock as the heavy solid curve and the interplanetary (solar wind) magnetic field as the thin solid lines. The orientation shown is the average configuration of the interplanetary field at the orbit of Earth determined by the competition between solar rotation and solar wind convection. The upstream field orientation can often be quite different from that illustrated.

Particles leaving the bow shock for the upstream region are constrained to follow the magnetic field lines and thus, unless cross-field diffusion is very strong, will not be found sunward of the magnetic field line which is just tangent to the bow shock. Moreover, all the while they are streaming back along the field lines, they are subject to the convective ($\tilde{E}\times\tilde{B}$) drift of the solar wind. The guiding centers of such backstreaming particles will thus follow trajectories as indicated by the dashed lines in the figure. The larger the velocity parallel to the magnetic field, the closer the trajectory will lie to the actual field direction. For a given parallel velocity, the particle trajectory which leaves the point of magnetic tangency defines the region sunward of which no such particle will be found. This trajectory is called the foreshock boundary and is shown as the heavy dashed line in Fig. 6. Very fast particles have a foreshock boundary near the tangent field line, while slower particles are found only deeper in the foreshock region.

A. Electrons

Figure 7 illustrates the suprathermal electron distributions which are observed in the Earth's electron foreshock. Below the diagram of the foreshock geometry are three schematic plots of contours of constant electron phase space density in two-

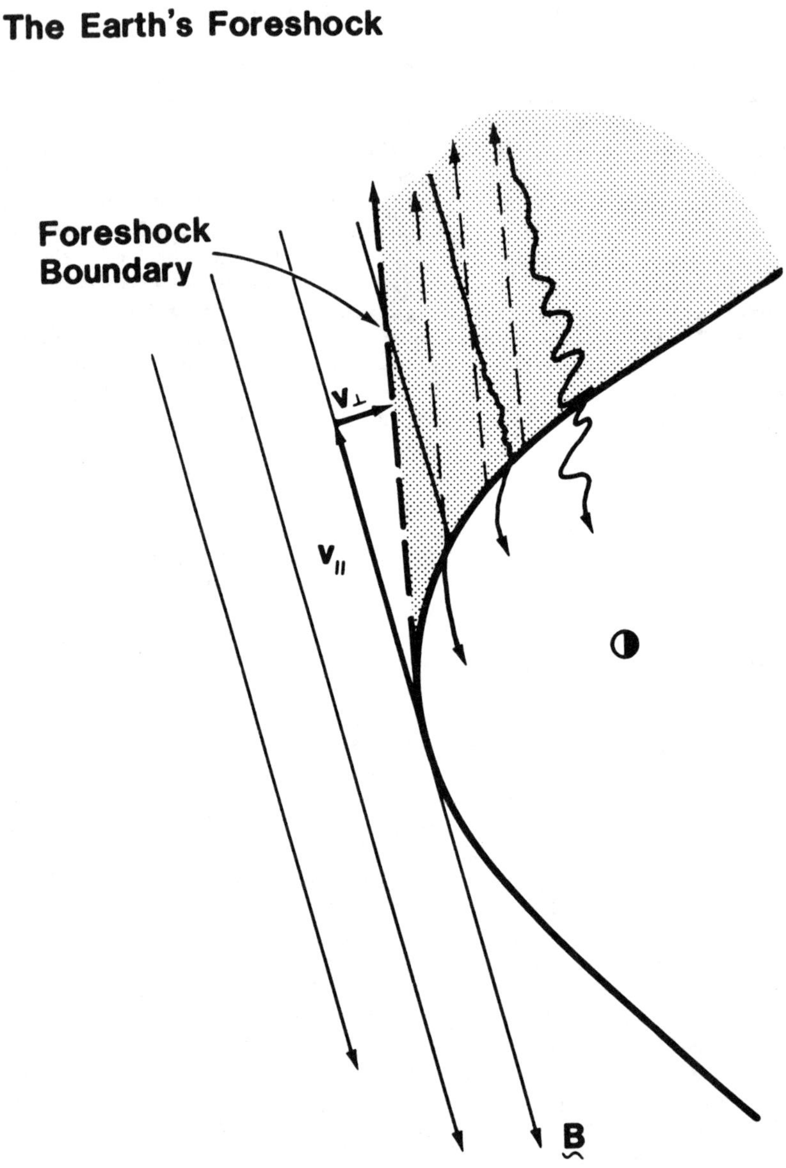

Fig. 6. Schematic illustration of the geometry of the Earth's foreshock for a nominal "garden hose" interplanetary magnetic field orientation. Particles leaving the shock travel along the magnetic field and at the same time are convected along with the solar wind, following trajectories represented by the dashed lines. Such particles will not be found sunward of the stippled foreshock region.

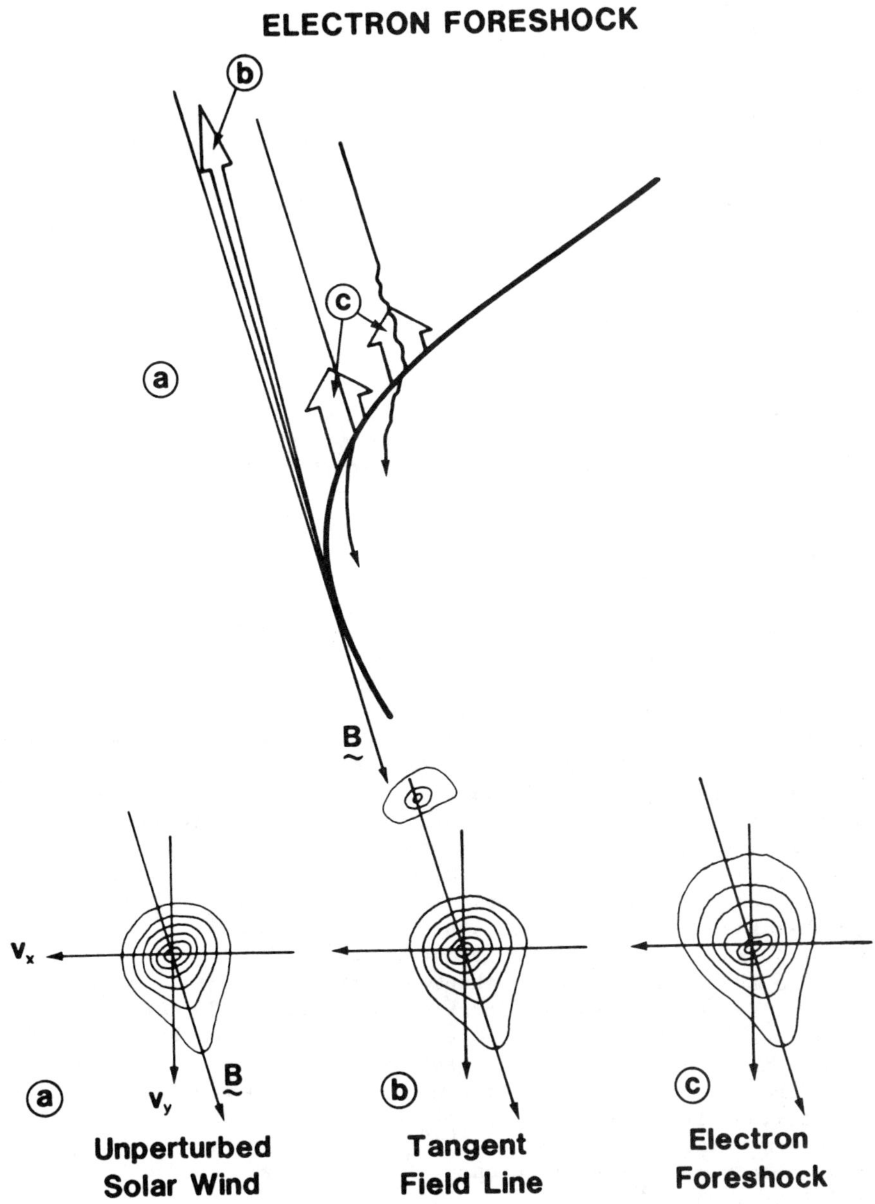

Fig. 7. Illustration of the electron distributions typically observed in different parts of the electron foreshock. The distributions are illustrated by schematic plots of contours of constant phase space density in two-dimensional velocity space as seen a) upstream of the foreshock boundary; b) at the foreshock boundary, which for electrons is near the tangent field line; and c) well downstream of the foreshock boundary.

dimensional velocity space as one would expect to observe them at three different locations upstream of the shock: a) In the unperturbed solar wind, b) just downstream of the tangent field line, and c) deeper in the foreshock.

Upstream of the tangent field line, in the unperturbed solar wind (region a), the electrons essentially consist of a Maxwellian (or two-temperature Maxwellian) distribution with a field-aligned heat flux away from the sun, carried by suprathermal electrons from the solar corona. Within the foreshock but very near the tangent field line (region b) one frequently observes superimposed on the solar wind distribution field-aligned beams of fairly energetic electrons ($E_e \lesssim 100$ keV) streaming away from the shock (e.g., see Anderson[33] and references therein). The energy of these suprathermal electrons decreases deeper in the foreshock region, and it appears that the most energetic of the electrons are produced only in a narrow region of the shock near the point of tangency.[34] These observations are consistent with a model in which the energetic electrons on nearly tangent field lines are produced by the adiabatic (magnetic moment-conserving) mirror reflection of solar wind electrons at the bow shock.[35,36] Evidence for such magnetic mirroring is present as well at lower energies ($E_e \sim 20$ eV) deeper in the foreshock.[7]

Well within the electron foreshock (region c in Figure 7), the very energetic electron component is not seen. Instead, hot, suprathermal electrons of bow shock origin are observed over essentially the entire backstreaming hemisphere.[7,12] These electrons carry a heat flux away from the bow shock which generally exceeds the heat flux carried toward the shock by the solar wind suprathermal electrons.[11] At higher energies ($E_e \gtrsim 40$ eV), this backstreaming population is nearly isotropic over the hemisphere away from the shock and has a power-law energy spectrum quite similar to the spectrum at similar energies in the region downstream of the shock. Indeed, as mentioned earlier, this population appears to be simply the higher-energy portion of the shock-heated electron distribution which is able to surmount the shock potential barrier and escape upstream.[7]

In addition to these shock-generated electron populations, bursts of very energetic electrons ($E_e \gtrsim 220$ keV) are also occasionally observed within the foreshock. These, however, appear to originate within the magnetosphere rather than at the shock.[37]

The suprathermal electrons in the foreshock region are associated with several types of waves, including electrostatic electron plasma waves,[38,39,40,41] electromagnetic waves at twice the electron plasma frequency,[42] and lower frequency (~1 Hz) whistler waves.[43,44]

B. Ions

The ion foreshock region is populated by a number of different types of suprathermal distributions. The properties of these various populations have been studied extensively (see especially the ISEE Upstream Waves and Particles special issue of J. Geophys. Res., June 1981).

The different types of suprathermal ion distributions and the foreshock regions in which they typically are found are illustrated schematically in Figure 8. The velocity space contour plots have the same format as those shown for electrons in Fig. 7. Near the foreshock boundary, in regions where the local magnetic field intersects the shock in the quasi-perpendicular ($\theta_{Bn} \gtrsim 45^o$) geometry (region 1), the most commonly observed suprathermal ion distribution is a field-aligned beam directed away from the bow shock. The beam energy is typically approximately a few keV, but on occasions has been found to exceed 40 keV. The beam temperature is typically ~ few x 10^{6o}K and is anisotropic, with $T_\perp/T_\parallel \sim 4$-9.[45]

Deeper in the ion foreshock (region 2) the suprathermal ion distributions are often of the type labelled "intermediate."[46] Intermediate ion distributions are similar to the field-aligned beams except that they extend to much larger pitch angles (appearing crescent-shaped in velocity space, as illustrated in Fig. 8), and they are somewhat hotter and slower on average.

Filling the rest of the deep foreshock (region 3), where the local magnetic field typically intersects the shock in the quasi-parallel geometry ($\theta_{Bn} \lesssim 45^o$), are suprathermal ion distributions known as "diffuse" ions.[47] These are characterized by broad, nearly isotropic angular distributions and by relatively flat energy spectra extending up to ~150 keV. Like the field-aligned beams and intermediate ions, they have a net bulk flow velocity away from the shock, but it is typically much smaller than for the other two populations. Although not indicated in Fig. 8, the diffuse ions are also commonly observed downstream of the shock within the magnetosheath; their net bulk flow velocity there is toward the downstream region, again away from the shock.[47,48] The energy spectra of the diffuse events frequently exhibit a peak at an energy of ~ 1-few keV, below which the phase space density drops off fairly rapidly.[47] At higher energies ($\gtrsim$ 30 keV) the energy spectra are generally well-described by either an exponential or Maxwellian in energy per charge.[49]

More recently an additional category of upstream suprathermal ion distributions has been identified.[50,51] These have been called "gyrating ions"[26] because the primary characteristic of the distribution is what one might call a "bulk gyromotion": The peak of the distribution does not lie along the magnetic field but rather at a nonzero pitch angle relative to the field (see Figure 8). Gyrating ion distributions often, but not always, exhibit confinement to a relatively narrow range of gyrophase; i.e., they are "non-gyrotropic" or "gyrophase bunched." The

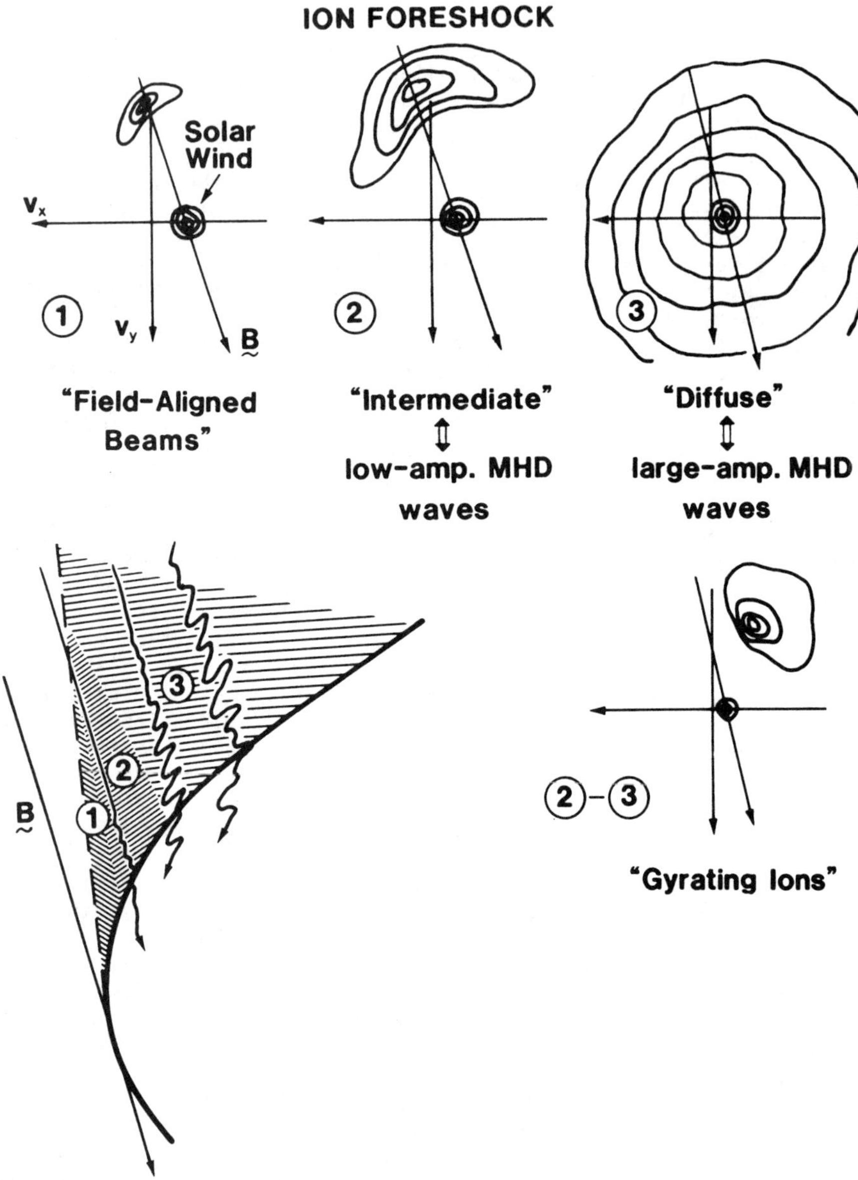

Fig. 8. Schematic illustration of the ion distributions typically observed in different parts of the ion foreshock. Same format as Figure 7.

occurrence statistics for gyrating ions in the foreshock region are not yet known, but they have been observed generally within regions 2 and 3 of Figure 8.

A number of production mechanisms have been proposed for these various types of suprathermal ion distributions, and evidence has accumulated that more than one process may be responsible for each type. The field-aligned beams were originally hypothesized to arise from a process of reflection at the shock, accompanied by drift parallel to the interplanetary electric field.[52] This suggestion has been found to be in good agreement with observed beam energies, especially the most energetic,[53,54] and numerical simulations and test particle trajectory calculations suggest that the necessary drift can be accomplished by means of multiple reflections along the shock front.[55,56] An alternative hypothesis is that the beams originate in the leakage of shock-heated magnetosheath ions back into the upstream region.[57,58,59] This mechanism is also consistent with some observed beam energies although it does not seem capable of producing the highest energies.[60] In addition, recent high time resolution observations of the evolution of several ion beam distributions adjacent to and through the bow shock are consistent with the leakage hypothesis.[61] It thus seems probable that some field-aligned ion beams arise from the leakage of magnetosheath ions while others (especially the most energetic) arise from multiple reflection at the shock. The relative importance of these two sources has yet to be determined.

The gyrating ions also probably have more than one source. The original suggestion was that they originate through a reflection process at the shock.[50] Subsequent quantitative analysis showed that in at least one case the backstreaming gyrating ions observed upstream of a quasi-parallel shock probably were produced by a specular reflection process at the shock analogous to that which produces the downstream heating in the quasi-perpendicular geometry (see above).[28] This process produces gyrophase-bunched ions at the shock, as observed. However, such a distribution should fairly rapidly become phase-mixed through kinematic effects as it propagates upstream away from the shock.[56,62] Hence, the observation of gyrophase-bunched gyrating ions at considerable distances upstream of the shock[50,63] suggests that other processes within the upstream region itself can either generate or maintain nongyrotropic ion distributions. One possibility is that such distributions are produced during one stage of the instability and disruption of a field-aligned ion beam.[64]

Instability and subsequent non-linear evolution of field-aligned ion beams is also thought to be the source of many of the intermediate ion distributions.[46] Calculations indicate that beam distributions of the type observed in the ion foreshock are unstable to the electromagnetic ion beam instability.[65,66,67] The instability should give rise to long-period MHD-like waves which in turn should scatter the beam ions in pitch-angle, initially producing a crescent-like distribution (the intermediate ions) and ultimately a more nearly isotropic distribution (the diffuse ions). Considerable observational evidence has accumulated in support of this scenario, including the presence of the long-period waves in direct association with intermediate and diffuse distributions.[46,68,69,70,71,72] Numerical simulations also confirm the basic elements of the beam-disruption hypothesis, including the linear growth stage, the simultaneous appearance of significant wave amplitudes and crescent-shaped distributions, and the ultimate production of diffuse-like distributions and large-amplitude magnetic turbulence.[73] Other evidence supporting the beam-disruption hypothesis has been summarized by Thomsen.[74]

The process of beam disruption described above applies to gyrating ion distributions as well as to field-aligned beams.[75] The ultimate non-linear result of the instability is again an intermediate-like and then diffuse-like distribution.

Finally, while the disruption of beams or gyrating ions seems well established as a source for intermediate ion distributions, the observation of field-aligned beams escaping from the magnetosheath[61] strongly suggests that magnetosheath leakage may also produce upstream intermediate distributions, as suggested by Edmiston et al.[58]

Since the diffuse suprathermal ion distributions observed in the Earth's foreshock include nearly as many particles moving toward the shock as away from it, it seems clear that scattering processes in the up-and downstream media are an essential element of the formation of diffuse distributions. The most likely origins of diffuse ions in the foreshock are 1) the disruption of field-aligned beams, intermediate distributions, or gyrating ion distributions, of whatever source, through scattering off of waves self-generated by those distributions, and 2) the "parasitic" disruption of such distributions by waves generated further upstream and convected back towards the shock by the solar wind. Backstreaming ions entering the foreshock near its forward boundary (in the quasi-perpendicular regime) are probably subject to the former process, whereas ions entering deeper in are probably subject to the latter process and are rapidly scattered before they can propagate very far from the shock.

INTERPLANETARY SHOCKS

The discussion above has been based primarily on observations made at and near the Earth's bow shock. Similar, though less comprehensive, observations are also available from interplanetary shocks, and they reveal some interesting similarities and differences with respect to the bow shock. The primary parametric differences between interplanetary shocks and the bow shock are that the interplanetary shocks are far less curved and that they typically have considerably lower Mach numbers.

The electron behavior at interplanetary shocks seems to be in most respects the same as that observed at weaker bow shock crossings: The nature of the heating and evolution of the electron distributions across the shock is completely analogous to that at the bow shock.[5,6] Upstream of the shock the same sort of heat flux away from the shock is found, as is the mid-energy (~20-40 eV) signature of magnetostatic reflection.[6] In addition, field-aligned energetic electrons ($E_e \gtrsim 2$ keV) are frequently observed streaming away from the shock.[76]

In contrast to the electrons, the ion behavior seems to be somewhat different than that at the bow shock by virtue of the combined effects of lower Mach number and less curvature. For lower Mach numbers, ion reflection is much reduced and hence can not contribute much to the heating or to the suprathermal shoulder on the downstream distribution. This latter situation results in fewer downstream particles with sufficient energy to escape upstream. In addition, the multiple reflection process is likely to be much inhibited by the reduced number of ions making the initial, nearly specular reflection. One thus expects the injection rate of suprathermal ions into the region upstream of interplanetary shocks to be reduced, and indeed the intensities of upstream suprathermal ions are found to be much smaller at interplanetary shocks.[77,78]

In addition, the nearly planar nature of the interplanetary shocks rules out the clearly demarcated foreshock structure of the bow shock. Specifically, the field-aligned beams observed in the forward part of the ion foreshock are essentially transient phenomena, present by virtue of the limited time of magnetic connection to the bow shock. At interplanetary shocks, suprathermal ions leaving the shock should immediately encounter and be scattered by magnetic perturbations generated further upstream. One might thus expect only diffuse-like suprathermal distributions upstream of these more "mature" interplanetary shocks. This is indeed the case: At the level of sensitivity of present instrumentation, suprathermal particles are sometimes seen in low intensities upstream of interplanetary shocks, but never in a beam-like configuration.[77,78]

The absence of field-aligned ion beams upstream of interplanetary shocks suggests that the magnetic turbulence there (and perhaps also in the deeper part of the Earth's foreshock) may be due to the "hot ion beam" version of the electromagnetic ion beam instability.[66,79] Alternatively, or perhaps concurrently, the waves may be generated by the more energetic ions typically observed fairly far upstream of such shocks.[80] This possibility is supported by the observation that the magnetic wave activity extends quite far upstream of the shock,[81] much further than the detectable fluxes of ions with energies on the order of several keV.

Another difference between interplanetary shocks and the Earth's bow shock is that interplanetary shocks are typically immersed in a "sea" of superthermal particles produced originally by the solar disturbance which gave rise to the shock (see, e.g., Tsurutani and Lin[82]). These pre-existing particles provide a natural seed population for acceleration to higher energies and may account for the fact that the intensities of more energetic ions (many keV) are typically higher at interplanetary shocks than at the bow shock, even though the injection rate of solar wind ions into the suprathermal seed population is probably lower, as discussed above.

SUMMARY

- Collisionless plasma shocks are quite effective at heating ions.
- They are not nearly as effective at heating electrons, possibly because of the strong heat flux carried by escaping suprathermal electrons.
- Downstream, both ion and electron distributions are non-Maxwellian.
- In general the downstream ion and electron temperatures are not equal.
- Collisionless shocks can accelerate both ions and electrons from upstream thermal energies to many tens of keV, with or without a pre-existing sea of suprathermals.

To these points should be added an important caveat: As pointed out above, the shocks available for study within the heliosphere have a limited range of Mach numbers. Because we now know that even over this fairly small range the dominant physical processes vary, one should be somewhat wary of uncritically extrapolating the present understanding to Mach numbers much outside of this range. Indeed, one should be prepared for the possibility that the physics at much higher Mach numbers may be completely different. Recent numerical simulations, as discussed by Quest at this conference, do indeed suggest that this is the case. It seems likely that in the near future such numerical simulations are the most promising means of extending our understanding of the physics of collisionless shocks to high Mach numbers.

ACKNOWLEDGMENTS

I am particularly indebted to W. C. Feldman and J. T. Gosling for their wisdom, insight, and guidance on the subjects discussed here. I am also very grateful to S. P. Gary, S. J. Schwartz, D. Winske, and K. Quest for numerous enlightening conversations. This review was written under the auspices of the U. S. Department of Energy and was funded in part by NASA through grants S-04039D and 10-23726.

REFERENCES

1. J. W. M. Paul, Review of experimental studies of collisionless shocks propagating perpendicular to a magnetic field, in Collision-Free Shocks in the Laboratory and Space, Proc. of a study group held at the European Space Research Institute (ESRIN), Frascati, 11-20 June 1969.
2. M. D. Montgomery, J. R. Asbridge, and S. J. Bame, J. Geophys. Res. 75, 1217 (1970).
3. J. D. Scudder, D. L. Lind, and K. W. Ogilvie, J. Geophys. Res. 78, 6535 (1973).
4. S. J. Bame, J. R. Asbridge, J. T. Gosling, M. Halbig, G. Paschmann, N. Sckopke, and H. Rosenbauer, Space Sci. Rev. 23, 75 (1979).
5. W. C. Feldman, J. R. Asbridge, S. J. Bame, J. T. Gosling, and R. D. Zwickl, Solar Wind Five, (NASA Conference Publication 2280, 1983), p. 403.
6. W. C. Feldman, R. C. Anderson, S. J. Bame, J. T. Gosling, R. D. Zwickl, and E. J. Smith, J. Geophys. Res. 88, 9949 (1983).
7. W. C. Feldman, R. C. Anderson, S. J. Bame, S. P. Gary, J. T. Gosling, D. J. McComas, M. F. Thomsen, G. Paschmann, and M. M. Hoppe, J. Geophys. Res. 88, 96 (1983).
8. W. C. Feldman, S. J. Bame, S. P. Gary, J. T. Gosling, D. J. McComas, M. F. Thomsen, G. Paschmann, N. Sckopke, M. M. Hoppe, and C. T. Russell, Phys. Rev. Lett. 49, 199 (1982).
9. C. C. Goodrich and J. D. Scudder, J. Geophys. Res. 89, 6654 (1984).
10. K. W. Ogilvie, J. D. Scudder, and M. Sugiura, J. Geophys. Res. 76, 8165 (1971).
11. W. C. Feldman, J. R. Asbridge, S. J. Bame, and M. D. Montgomery, J. Geophys. Res. 78, 3697 (1973).
12. W. C. Feldman, R. C. Anderson, J. R. Asbridge, S. J. Bame, J. T. Gosling, and R. D. Zwickl, J. Geophys. Res. 87, 632 (19[illegible]).
13. J. W. M. Paul, G. C. Goldenbaum, A. Iiyoshi, L. S. Holmes, and R. A. Hardcastle, Nature 216, 363 (1967).
14. M. Keilhacker, M. Kornherr, and K.-H. Steuer, Z. Physik 223, 385 (1969).
15. M. Kornherr, Z. Physik 233, 37 (1970).
16. S. E. Segre and M. Martone, Plasma Phys. 13, 113 (1971).
17. P. E. Phillips and A. E. Robson, Phys. Rev. Lett. 29, 154 (1972).
18. L. C. Woods, Plasma Phys. 11, 25 (1969).
19. D. A. Tidman and N. A. Krall, Shock Waves in Collisionless Plasmas (Wiley-Interscience 1971).
20. D. Biskamp, Nucl. Fusion 13, 719 (1973).
21. J. P. Edmiston and C. F. Kennel, J. Plasma Phys. 32, 429 (1984).

22. C. T. Russell, M. M. Hoppe, W. A. Livesey, J. T. Gosling, and S. J. Bame, Geophys. Res. Lett. 9, 1171 (1982).
23. M. F. Thomsen, J. T. Gosling, S. J. Bame, and M. M. Mellott, J. Geophys. Res. 90, 137 (1985).
24. M. M. Leroy, C. C. Goodrich, D. Winske, C. S. Wu, and K. Papadopoulos, Geophys. Res. Lett. 8, 1269 (1981).
25. M. M. Leroy, D. Winske, C. C. Goodrich, C. S. Wu, and K. Papadopoulos, J. Geophys. Res. 87, 5081 (1982).
26. G. Paschmann, N. Sckopke, S. J. Bame, and J. T. Gosling, Geophys. Res. Lett. 9, 881 (1982).
27. N. Sckopke, G. Paschmann, S. J. Bame, J. T. Gosling, and C. T. Russell, J. Geophys. Res. 88, 6121 (1983).
28. J. T. Gosling, M. F. Thomsen, S. J. Bame, W. C. Feldman, G. Paschmann, and N. Sckopke, Geophys. Res. Lett. 9, 1333 (1982).
29. W. C. Feldman, J. T. Gosling, M. F. Thomsen, S. J. Bame, G. Paschmann, M. Mellott, C. T. Russell, EOS Trans. Am. Geophys. Union 64, 824 (1983).
30. J. T. Gosling, M. F. Thomsen, S. J. Bame, and M. M. Mellott, AGU Chapman Conference on Collisionless Shock Waves in the Heliosphere (Napa, CA, Feb. 1984).
31. J. T. Gosling and A. E. Robson, Proc. AGU Chapman Conference on Collisionless Shock Waves in the Heliosphere (American Geophysical Union, Washington, in press, 1985).
32. E. W. Greenstadt, Proc. AGU Chapman Conference on Collisionless Shock Waves in the Heliosphere (American Geophysical Union, Washington, in press, 1985).
33. K. A. Anderson, J. Geophys. Res. 86, 4445 (1981).
34. K. A. Anderson, R. P. Lin, F. Martle, C. S. Lin, G. K. Parks, and H. Reme, Geophys. Res. Lett. 6, 401 (1979).
35. M. M. Leroy and A. Mangeney, Ann. Geophys. 2 449 (1984).
36. C. S. Wu, J. Geophys. Res. 89, 8857 (1984).
37. S. M. Krimigis, D. Venkatesan, J. C. Barichello, and E. T. Sarris, Geophys. Res. Lett. 5, 961 (1978).
38. F. L. Scarf, R. W. Fredericks, L. A. Frank, and M. Neugebauer, J. Geophys. Res. 76, 5162 (1971).
39. P. C. Filbert and P. J. Kellogg, J. Geophys. Res. 84, 1369 (1979).
40. R. R. Anderson, G. K. Parks, T. E. Eastman, D. A. Gurnett, and L. A. Frank, J. Geophys. Res. 86, 4493 (1981).
41. J. Etcheto and M. Faucheux, J. Geophys. Res. 89, 6631 (1984).
42. S. Hoang, J. Fainberg, J. L. Steinberg, R. G. Stone, and R. D. Zwickl, J. Geophys. Res. 86, 4531 (1981).
43. M. M. Hoppe, C. T. Russell, T. E. Eastman, and L. H. Frank, J. Geophys. Res. 87, 643 (1982).
44. D. D. Sentman, M. F. Thomsen, S. P. Gary, W. C. Feldman, and M. M. Hoppe, J. Geophys. Res. 88, 2048 (1983).

45. G. Paschmann, N. Sckopke, I. Papamastorakis, J. R. Asbridge, S. J. Bame, and J. T. Gosling, J. Geophys. Res. 86, 4355 (1981).
46. G. Paschmann, N. Sckopke, S. J. Bame, J. R. Asbridge, J. T. Gosling, C. T. Russell, and E. W. Greenstadt, Geophys. Res. Lett. 6, 209 (1979).
47. J. T. Gosling, J. R. Asbridge, S. J. Bame, G. Paschmann, and N. Sckopke, Geophys. Res. Lett. 5, 957 (1978).
48. J. R. Asbridge, S. J. Bame, J. T. Gosling, G. Paschmann, and N. Sckopke, Geophys. Res. Lett. 5, 953 (1978).
49. F. M. Ipavich, A. B. Galvin, G. Gloeckler, M. Scholer, and D. Hovestadt, J. Geophys. Res. 86, 4337 (1981).
50. C. Gurgiolo, G. K. Parks, B. H. Mauk, C. S.Lin, K. A. Anderson, R. P. Lin, and H. Reme, J. Geophys. Res. 86, 4415 (1981).
51. T. E. Eastman, R. R. Anderson, L. A. Frank, and G. K. Parks, J. Geophys. Res. 86, 4379 (1981).
52. B. U. Ö. Sonnerup, J. Geophys. Res. 74, 1301 (1969).
53. G. Paschmann, N. Sckopke, J. R. Asbridge, S. J. Bame, and J. T. Gosling, J. Geophys. Res. 85, 4689 (1980).
54. C. Bonifazi, G. Moreno, and C. T. Russell, J. Geophys. Res. 88, 7853 (1983).
55. M. M. Leroy and D. Winske, Ann. Geophys. 1, 527 (1983).
56. D. Burgess and S. J. Schwartz, J. Geophys. Res. 89, 7407 (1984).
57. D. Eichler, Astrophys. J. 229, 419 (1979).
58. J. P. Edmiston, C. F. Kennel, and D. Eichler, Geophys. Res. Lett. 9, 531 (1982).
59. M. Tanaka, C. C. Goodrich, D. Winske, and K. Papadopoulos, J. Geophys. Res. 88, 3046 (1983).
60. M. F. Thomsen, S. J. Schwartz, and J. T. Gosling, J. Geophys. Res. 88, 7843 (1983).
61. M. F. Thomsen, J. T. Gosling, S. J. Bame, W. C. Feldman, G. Paschmann, and N. Sckopke, Geophys. Res. Lett. 10, 1207 (1983).
62. C. Gurgiolo, G. K. Parks, and B. H. Mauk, J. Geophys. Res. 88, 9093 (1983).
63. M. F. Thomsen, J. T. Gosling, S. J. Bame, and C. T. Russell, J. Geophys. Res. 90, 267 (1985).
64. M. Hoshino and T. Terasawa, J. Geophys. Res. 90, 57 (1985).
65. A. Barnes, Cosmic Electrodyn. 1, 90 (1970).
66. D. D. Sentman, J. P. Edmiston, and L. A. Frank, J. Geophys. Res. 86, 7487 (1981).
67. S. P. Gary, J. T. Gosling, and D. W. Forslund, J. Geophys. Res. 86, 6691 (1981).
68. E. W. Greenstadt, I. M. Green, G. T. Inouye, A. J. Hundhausen, S. J. Bame, and I. B. Strong, J. Geophys. Res. 73, 51 (1968).

69. D. H. Fairfield, J. Geophys. Res. 74, 3541 (1969).
70. F. L. Scarf, R. W. Fredericks, L. A. Frank, C. T. Russell, P. J. Coleman, Jr., and M. Neugebauer, J. Geophys. Res. 75, 7316 (1970).
71. M. M. Hoppe, C. T. Russell, L. A. Frank, T. E. Eastman, and E. W. Greenstadt, J. Geophys. Res. 86, 4471 (1981).
72. M. M. Hoppe and C. T. Russell, J. Geophys. Res. 88, 2021 (1983).
73. D. Winske and M. M. Leroy, J. Geophys. Res. 89, 2673 (1984).
74. M. F. Thomsen, Proc. AGU Chapman Conference on Collisionless Shock Waves in the Heliosphere (American Geophysical Union, Washington, in press, 1985).
75. D. Winske, C. S. Wu, Y. Y. Li, and G. S. Zhou, J. Geophys. Res. 89, 7327 (1984).
76. D. W. Potter, J. Geophys. Res. 86, 11,111 (1981).
77. J. T. Gosling, Space Sci. Rev. 34, 113 (1983).
78. J. T. Gosling, S. J. Bame, W. C. Feldman, G. Paschmann, N. Sckopke, and C. T. Russell, J. Geophys. Res. 89, 5409 (1984).
79. S. P. Gary, Astrophys. J. 288, 342 (1984).
80. T. R. Sanderson, P. Van Nes, R. Reinhard, K.-P. Wenzel, B. T. Tsurutani, and E. J. Smith, AGU Chapman Conference on Collisionless Shock Waves in the Heliosphere (Napa, CA, Feb. 1984).
81. B. T. Tsurutani, E. J. Smith, and D. E. Jones, J. Geophys. Res. 88, 5645 (1983).
82. B. T. Tsurutani and R. P. Lin, J. Geophys. Res. 90, 1 (1985).
83. W. C. Feldman, Proc. AGU Chapman Conference on Collisionless Shock Waves in the Heliosphere (American Geophysical Union, Washington, in press, 1985).

SUBSTORMS IN THE EARTH'S MAGNETOSPHERE

D. N. Baker
University of California, Los Alamos National Laboratory
Los Alamos, NM 87545

ABSTRACT

Magnetospheres are plasma regions of large scale in space dominated by magnetic field effects. The Earth, and many planets in our solar system, are known to have magnetospheric regions around them. Magnetospheric substorms represent the intense, rapid dissipation of energy that has been extracted from the solar wind and stored temporarily in the terrestrial magnetotail. In this paper a widely, but not universally, accepted model of substorms is described. The energy budgets, time scales, and conversion efficiencies for substorms are presented. The primary forms of substorm energy dissipation are given along with the average levels of the dissipation. Aspects of particle acceleration and precipitation, Joule heating mechanisms, ring current formation, and plasmoid escape are illustrated based on in situ observations taken from the large available data base. A brief description is given of possible analogues of substorm-like behavior in other astrophysical systems.

INTRODUCTION

A magnetosphere is a relatively self-contained region in space whose global topology is organized by the magnetic field associated with the parent compact object. A number of the major planets (e.g., Mercury, Earth, Jupiter, and Saturn) are known to have intrinsic magnetic fields and have magnetospheric regions around them. These differ from one another in very significant, and interesting, ways. Indeed, the Sun itself may be viewed as having a magnetosphere (the "heliosphere") within which all of the other magnetospheres are embedded. It is the continual outflow of hot coronal gas from the sun which gives rise to the heliosphere and this supersonic, superalfvenic gas flow is called the solar wind (Fig. 1). The solar wind compresses, distorts, and confines the planetary magnetospheres[1] and imparts to them much (if not most) of their available free energy.

In addition to planetary and solar-system scale magnetospheres, there also appear to be magnetospheres of galactic proportions.[2] In each of these cases there are analogous features as well as distinctly different features (Fig. 2). By far the most thoroughly explored magnetosphere is that of the Earth and it is this physical system which will be discussed in detail in this paper.

The intrinsic magnetic field of the Earth arises from a complex dynamo action in the molten metal core of the Earth and may be well-represented for most purposes by an Earth-centered dipole of

0094-243X/86/1440184-24$3.00

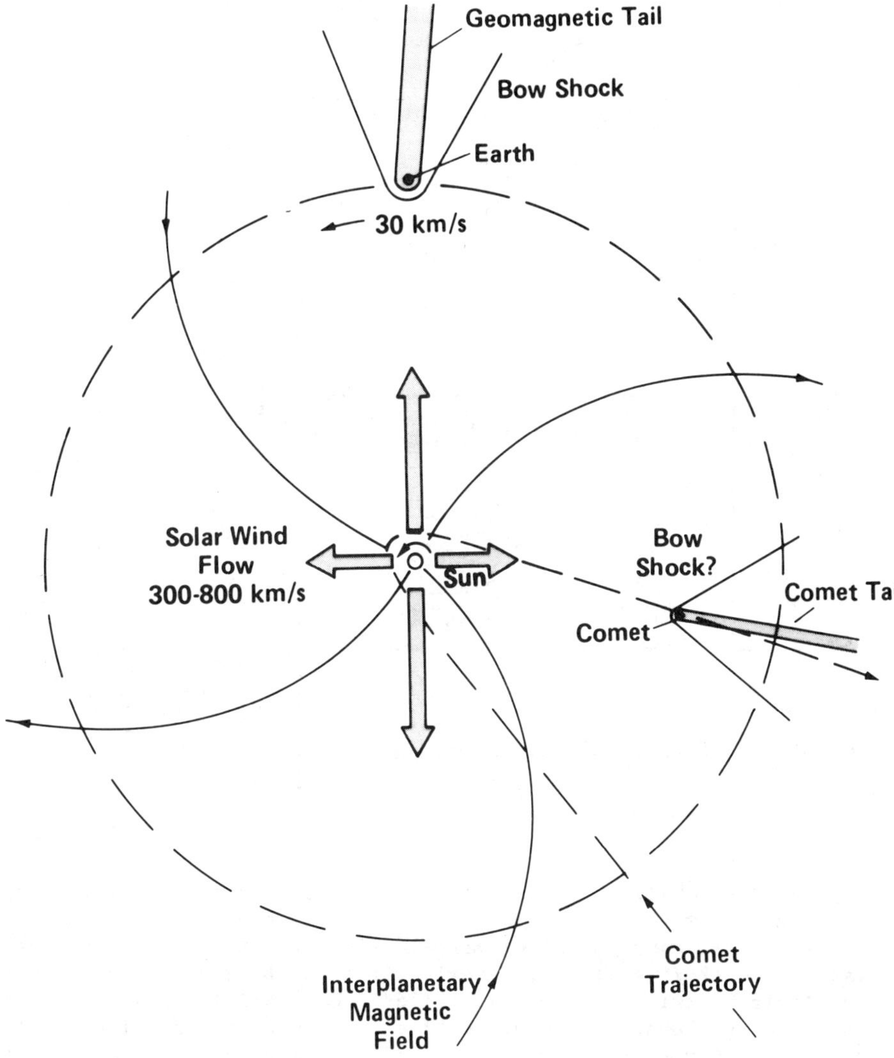

Fig. 1. The solar system is filled with a supersonic solar wind blowing nearly radially outward from the sun. Embedded in the flow is a remnant of the solar magnetic field, which, however, is not radial but is bent into an Archimedean spiral by the rotation of the sun. The flow of the solar wind past the Earth produces a stretching of the Earth's magnetic field into a long, tail-like structure on the nightside and causes a detached bow shock to form on the dayside. This geomagnetic tail is similar in some respects to the ionic tail of a comet, which results from the interaction of the solar wind with gases emitted from the head of the comet.[1]

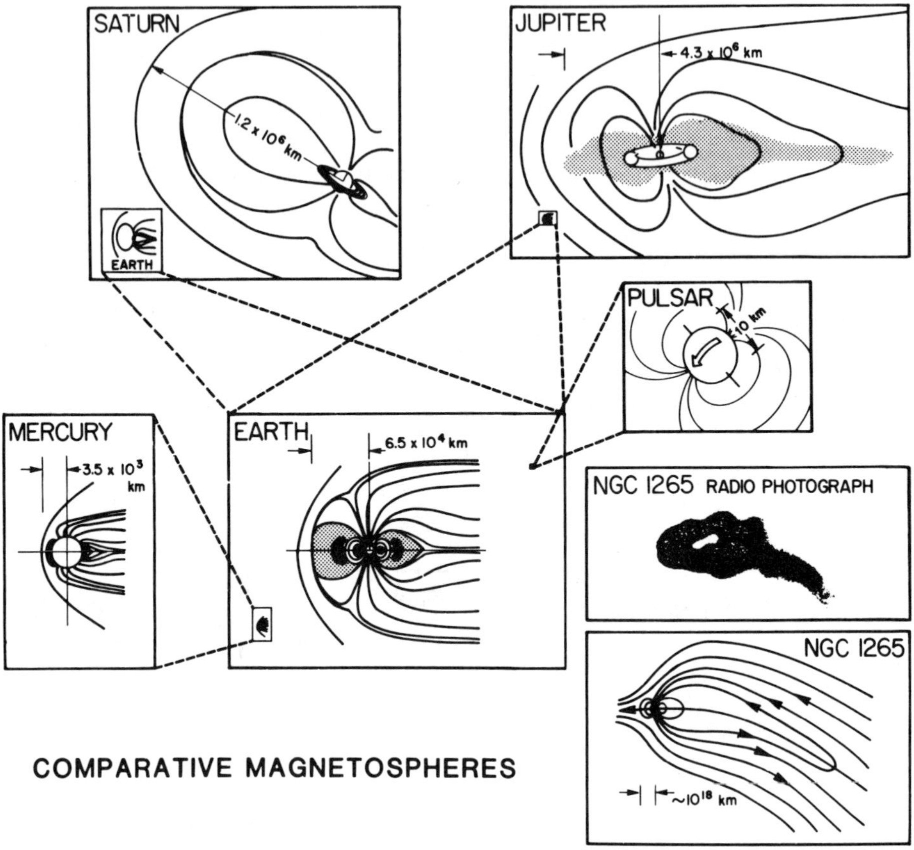

Fig. 2. Comparative magnetospheres. Fundamental similarities characterize the magnetospheric configurations of the planets in the solar system and some other celestial objects in the Universe at large, but their scales are vastly different because of differences in their intrinsic magnetic fields and associated plasmas. Among the planets, Mercury's magnetosphere is tiny compared with Earth's while Jupiter's is enormous - an order of magnitude larger than the Sun itself. The size of the magnetosphere of a typical pulsar is of the same order of magnitude as the Earth's, but its magnetic field is a trillion times stronger. Plasma is locked into its magnetic field until it is spun up to nearly the speed of light. As the radio galacy NGC 1265 (shown here with a map of radio emission for comparison) plows through the intergalactic gas, the ram pressure creates a magnetospheric tail stretching millions of light years into space. The distance of a million trillion kilometers shown by the marks is the equivalent of a hundred thousand light years (from p. 118, NRC: <u>Solar-Terrestrial Research for the 1980's</u>, 1981).

magnetic moment M = 8.06 x 10^{25} G-cm^3. This field extends far into space and serves to deflect the on-rushing solar wind plasma. The stand-off distance at the subsolar point is highly variable (depending on solar wind pressure), but is commonly about 10 R_E (1 R_E = 1 Earth radius = 6375 km). The flowing solar wind applies tangential stresses to the outer reaches of the Earth's intrinsic field and sets up a system of currents in the boundary regions. The $\vec{j}\times\vec{B}$ forces due to these currents act to distort the magnetic field and field lines are dragged downstream to form a very elongated magnetotail (Fig. 3).

Through over two decades of experimental in situ probing and theoretical modeling, many distinct and identifiable plasma regimes in the Earth's magnetosphere have become recognized. The complexity of these plasma regimes is illustrated in Figure 4. Although these regions are very important to the detailed workings of the terrestrial magnetosphere, it is possible to consider the behavior of the system in a much more simplified sense, much more akin to the diagram shown in Fig. 3. This point is especially important when one considers that there is little hope of ever achieving such a detailed mapping of remote astrophysical systems as is shown in Fig. 4.

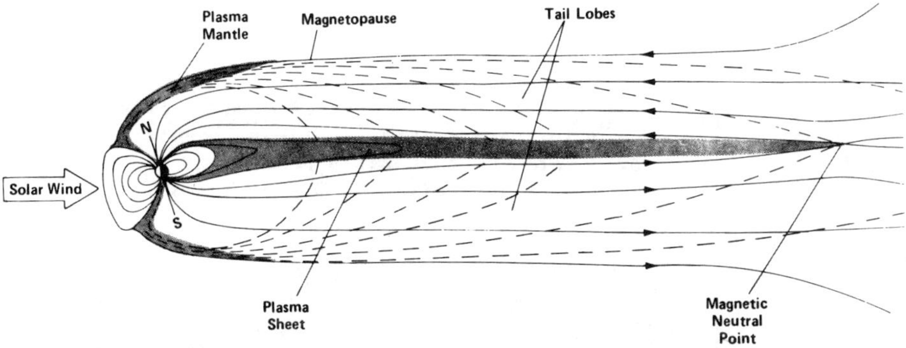

Fig. 3. A modern simplified view of the magnetosphere. Solar wind plasma crosses the magnetopause (the surface of the magnetosphere) on reconnected field lines on the dayside to form the plasma mantle and drifts along the dashed lines through the plasma-poor tail lobes to populate the flattened, horizontal plasma sheet. A distant magnetic neutral line, or neutral point, is the site of continual magnetic reconnection and replenishment of the plasma sheet.[1]

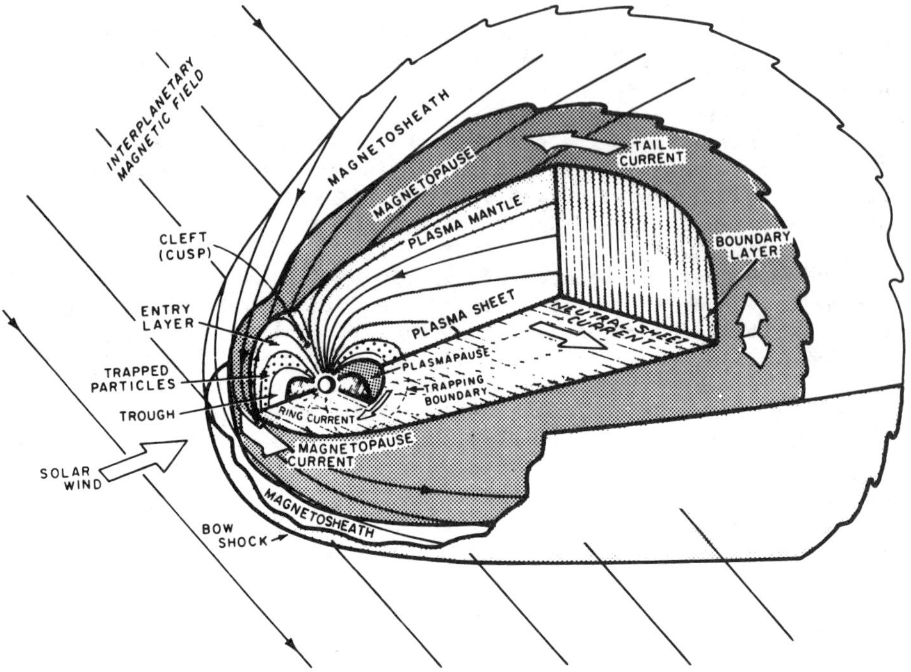

Fig. 4. A cutaway view of the Earth's magnetosphere illustrating the complex array of plasma regimes that have been revealed by in situ probing of this system by near-Earth spacecraft during the past two decades (courtesy W. J. Heikkila, U. of Texas).

The Earth's magnetosphere extends from the ionosphere to altitudes of $\sim 10\ R_E$ on the dayside, and perhaps to 1000 R_E on the nightside. The solar wind flows continually over, around, and into the terrestrial magnetosphere and in so doing it continually imparts mass, momentum, and energy to the system. This transfer, however, occurs with great variability. When the added amount of energy is high, the magnetosphere moves far out of its equilibrium "ground state." The added energy must then be dissipated either continuously or sporadically. It has been found, in fact, that the dissipation of added solar wind energy occurs in a quite sporadic way and the sudden occurrence of this magnetospheric dissipation is a major feature of the collection of physical processes that we call a magnetospheric substorm.

Nearly all of the energy dissipated in substorms comes from the solar wind. As shown in Fig. 5, substorm processes are a mixture of "driven" processes and "unloading" processes. In general, part of

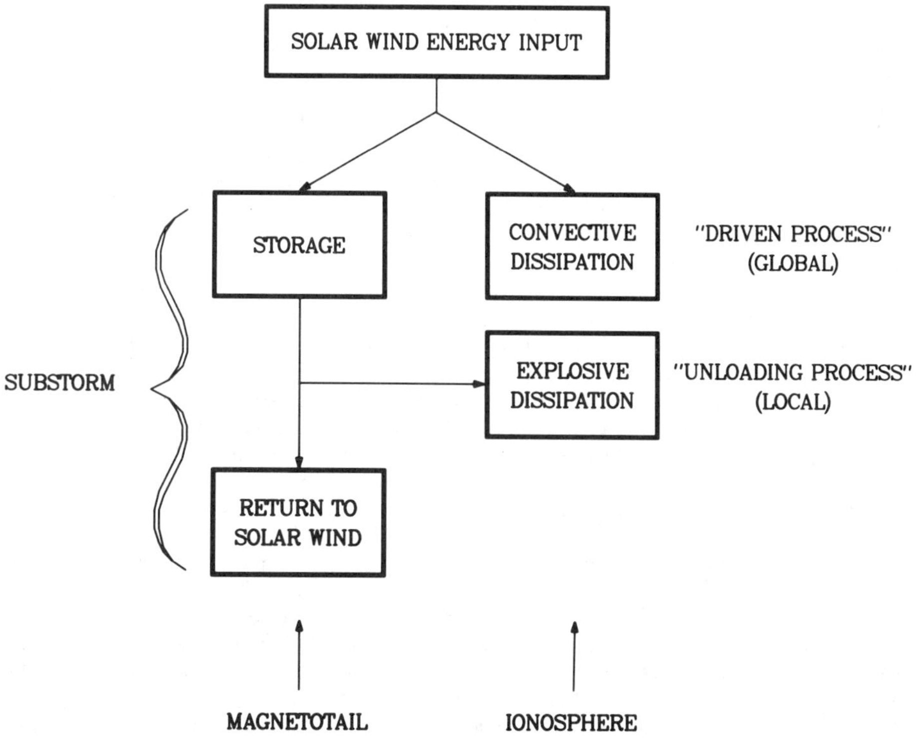

Fig. 5. A summary of the substorm sequence showing the roles of the magnetotail and the ionosphere and also showing the aspects of substorms which are regarded as driven and unloading processes (adapted from Baker et al., 1984).

the energy input from the solar wind to the magnetosphere goes into convective dissipation in the ionosphere and this has been termed the driven aspect of substorm dissipation. A major part of the solar wind energy input also goes into the form of stored energy in the Earth's magnetotail. To be a substorm there has to be at least one episode of explosive dissipation. This explosive dissipation phase represents the "unloading" of energy that was previously stored in the magnetotail. Some of the unloaded energy eventually appears in the inner magnetosphere/ionosphere in the form of ring current enhancement, ionosphere Joule heating, and auroral particle precipitation, while the rest of it is returned to the solar wind, probably in the form of a severed portion of the magnetotail (i.e., the "plasmoid" to be discussed further below). In the remainder of this paper we illustrate a model of substorms and discuss aspects of plasma heating, energetic particle acceleration, and energy conversion efficiencies that may be of significance in the astrophysical context.

The Substorm Model

The model of substorms alluded to above and which will be described in more detail here is the product of over two decades of research. It has much of its basis in the observational work of McPherron, Russell, and coworkers at UCLA [3,4] and in work at Los Alamos.[5,6] There has also been a strong interplay of theory with observation in order to develop the model.[7,8] Most aspects of substorm studies prior to 1983 have been thoroughly reviewed in a comprehensive document[9] to which the interested reader is referred.

Figure 6 presents a noon-midnight meridional cross-section of the near-Earth magnetosphere. The figure is drawn roughly to scale, although the Earth's size is somewhat exaggerated. In the figure we show (at the left) the incoming solar wind plasma. This plasma is collisionless, and its flow speed is typically in the range of 300 to 800 km/s in the Earth's frame of reference. Since upstream sound speeds are typically ~50 km/s and Alfvén speeds are typically $\lesssim$100 km/s, the solar wind flow is highly supersonic and superalfvénic (Alfvén mach numbers M_A = 3-10). The magnetospheric field (and plasma) constitutes a blunt obstacle in the solar wind flow, and thus a collisionless bow shock wave forms upstream of the magnetosphere. The shock slows down and heats the inflowing solar wind, which is then largely diverted around the magnetosphere to form a plasma sheath region (the "magnetosheath"). It is, therefore, shocked solar wind plasma and the embedded magnetic field that actually makes contact with the magnetospheric surface (the "magnetopause").

A broad range of observational data suggest that the Earth's magnetic field rather effectively excludes solar wind plasma unless the interplanetary magnetic field (IMF) is at least somewhat antiparallel to the Earth's intrinsic northward-directed field. Under such conditions there appears to be significant reconnection between the solar wind and terrestrial magnetic fields.

The substorm model assumes that such field line cutting and "reconnection" between solar wind field lines and terrestrial field lines in fact occurs. As shown in Fig. 6, reconnected field lines are dragged, by virtue of continued solar wind flow, to the nightside, or magnetotail, portion of the system. If one prefers to avoid the concept of reconnected magnetic field lines, one can equivalently talk about electric fields and currents in the plasmas. In either case, the net effect of dayside reconnection is to transfer mass, momentum, and energy from the solar wind to the magnetosphere. The added energy appears as an increased magnetic energy density ($B^2/8\pi$) in the magnetotail lobes. This increased magnetic pressure gradually "squeezes" and distorts the plasma sheet (which carries the cross-tail currents).

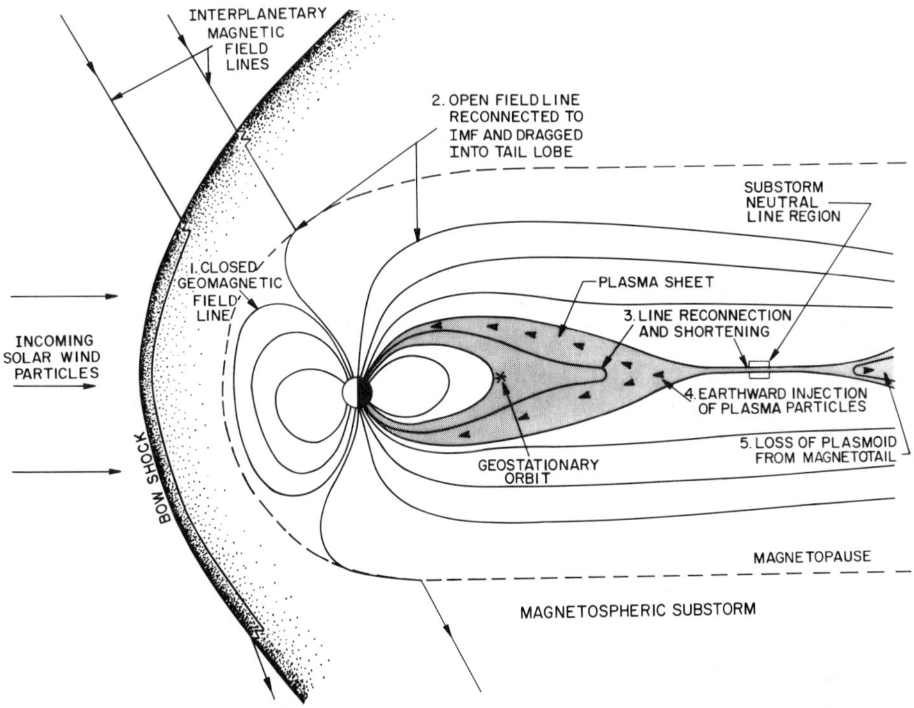

Fig. 6. Features of magnetospheric substorms inferred from many years of near-Earth observations. Interplanetary field lines interconnect with terrestrial field lines near the front of the magnetosphere and are dragged back to form the tail lobes, carrying plasma and energy with them. Far down the tail (off the figure to the right, see Fig. 3), lobe field lines reconnect to form the plasma sheet. When the midsection of the plasma sheet becomes thin enough (due to increased magnetic pressure from the tail lobes), field lines reconnect there also, forming the magnetic neutral region shown. The release of energy that results from the reconnetion heats and accelerates magnetotail plasma, driving part of the plasma back along field lines into the Earth's atmosphere at the poles. Another blob of plasma is driven to the right, flowing at great velocity down the tail. Before ISEE-3's journeys into the magnetotail, one could only speculate about what occurred beyond the right-hand side of the figure.

Under ordinary (i.e., near equilibrium) conditions, only a small amount of reconnection occurs on the dayside, and this is counterbalanced by reconnection at the distant neutral line ~100 R_E

down the tail (see Fig. 3). When energy input is greatly increased, however, the distant tail reconnection rate (determined basically by the local Alfvên speed at that distance) is too slow to dissipate the added energy effectively.

The magnetospheric system responds by forming a new, near-Earth reconnection region at ~10-20 R_E in the magnetotail. Such a response is energetically favorable for several reasons. First, in the near-Earth region the Alfvên speed ($B/\sqrt{4\pi\rho}$) is substantially higher since B is several times greater than in the deep tail, and ρ (the mass density) is only slightly larger. Thus, the intrinsic reconnection rate (dependent on V_A) is increased. Secondly, with reconnection occurring in the near tail, plasma need not convect all the way from ~100 R_E on the nightside to the dayside in order to return flux to the dayside magnetosphere. Rather, flux can be quickly reconnected in the near-tail, and this stored energy can be rapidly dissipated. A third advantage of near-Earth reconnection is that much of the plasma sheet and the associated currents can simply be severed (as a plasmoid) and allowed to leave the magnetosphere. Since the severed part of the plasma sheet contains a large amount of energy in the form of hot plasmas and magnetic field loops, the loss of the plasmoid is one of the most efficient possible ways for the magnetosphere to return toward its ground state.

The sequence of events shown in Fig. 6, which constitute the substorm sequence, may thus be summarized as follows. First, a southward turning of the interplanetary magnetic field (IMF) initiates enhanced solar wind field reconnection with the terrestrial field. Dayside magnetospheric flux is eroded, and field lines are dragged across the polar caps into the nightside tail lobes. As flux is continually, and rapidly, added to the lobes, the plasma sheet gradually thins. A plasma sheet instability eventually occurs, which causes a new neutral line to form in the near-Earth tail region. There is then greatly enhanced reconnection of plasma sheet and, eventually, lobe field lines which rapidly dissipates stored magnetic energy. When reconnection has proceeded sufficiently, a large segment of the plasma sheet is severed and lost in the form of a plasmoid which then rapidly carries away great amounts of the stored energy in the tail. The plasma sheet instability that gives rise to the new substorm-associated neutral line is probably an ion tearing mode instability.[8] A substantial amount of the recent observational and theoretical work on this aspect is reviewed in Ref. 9.

The model of Fig 6 has a firm observational basis. It is very well established, for example, that energy coupling between the solar wind and the magnetosphere is strongly dependent on the IMF direction,[10] and dayside reconnection has been observed directly.[11] Further, the storage of energy in the magnetotail lobes prior to substorm onsets has also been observed many times (often at geostationary orbit, 6.6 R_E).[6,12] Plasma sheet thinning has been observed for many substorms[5] as has dayside flux erosion.[13] Finally, plasmoids released during substorms have been observed in the deep geomagnetic tail by ISEE-3.[14] Figure 7 is adapted from a study of ISEE-3 data when the spacecraft was within the tail at ~220 R_E geocentric distance[14] and illustrates the formation and loss of a plasmoid from the tail.

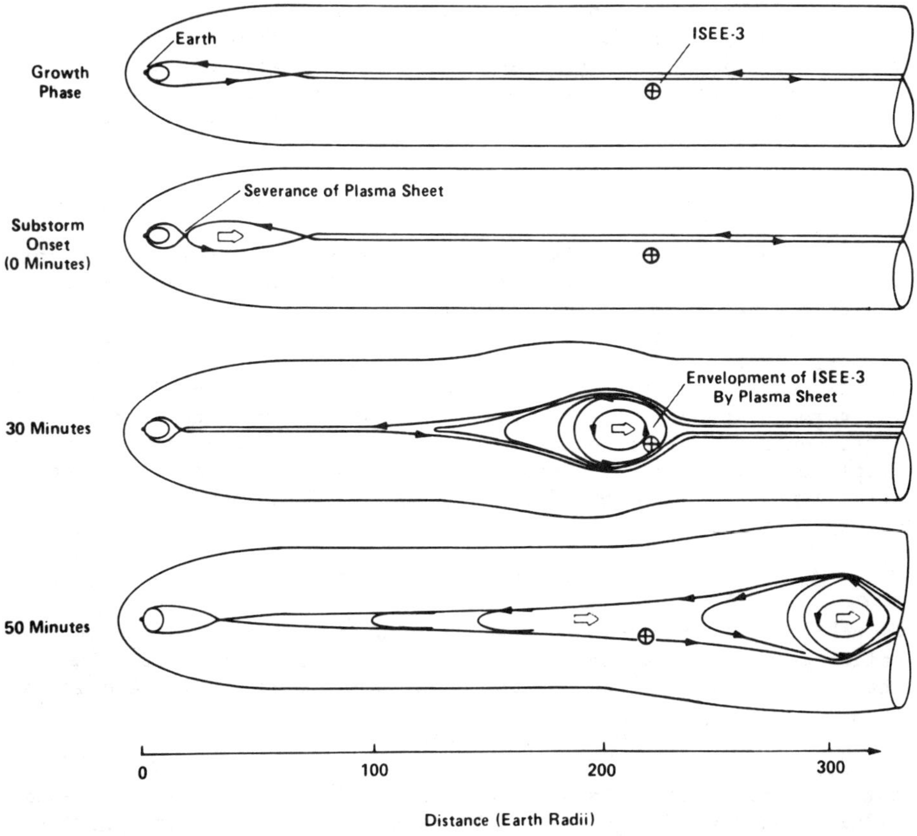

Fig. 7. After plasma and energy have built up in the magnetotail during the growth phase, a portion of the plasma sheet is severed by reconnection close to Earth, causing substorm onset. The plasmoid thus formed hurtles down the tail, eventually enveloping ISEE-3 about thirty minutes later at its vantage point nearly 1,400,000 kilometers from Earth. The envelopment is seen by instruments in the spacecraft as a transition into the plasma sheet. White arrows indicate plasma flow; black arrows indicate the direction of the magnetic field (adapted from Ref. (14)).

The Energy Budget for Solar Wind-Magnetosphere Coupling

In order to place substorms into proper context, it is advisable to illustrate the kinds of energy dissipation being discussed, the time scales on which substorms occur, and efficiencies of energy conversion. In this section, these energy budgets and time scales will be addressed.

The dominant type of energy output from the sun is in the form of electromagnetic radiation, and this is strongly peaked in the visible portion of the spectrum. The "solar constant" at 1 astronomical unit (A.U.) is 1.35×10^6 ergs/cm^2-sec. If we consider a disk of radius 1 R_E at the Earth's orbit, then the Earth intercepts approximately

$$W_s = 1.74 \times 10^{24} \text{ ergs/sec.}$$

This is the amount of radiant energy from the sun possibly available for atmospheric heating, etc. (although, of course, much is reradiated and is not geoeffective).

The solar wind kinetic motion relative to the Earth constitutes the primary form of plasma energy available at 1 A.U. The majority of energy in the solar wind plasma is in this bulk flow (rather than in thermal energy). The bulk flow of several hundred km/sec corresponds to ~1 keV per solar wind ion. If we consider the range of plasma flow speeds (300-800 km/sec) and densities (1-100 cm^{-3}) actually observed in the solar wind, we find an approximate kinetic energy flux range of

$$w_k \sim 0.05\text{-}20 \text{ ergs/cm}^2\text{-sec}$$

Since this plasma interacts with the projected cross-section of the entire magnetosphere (which we can take as a disk of radius ~20 R_E), the total power intercepted due to the solar wind kinetic energy is

$$W_k \sim 5 \times 10^{19}\text{-}10^{22} \text{ ergs/sec}$$

Note that although w_k is only perhaps 10^{-6} or so of the solar radiant energy, W_k is $\sim 10^{-3}$ W_s because of the relatively large magnetospheric "collection" area.

As described above, however, it is not the solar wind kinetic energy flux which seems to control geomagnetic activity, but rather the embedded solar wind magnetic field. Typically at 1 A.U. this field ranges from $\sim 5 \times 10^{-5}$ gauss to $\sim 30 \times 10^{-5}$ gauss. Thus, the IMF energy density, $B^2/8\pi$, ranges from about 10^{-10} to 4×10^{-9} ergs/cm^3. Taking into account the possible range of solar wind convective speeds, the magnetic energy flux is in the approximate range

$$w_B \sim 0.003\text{-}0.3 \text{ ergs/cm}^2\text{-sec.}$$

Again using a magnetospheric collection cross-section of radius 20 R_E gives a total intercepted magnetic power of

$$W_B \sim 10^{18}\text{-}10^{20} \text{ ergs/sec.} \quad (1)$$

This value is approximately 1% of the kinetic energy power of the solar wind and is of order 10^{-5} of the solar radiant energy rate striking the Earth.

Much effort has gone into empirical determinations of how the solar wind magnetic energy couples into the magnetospheric system. The geoeffective magnetic energy, i.e., the solar wind input energy is given by a product of the form

$$W_{in} = k_o \, V \cdot \frac{B^2}{8\pi} F(\theta) \quad (2)$$

In this equation, V is the solar wind speed, B is the IMF strength, k_o is a geometric parameter, and $F(\theta)$ is a factor which accounts for the strong dependence of the energy coupling upon the polar angle θ between the IMF and the Earth's dipole axis. For $0 < \theta \lesssim \frac{\pi}{2}$, very little energy coupling (i.e., dayside reconnection) occurs. For $\theta \sim \pi$ the coupling (dayside reconnection) is very efficient. Roughly speaking, a maximum energy coupling occurs when for the case of $F(\theta) = 1$ and $k_o = \pi r^2 = \pi(20\ R_E)^2$. A more refined analysis shows that a better approximation may be

$$F(\theta) = \sin^4(\frac{\theta}{2})$$

with the definition

$$\frac{k_o}{8\pi} \sim \ell_o{}^2$$

and $\ell_o \sim 7\ R_E$. Thus, a popular energy coupling relation[15] in current use is

$$W_{in} \sim \ell_o{}^2 \, V \, B^2 \sin^4 (\frac{\theta}{2}) \quad (3)$$

but better estimates of the coupling strength for substorms have also been found.[10]

The substorm model described in the previous section predicts a repeatable sequence of events in terms of solar wind energy input and storage and energy release from the magnetotail. An idealization of this input-output relation is shown in Figure 8. For approximately one hour there is strong energy input to the magnetosphere at a level typically between 10^{18} and 10^{19} ergs/sec. (Note that, in general, energy <u>input</u> need not decrease prior to the beginning of tail energy <u>output</u> as is schematically illustrated in Fig. 8.) With a short time delay from the beginning of the energy input, ($\lesssim$20 min) there begins to be a weak resistive dissipation in the ionosphere (typically $\lesssim 10^{18}$ ergs/sec) due to the dragging of field lines through the polar cap ionosphere. Concurrently 10^{22} to

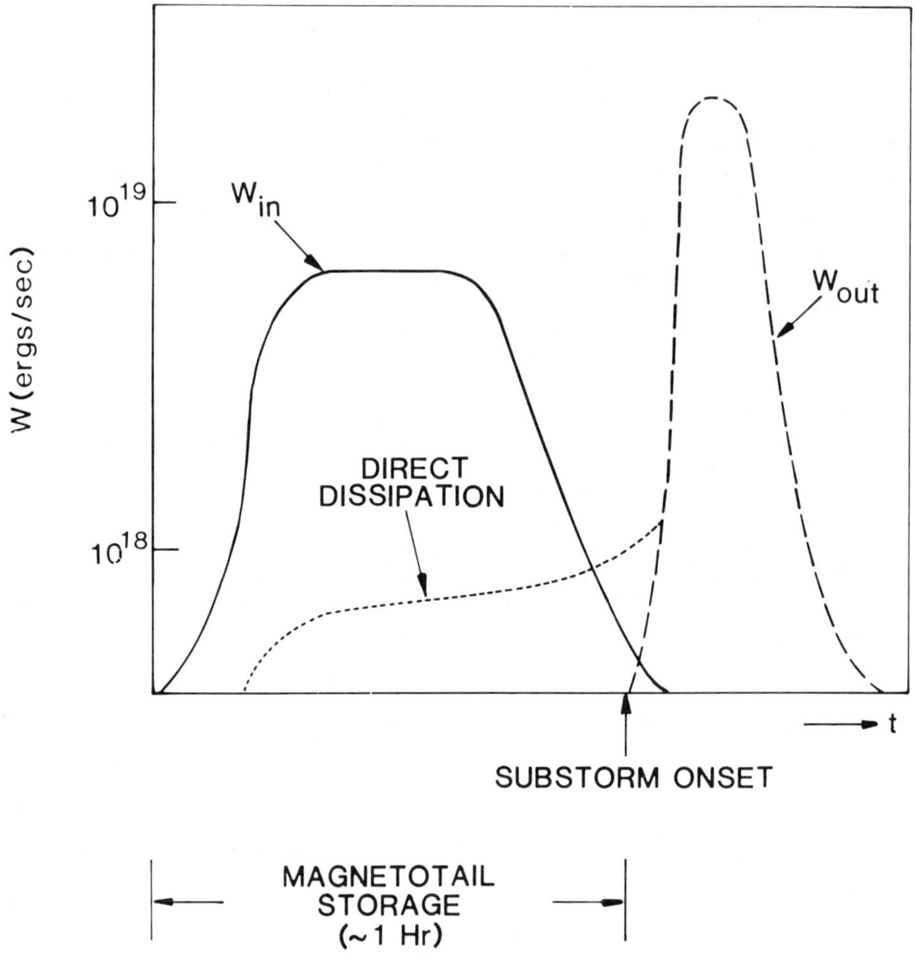

Fig. 8. An idealized illustration of the relationship between magnetospheric energy input (W_{in}) and energy output (W_{out}) during substorms. Energy is typically stored for ~1 hr (with concomitant weak dissipation) after which the stored energy is strongly and rapidly dissipated.[19]

10^{23} ergs ($\int W_{in}dt$) is stored in the magnetotail lobes. This latter energy is released primarily at substorm onset and dissipation becomes very intense then. The peak of W_{out} is usually greater than W_{in}, but we ordinarily find that

$$\int W_{in}\, dt > \int W_{out}\, dt \qquad (4)$$

This is true because W_{out} is usually estimated only from near-Earth and ground-based data sets: it does not include plasmoid energy-loss estimates. As will be discussed below, plasmoids may carry away half, or more, of the stored tail energy.

Forms and Levels of Substorm Energy Dissipation

The rate and form of energy input to the magnetosphere is fairly well-established and agreed upon.[16] With energy transfer as described above, substorm activity always follows and, thus, the substorm may be viewed as the fundamental energy release element permitting the return of the magnetosphere to its ground state. Examination of W_{out} traditionally concentrated upon dissipation detectable in some sense from the Earth's surface. Thus, three primary forms of substorm energy dissipation were considered.

(1) Auroral particle precipitation;
(2) Auroral Joule heating; and
(3) Ring current flows.

During substorms there are large auroral displays which are caused by energized electrons bombarding the upper atmosphere, colliding with atmospheric constituents, and thus dissipating their energy. From the brightness and area of such displays, one can estimate the precipitation energy dissipation involved. More recently, polar orbiting spacecraft have been able to measure directly the precipitating energy fluxes. Globally-integrated energies for this dissipative component are of the order of 0.5 - 1.0 X 10^{18} ergs/sec during most substorms.[17]

Substorms also have long been detected by virtue of the intense magnetic disturbances that they cause in the auroral regions. These magnetic effects are due to strong field-aligned (Birkeland) currents that flow in the auroral zones which close (and dissipate energy resistively) in the upper ionosphere. This Joule heating associated with substorm currents can also be monitored from Earth (through arrays of magnetometers) and ionospheric conductivity models can be employed to convert measured currents to ohmic (I^2R) dissipation. Global numbers for this component of dissipation are typically 1-5 X 10^{18} ergs/sec during most substorms.[15]

Finally, there are always trapped particles drifting in the Earth's inner magnetospheric region. These particles constitute what is known as the extraterrestrial ring current. A large enhancement of this ring current particle population is one of the primary manifestations of a geomagnetic storm and during such enhancements the ring current can cause large magnetic disturbances in the equatorial magnetic field at Earth's surface. During many storm intervals, accelerated particles and hot plasma are injected from the tail into the ring current.[9,18] There the energy of these particles is gradually (over hours or days) lost due to precipitation and charge exchange processes. Hence, the ring current is a major sink of magnetospheric energy. During even quiet times, the ring current contains an energy comparable to that stored in the magnetotail: $\sim 10^{22}$ ergs. During a typical substorm, the ring current energy can grow at a rate of $2\text{-}3 \times 10^{18}$ ergs/sec, while during intense storms it may grow by as much as 10^{19} ergs/sec.[15]

Therefore, a significant part of the energy dissipation during substorms can be assessed by examination of auroral and ring current terms. Over the years, indices of auroral disturbances (e.g., the AE index) and ring current disturbances (e.g., the Dst index) have been formulated. These are basically measures of levels of magnetic disturbances as measured on the Earth's surface and they "calibrate" the disturbance level. In turn, it has been possible to calibrate magnetospheric energy losses in terms of these indices. If one calls the total energy dissipation W_{out}, then

$$W_{out} = W_{RC} + W_A + W_J \tag{5}$$

where W_{RC} is ring current loss, W_A is auroral precipitation, and W_J is Joule heating loss.

It has been suggested[15] that W_A and W_J scale directly with AE (in nT):

$$W_A = 1 \times 10^{15} \text{ AE}$$

and (6)

$$W_J = 2 \times 10^{15} \text{ AE}$$

The W_{RC} term is estimated to be related to Dst and its time derivative:

$$W_{RC} = 4 \times 10^{20} \left(\frac{\partial \text{Dst}}{\partial t} + \frac{\text{Dst}}{\tau}\right) \tag{7}$$

with τ being a charge-exchange lifetime (several hours).[15,16]

Figure 9 is a specific study of a substorm that occurred on 22 March 1979, with onset at ~1100 UT.[19] The dashed line shows (scale to the right) the estimated energy input rate (W_{in}) from solar wind measurements. A large value was estimated, viz., $3\text{-}4 \times 10^{19}$ ergs/sec. An excellent determination of W_{out} was obtained in this study. In Fig. 9 we show one component of W_{out}, namely the Joule heat rate W_J. Peak values of W_J were of order 5×10^{18} ergs/sec.

The W_{in} curve of Fig. 9 compares favorably with the "model" curves of Fig. 8. During ~1 hour of energy storage (~1010-1110 UT) it is seen that W_J is increased by a relatively small amount. At substorm onset W_J increased substantially. Since this large increase in W_J occurred when W_{in} was essentially zero, the dissipated energy (W_J) must have come from energy stored in the tail.[19]

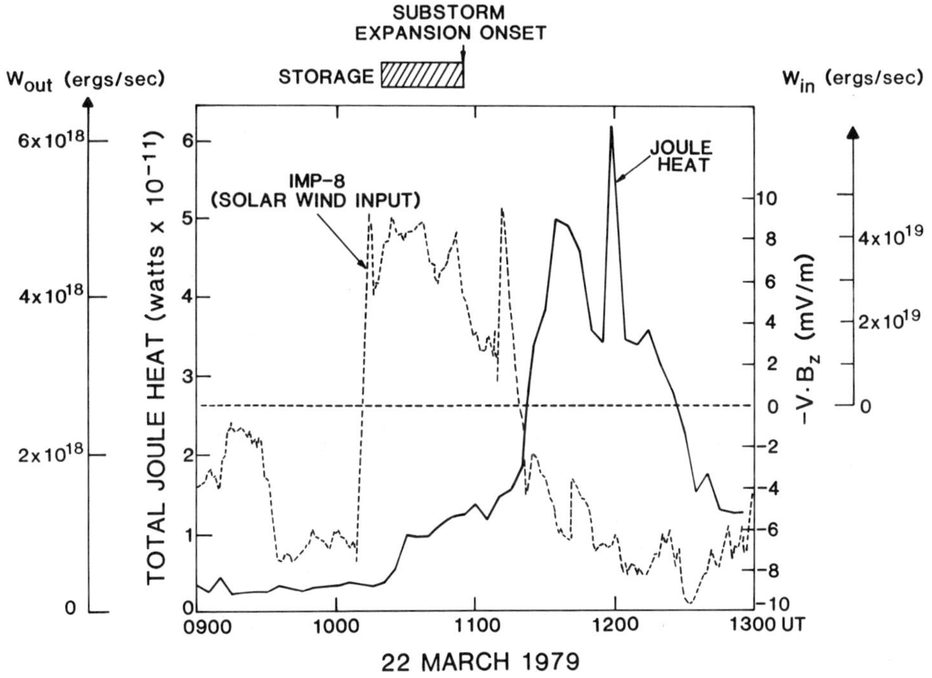

Fig. 9. A comparison of solar wind energy input to the magnetosphere (dashed line) compared to a measure of magnetospheric substorm energy output (Joule heat rate; solid line) for a particular event observed on 22 March 1979.[19] The parameter $-VB_z$ is the dawn-to-dusk component of the solar wind electric field and is a measure of solar wind-magnetosphere coupling. W_{in} is positive for $-VB_z > 0$ and this energy input is estimated by the scale at the far right. W_{out} has several components (see text) but a major one is the Joule heating rate scaled to the far left. These actual results are very similar to the idealization shown in Fig. 8.

In studies of Earth's magnetosphere we need not be content with only remote (i.e., ground-based) estimates of energy conversion processes--often we can observe magnetotail storage and dissipation directly. An example taken from direct spacecraft observations is shown[6] in Fig. 10. The figure combines particle measurements at geostationary orbit (6.6 R_E) with magnetic field data from IMP-8 positioned in the middle of the magnetotail lobe at ~35 R_E geocentric distance. The small inset in the middle panel shows the spacecraft positions projected onto the equatorial plane.

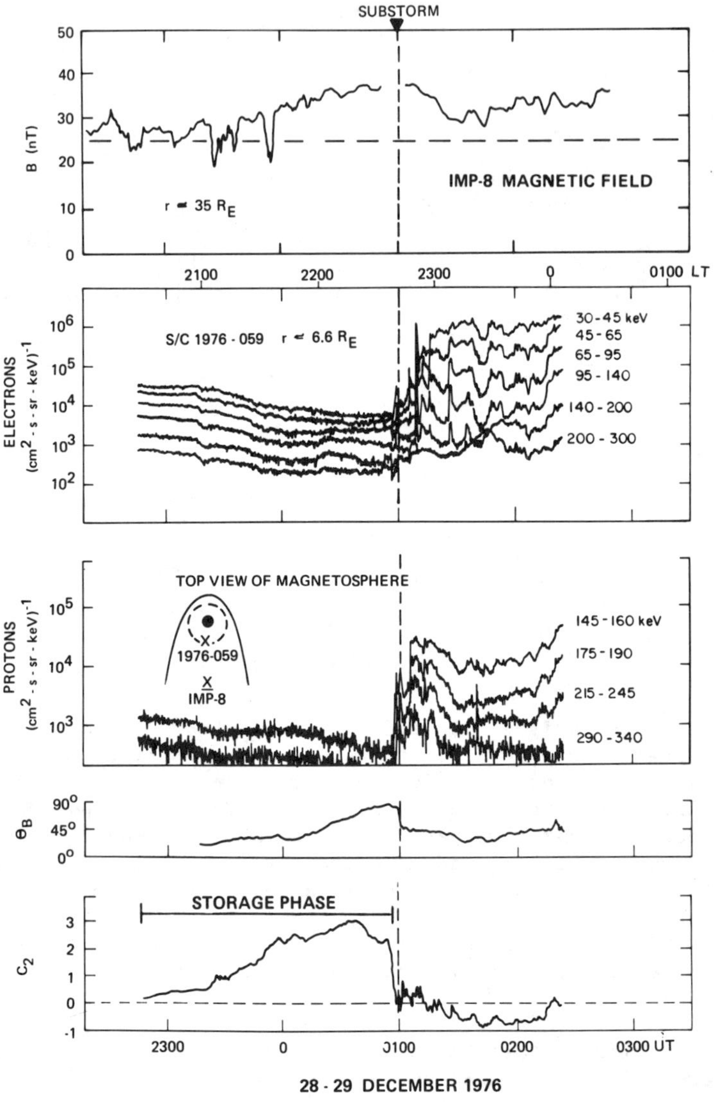

Fig. 10. The upper panel shows the total magnetic field strength measured by IMP-8 in the Earth's magnetotail lobe at ~35 R_E geocentric distance. The lower four panels show data from geostationary spacecraft 1976-059 at 6.6 R_E radial distance. As described in the text, the data provide strong evidence for storage of energy in the tail between 2300 UT and 0100 UT and for rapid dissipation of this energy after the 0100 UT substorm onset.

Between 2300 and 0100 UT the IMP-8 tail field strength increased (25 → 40 nT). Therefore, the magnetic energy density in the tail increased by

$$\delta(B^2/8\pi) = 2B\delta B/8\pi$$

Using an average B ~32 nT and δB ~15 nT gave

$$\delta(B^2/8\pi) \simeq 4 \times 10^{-9} \text{ ergs/cm}^3$$

Assuming that this energy density was characteristic of the whole lobe region allows an estimate of the increase in stored tail energy for this event. For an effective tail length of ~100 R_E and a constant (and conservative) 15 R_E tail radius the volume of the tail is

$$V \sim 2 \times 10^{31} \text{ cm}^3$$

Thus the excess free energy in the tail was estimated to be

$$\delta W = V \cdot \delta(B^2/8\pi)$$

$$\sim 8 \times 10^{22} \text{ ergs}$$

Further, by taking a characteristic time δt for the substorm energy dissipation time (see Fig. 10) as ~40 min (i.e., ~0100 to ~0140 UT), then

$$\delta W/\delta t \simeq 8 \times 10^{22}/2400 \text{ s} \sim 3 \times 10^{19} \text{ ergs/s}$$

As noted for the substorm model, this suggests a large, relatively short duration energy dissipation process. Such results provide direct <u>in situ</u> evidence that large amounts of energy are accumulated in the magnetosphere before substorms and this energy is rapidly converted to other forms (e.g., energetic plasma) during the substorm itself.

We estimate the total energy contained in the energetic electron component in the magnetosphere (cf. Fig. 10) in this case to be

$$W_e \sim 6 \times 10^{20} \text{ergs}$$

Similar, or larger, estimates were expected to apply to the total substorm-generated energetic proton energy. Thus, of order 10^{21} ergs, or greater, appeared in the energetic particle component during this substorm, and this could be readily accounted for by the 8×10^{22} ergs of dissipated energy suggested by IMP 8 observations. Furthermore, if we assume the $\gtrsim 10^{21}$ ergs of energetic particle energy was produced on a timescale of ~1/2 hour we get

$$W_{particles} \gtrsim 6 \times 10^{17} \text{ ergs/s}$$

This estimate represents a significant fraction (~ few percent) of the total energy dissipation rate suggested by IMP 8 data and it represents over 10% of typical total substorm energy dissipation rates. Thus the energetic particle increase at ~0100 UT in Fig. 10, although it represents the tail of a plasma distribution, nonetheless corresponds to a significant part of the substorm energy output.

Results of this sort show that the total energy dissipated to the Earth and to the inner magnetosphere during substorms is often less than half of the energy that was stored and then released from the magnetotail. This raises the question of what becomes of the other very substantial fraction of the energy. It seems likely that it is released directly to the down-stream solar wind in the form, first, of a plasmoid and then as energized plasma flowing tailward directly from the reconnection region.

Continuing studies of the near-Earth substorm neutral line model give a good indication of the nature of reconnection, and the associated energy conversion, during substorms.[5] The general character of this model is illustrated in Fig. 11. Relatively cool, low β plasma flows from the lobes into the region of the near-Earth neutral line; for a large part of the substorm, reconnection proceeds on lobe field lines. The incoming plasma is heated and accelerated both earthward and tailward. This acceleration is manifested as strong ($\gtrsim$ 1000 km/s) jetting of plasma away from the neutral line.[5,9] This is the source of most of the plasma seen in the plasma sheet during substorms, and these plasmas represent much of the converted magnetic energy formerly in the lobes.

In most substorms a variety of mechanisms, largely driven by magnetotail energy dissipation, are found to contribute to acceleration of ions and electrons to moderate (< 100 keV) energies. A list of these includes:

- Parallel electric fields above the auroral oval accelerate particles to 1-10 keV.
- Electrostatic ion-cyclotron waves probably produce ion beams over the auroral zone.
- Betatron acceleration of particles convecting from the plasma sheet to 6.6 R_e can account for a part of the low-energy component of particle injections (see Fig. 10).
- Acceleration of particles in the distant (~100 R_E) neutral sheet may account for the characteristics of ion beams in the plasma sheet boundary layer.
- Acceleration in substorm compression waves may occur.

The energization mechanisms described above do not appear capable of producing the high energy (>100 keV) component of substorm-related particle enhancements. Such particles could, however, be rapidly accelerated in the parallel electric field which exists along a near-Earth neutral line. We note that induction effects related to dynamic reconnection can raise the total

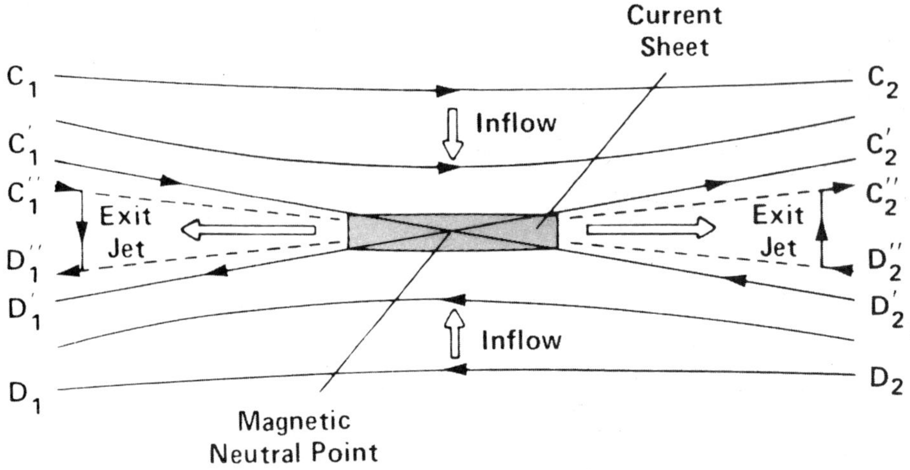

Fig. 11. Magnetic reconnection induced by flow of plasma (arrows). Carrying magnetic field lines of opposite orientation (unprimed) with it, plasma moves slowly from top and bottom toward the neutral point. Here the field lines (primed) break and reconnect. Tension within the reconnected field lines produces fast plasma jets moving to the left and right that carry the reconnected field lines (double printed) with them. (Adapted from a figure on page 75 in Space Science Physics: The Study of Solar-System Plasmas (National Academy of Sciences, Washington, D.C., 1978).)

potential drop along the neutral line far above its expected steady-state value.[20]

Enhancements of the fluxes of particles having energies of several hundred keV are commonly observed at synchronous orbit (Fig. 10) and in the magnetotail during geomagnetically active periods.[21] A comprehensive model for the morphology of energetic ion enhancements is illustrated in Figure 12. This model[22] suggests that after acceleration at the X-line, the particles stream both sunward and tailward. Those reaching the synchronous orbit region are transported westward around Earth via curvature and grad-B drifts and may be identifiable as distinct particle bunches ('drift echoes') through several circulations.[22]

The tailward-streaming particles appear as 'impulsive bursts.'[21] The inverse velocity dispersion (i.e., observation of slower particles before faster ones) exhibited by these bursts is supportive of their hypothesized origin at a magnetic X-line.[23] As suggested by the inset at the bottom of Figure 12, a spacecraft in

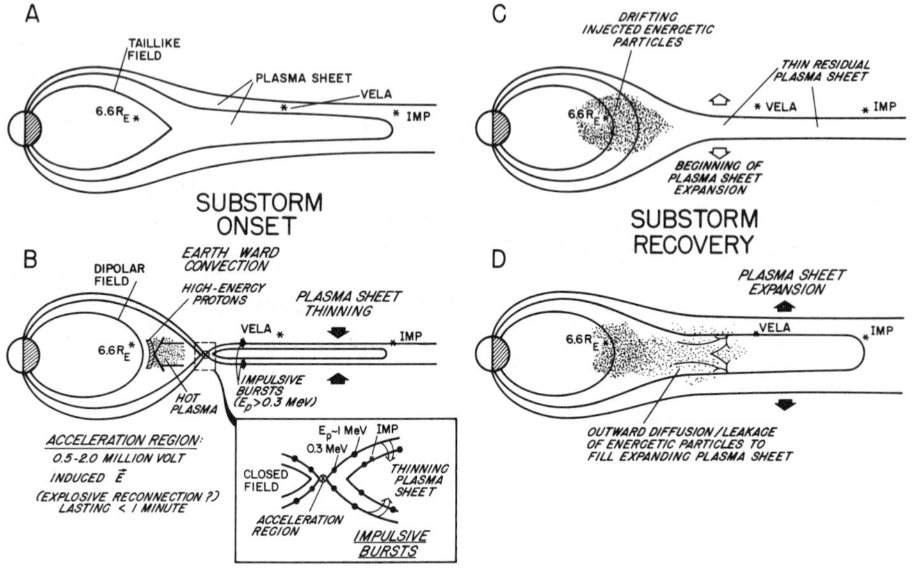

Fig. 12. Schematic depicting the sequence of energetic particle events predicted by the model of Baker and coworkers.[22] (A) The inner magnetosphere just prior to substorm onset showing the buildup of stress evidenced by the tail-like field. (B) The magnetosphere just after onset showing a dipolar field configuration and the acelerated proton bunches streaming sunward toward the trapped radiation zones and antisunward along the thinning plasma sheet. (C) Conditions just prior to substorm recovery and the beginning of the plasma sheet expansion. (D) Expansion of the sheet and the subsequent filling of the expanding sheet with energetic protons diffusing out of the trapping region. Note in the small inset that very large, impulsive bursts of energized particles must be related to internal induced electric fields at the near-Earth substorm neutral line.

the thinning plasma sheet successively samples field lines that have reconnected more and more recently at the X-line. These field lines contain distributions that are less depleted at the high energy end of the spectrum by escape of the faster particles. Finally, just as the spacecraft enters the lobe, it samples preferentially the fastest particles streaming along field lines connected directly to the X-line source.

As noted above, as much as 10% of the available dissipated substorm energy goes into the (suprathermal) energetic particle population. The ion spectra are quite steep[21,22] going as E^{-5} to

E^{-7}. Electron spectra are typically considerably harder, going as $\sim E^{-3}$. Of course, much of the energy which appears in these energetic particle populations goes eventually into the auroral precipitation and ring current particles. In a final tallying of energy budgets, therefore, one must be careful not to count energies twice.

From the above discussion one sees that substorms represent the extraction, storage, and delayed dissipation of energy from the flowing solar wind. From a total solar wind power of order 10^{21} ergs/s, 10^{18}-10^{19} ergs/s typically enter the magnetosphere. Of this entering energy perhaps 50% is converted to geoeffective forms while the remainder is released back into the downstream solar wind.

The major form of substorm energy conversion (and dissipation) is to generate hot plasma and energetic particles from stored magnetic field. Considering auroral and ring current particles, typically 60-70% of the energy goes into this form.[24] Remarkably, the efficiency of energetic particle production is quite high with $\geqslant$ 10% of the available energy appearing in the suprathermal population. Eventually, plasma particles may further convert their energy to radiation modes such as auroral kilometric radiation ($\sim 10^{15}$ ergs/s) and ULF/VLF waves ($\leqslant 10^{16}$ ergs/s).[24] Thus, of order 1% of the incoming energy in the magnetosphere may eventually be emitted as electromagnetic radiation detectable at large distances.

Substorms as Astrophysical Analogues

Some solar flares have long been recognized to bear similarity to magnetospheric substorms. In these classes of flares, magnetic energy appears to build up in coronal structures for periods of hours and then there appears to be an explosive dissipation of the magnetic energy.[25] As in substorms, much of the energy appears to go into suprathermal ions and electrons, many of which convert their energy (eventually) into X-rays, radio emissions, and other forms.[25] Presumably, such flares also occur on the surfaces of other dwarf stars similar to the sun and give rise to observed X-ray emissions, for example.

Another interesting class of astrophysical object is the X-ray burster. This is probably a compact object (e.g., neutron star) having a strong magnetic field and a well-developed magnetosphere.[26] As in the case of the Earth's magnetosphere, such an object (for example, Sco X-1) may accrete material from a companion object, "store" the material in its outer magnetosphere for $\sim 10^2$-10^3s, and then due to large-scale instabilities this stored accretion material may be dumped onto the stellar surface. This unloading of mass onto the surface presumably gives rise to the burst of X-rays observed. The analogy with the substorm model presented here has been noted previously.[27]

A final analogy is that of radio jets.[2] Given the (at least superficial) similarity of radio jet structures and the Earth's magnetotail, we might expect substorm-like behavior in these remote objects. In radio jets there is ostensibly supersonic flow of plasma from a central object with an embedded magnetic field in the plasma. If, in fact, there arises oppositely directed magnetic field in the jet structure due to large-scale current sheets, then as in the terrestrial magnetotail we would expect the occurrence of sporadic magnetic reconnection. This, in turn should produce (with high efficiency) large fluxes of energetic particles. Such particles, as in the Earth's case, ought then to produce intense radio waves both through synchrotron radiation and by collective plasma modes.

Acknowledgments. This work was done under the auspices of the U.S. Department of Energy. The author thanks J. T. Gosling and E. W. Hones, Jr. for several useful comments and C. F. Kennel for helpful discussions. Thanks are also extended to the Editors for many valuable suggestions concerning the contents of this paper.

References

1. J. T. Gosling, D. N. Baker, and E. W. Hones, Jr., Los Alamos Science, p. 32, Spring (1984).
2. M. C. Begelman, R. D. Blandford, and M. J. Rees, Rev. Mod. Phys, 56, 255 (1984).
3. R. L. McPherron, C. T. Russell, and M. P. Aubry, J. Geophys. Res., 78, 3131 (1973).
4. C. T. Russell and R. L. McPherron, Space Sci. Rev., 15 , 205 (1973).
5. E. W. Hones, Jr., Space Sci. Rev., 23, 393 (1979).
6. D. N. Baker, E. W. Hones, Jr., P. R. Higbie, R. D. Belian, and P. Stauning, J. Geophys. Res., 86, 8941 (1981).
7. F. V. Coroniti and C. F. Kennel, J. Geophys. Res., 77, 3361 (1972).
8. K. Schindler, J. Geophys. Res., 79, 2803 (1974).
9. D. N. Baker, S.-I. Akasofu, W. Baumjohann, J. W. Bieber, D. H. Fairfield, E. W. Hones, Jr., B. Mauk, and R. L. McPherron, Chapter 8 - Substorms in the Magnetosphere (in Solar Terrestrial Physics - Present and Future, D. Butler and K. Papadopoulos, Eds.), NASA (1984).
10. D. N. Baker, E. W. Hones, Jr., J. B. Payne, and W. C. Feldman, Geophys. Res. Lett., 8, 1979 (1981).
11. B. U. O. Sonnerup, G. Paschmann, I. Papamastorakis, N. Sckopke, G. Haerendel, S. J. Bame, J. R. Asbridge, J. T. Gosling, and C. T. Russell, J. Geophys. Res., 86, 10049 (1981).
12. D. H. Fairfield, R. P. Lepping, E. W. Hones, Jr., S. J. Bame, and J. R. Asbridge, J. Geophys. Res., 86, 1396 (1981).
13. R. E. Holzer and J. A. Slavin, J. Geophys. Res., 83, 3831 (1978).
14. E. W. Hones, Jr., D. N. Baker, S. J. Bame, W. C. Feldman, J. T. Gosling, D. J. McComas, R. D. Zwickl, J. Slavin,

E. J. Smith, and B. T. Tsurutani, Geophys. Res. Lett., 11, 5 (1984).

15. S.-I. Akasofu, Space Sci. Rev., 28, 121 (1981).
16. D. N. Baker, R. D. Zwickl, S. J. Bame, E. W. Hones, Jr., B. T. Tsurutani, E. J. Smith, and S. -I. Akasofu, J. Geophys. Res., 88, 6230 (1983).
17. P. Perrault and S. -I. Akasofu, Geophys. J. R. Astr. Soc., 54, 547 (1978).
18. T. E. Moore, R. L. Arnoldy, J. Feynman, and D. A., Hardy, J. Geophys. Res., 86, 6713 (1981).
19. D. N. Baker, T. A. Fritz, R. L. McPherron, D. H. Fairfield, Y. Kamide, and W. Baumjohann, J. Geophys. Res., 90, 1205 (1985).
20. D. N. Baker, T. A. Fritz, B. Wilken, P. R. Higbie, S. M. Kaye, M. G. Kivelson, T. E. Moore, W. Studemann, A. J. Masley, P. H. Smith, and A. L. Vampola, J. Geophys. Res., 87, 5917 (1982).
21. E. T. Sarris, S. M. Krimigis, and T. P. Armstrong, J. Geophys. Res., 81, 2341 (1976).
22. D. N. Baker, R. D. Belian, P. R. Higbie, and E. W. Hones, Jr., J. Geophys. Res., 84, 7138 (1979).
23. E. T. Sarris and W. I. Axford, Nature, 77, 460 (1979).
24. D. P. Stern, NASA Tech. Memorandum 86063, NASA Goddard SFC, Greenbelt, MD (1984).
25. Chap. 2 and Chap. 9, Solar-Terrestrial Physics - Present and Future (D. Butler and K. Papadopoulos, Eds.) NASA, Washington, DC (1984).
26. F. K. Lamb, A. C. Fabian, J. E. Pringle, and D. Q. Lamb, Ap. J, 217, 197 (1977).
27. M. Neugebauer and B. T. Tsurutani, Ap. J., 227, 494 (1978).

JOVIAN MAGNETOSPHERIC PROCESSES

C. K. Goertz
Dept. of Physics and Astronomy, Univ. of Iowa, Iowa City, IA 52242

ABSTRACT

Jupiter's rotational energy (6×10^{34} J) powers a large number of processes such as auroral UV emission, radio waves, and charged particle energization. We describe how the rotational energy may be dissipated by injection of plasma, magnetic pumping and field aligned electric fields. In addition, we describe energization by radial diffusion and plasma wave absorption. We also describe the generation of Alfven waves by the moon Io and their relation to the emission of the Jovian decametric (DAM) radio waves.

INTRODUCTION

Jupiter's huge magnetosphere may be of special interest to astrophysicists for several reasons.

1. Its structure is determined by the rapid rotation of Jupiter and strong internal plasma sources (Jovian ionosphere and the moon Io).
2. It produces very energetic charged particles (energies in excess of several 10's of MeV).
3. It is a strong source of radio waves.
4. It also emits UV and X-rays at levels which suggest non-thermal processes.

Jupiter has been likened to a pulsar[1] and to a binary star system.[2] Even though these analogies may seem somewhat artificial, we believe that processes occurring in the Jovian magnetosphere are of astrophysical interest. In this paper we will discuss the rotation of the Jovian magnetosphere and the energization of plasma by the rotation. The existence of MeV electrons and ions, however, requires additional energization mechanisms which may involve betatron and Fermi acceleration as well as magnetic pumping. We also discuss the interaction of Io with the corotating plasma which controls much of the Jovian radio emission. Finally, we discuss particle precipitation by wave-particle interactions and its relation to the auroral UV emissions.

Our knowledge of the Jovian magnetosphere has made great leaps following the two Pioneer (in the early 1970's) and the two Voyager fly-bys in 1979. Nevertheless, most of the Jovian magnetosphere has not been explored. For example, the high latitude region of the inner magnetosphere and the dusk to midnight region of the magnetotail have not been traversed by any of the spacecraft. Satellites have traversed the Io torus only at a few longitudes and we are still not certain whether it is longitudinally symmetric or not. They have flown close to Io only once and not through the region which carries the Io-driven Birkeland currents. We still have no measurements of plasma parameters in the source region of

0094-243X/86/1440208-16$3.00

DAM. Clearly, many of our ideas about the Jovian magnetosphere are still in the stage of enlightened speculation.

MAGNETOSPHERIC ROTATION

There are important differences between pulsar magnetospheres and the Jovian magnetosphere which make a comparison of limited value. In the local stress balance equation

$$\rho(d\vec{v}/dt - \vec{g}) + \nabla\cdot p = \vec{j} \times \vec{B} + \rho_e\vec{E} = \left(\frac{\nabla \times \vec{B}}{\mu_o}\right) \times \vec{B} + \varepsilon_o(\nabla\cdot\vec{E})\vec{E} \qquad (1)$$

the ratio R_1 of the two terms on the right-hand side is quite different for the two cases. Reducing we have:

$$R_1 = \frac{\mu_o\varepsilon_0(\nabla\cdot\vec{E})\vec{E}}{(\nabla \times \vec{B}) \times B} \approx \frac{\Omega^2 r^2}{c^2} \qquad (2)$$

(where $\vec{E} + (\vec{\Omega} \times \vec{r}) \times \vec{B} = 0$, has been assumed). R_1 is extremely small at Jupiter whereas it is comparable to 1 for pulsars. Thus for Jupiter we can neglect the charge density. In addition, we can safely neglect the displacement current $\varepsilon_o\ \partial\vec{E}/\partial t$ as compared to the conduction current. For pulsar magnetospheres the inertia terms on the left-hand side are often small, at Jupiter they are very important. The ratio of inertia current to the convection current $\rho_e\Omega r$ is

$$R_2 = \frac{\rho(d\vec{v}/dt - \vec{g}) + \nabla p}{B\ \varepsilon_o(\nabla\cdot E)E\ \Omega r} \approx \frac{\rho}{\varepsilon_o B^2}\left(1 - \frac{g}{\Omega^2 r} - \frac{kT}{m\Omega^2 r^2}\right) \quad . \qquad (3)$$

The ratio $\rho/\varepsilon_o B^2$ is equal to $c^2/v_A{}^2$, where v_A is the Alfvén speed. In the outer magnetosphere of Jupiter, the Alfvén speed is on the order of a few 100 km/s. Even near the surface where the magnetic field is large the ratio $c^2/v_A{}^2$ is greater than 10^{+2}. Thus the inertia current is most important at Jupiter.

If the field lines are equipotentials (i.e., if the parallel electric field $E_\parallel$ equals 0) the angular rotation rate Ω is a constant along field lines and for an azimuthally symmetric magnetic field only a function of L, the usual L-shell parameter. The total rotational energy of Jupiter is 6×10^{34} J and is believed to be the dominant energy source for processes such as enforcing (partial) corotation of plasma produced within the magnetosphere, plasma heating, auroral emissions, production of high energy charged particles and the system of radial transport in the outer magnetosphere.[1]

The simplest way of tapping the rotational energy involves the magnetic-field-aligned (Birkeland) current system shown in Figure 1

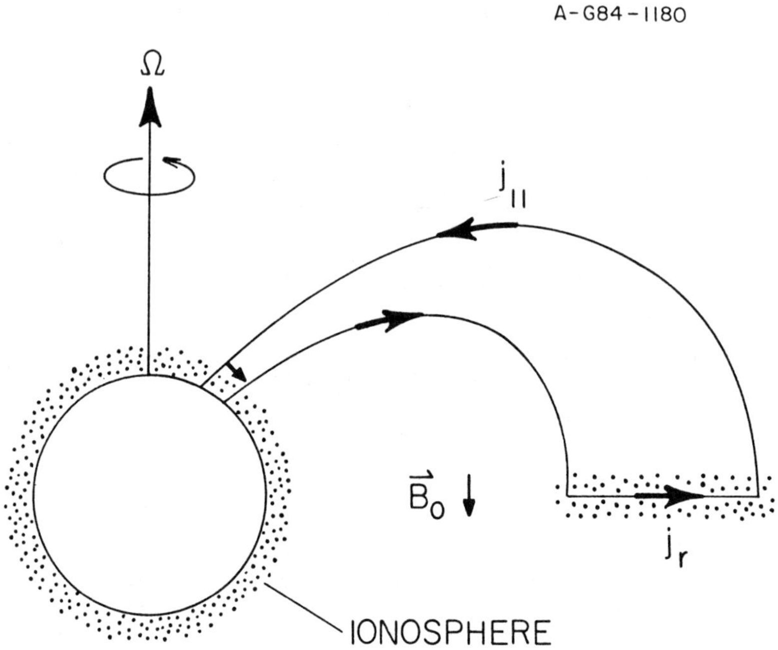

Fig. 1. The current system that transmits torque from Jupiter's ionosphere to the plasma in the equatorial region (e.g., the Io torus). The equatorial current spins up the plasma whereas the ionospheric current tends to spin down the ionosphere. Note that field aligned Birkeland currents must flow. These may become unstable to double-layer formation or other mechanisms supporting a parallel electric field.

which exerts the necessary torque to produce and maintain the corotation of plasma produced in, or injected into, the magnetosphere. Note that the torque in the ionosphere tends to spin down the ionosphere which is coupled to the neutral atmosphere by ion-neutral collisions. Regardless of where the new plasma particles are produced they gain an energy of about twice the local corotation energy $W_c = m\,\Omega^2 r^2/2$. If they are produced in the equatorial plane the corotation electric field increases their gyroenergy by $W_\perp = W_c + W_o - 2\sqrt{W_o W_c}$ where W_o is the kinetic energy of the neutral which is ionized ($W_o = mM_J G/r \ll W_c$ beyond about 2 R_J). In addition, they gain the corotational drift energy ($= W_c$). If the ions originate in the ionosphere and are pulled up by hot photoelectrons, the centrifugal sling-shot effect[3] increases their energy by $W_\parallel = W_c - mM_J G/r$ in addition to the corotational drift energy. For intermediate source positions the energy gain is distributed between parallel and perpendicular energy. For an injection rate of S (ions/s) the rate of energy extraction is thus 2 S W_c.

When the injected plasma is transported outward, even more investment of rotational energy is needed to maintain (partial) corotation.[4,5,6,7] The total power extracted is given by Hill et

$$P = S \left[\frac{1}{2}\,\Omega_J^2 (L_s R_J)^2 - M_J G/(L_s R_J) + \int_{L_s R_J}^{\infty} \omega^2 r dr\right] \qquad (4)$$

where L_s (usually 6 for injection from Io) is the L value of the field line on which the injection occurs. The rotation rate of the plasma in the magnetosphere, ω, may be different from the planetary rotation rate Ω. The first two terms represent the energy gained in the source and the third represents the energy gained as the ions are transported outward. Hill[8] has shown that corotation is maintained up to a critical distance

$$r_c/R_J = (\pi\, \Sigma\, B_J^2\, R_J^2/S)^{1/4} \qquad (5)$$

where Σ is the ionospheric Pedersen conductance (Σ ~ 0.1 mho). For an Io injection strength of S ~ 1700 kg/s this yields $r_c = 20\ R_J$ consistent with observations.[9] For these nominal values, the total rotational energy extracted is about 5×10^{13} W which is somewhat less than the power needed to maintain the auroral UV emission (10^{14} W).

The centrifugal force necessary to maintain the plasma in (partial) corotation is provided for by the $\vec{j} \times \vec{B}$ term in equation (1). The centrifugal current is in the azimuthal direction and corresponds to a slight difference in azimuthal drifts of the ions and electrons. It distorts the magnetic field into a disc-like configuration. Various theoretical models for this distortion have been discussed and are reviewed by Vasyliunas.[10]

The mode of outward transport is not clear. Kennel and Coroniti[11] have suggested that an axially symmetric stellar-wind-like transport occurs. The average radial velocity is non-zero. However, this does not really apply to Jupiter because a steady wind would require open field lines which do not seem to exist in the inner and middle magnetosphere. Their solution also lacks the day-night asymmetry imposed by the solar wind. Thus Hill et al.[12] have proposed that the wind only exists on the nightside where it is not constrained by the solar wind. Vasyliunas[10] has suggested that a neutral line exists in the nightside magnetotail beyond which the field lines are open and the wind solution can apply (Figure 2). The observations of Krimigis et al.[13] seem to confirm this picture.

The transport in the closed field regions (inner and middle magnetosphere) must occur by diffusion, i.e., a correlation between density and radial plasma flow. In this case the average radial velocity $\langle v_r \rangle = \langle E_\phi \rangle / B$ is zero, but $\langle \rho v_r \rangle$, the average radial mass flux is not zero. The mechanism responsible for diffusing low energy plasma may be different from that responsible for diffusing

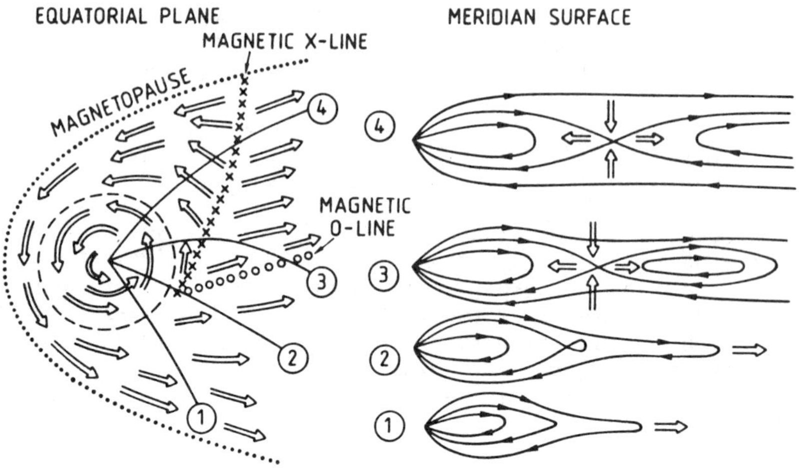

Fig. 2. A sketch of the equatorial plasma flow (left) and associated magnetic fields (right) according to Vasyliunas.[10]

high energy particles (see below). Siscoe and Summers[14] have pointed out that the inward plasma density gradient on the outer edge of the plasma torus would be unstable to the centrifugal drift instability. This instability is similar to the Rayleigh-Taylor instability. In a rotating plasma the effective force is outwards beyond synchronous orbit. Thus an inward density gradient corresponds to a heavy fluid on top of a light fluid which is an unstable situation with an excess of free energy. The flute mode ($k_{\parallel} \simeq 0$) would cause radial transport of plasma by interchange of magnetic flux tubes. The size of the flux tubes is not known, but a likely size seems to be 1 to 2 R_J.[14]

Independent of the mechanism of radial transport, a time-averaged radial outward transport requires a radial current to impart the torque on the plasma necessary for maintaining (partial) corotation. This radial current distorts the magnetic field into a spiral configuration. This has, indeed, been observed. The observed azimuthal magnetic field is consistent with an outward mass transport of 10^3 kg/s.[10,15]

ENERGIZATION MECHANISMS

The energies produced directly by the rotation of Jupiter are of the order of (60 eV) $(r/6\ R_J)^2\ (m/m_p)$ and thus considerably smaller than the MeV energies of some electrons and ions observed in the magnetosphere. The acceleration mechanisms which can produce such large energies can be classified according to the degree to which they violate the three adiabatic invariants. According to Hill, Dessler, and Goertz[1] processes which violate only the third adiabatic invariant (related to the magnetic flux enclosed by a drift orbit) are called "adiabatic". Those which violate one or both of the other two (related to the magnetic moment of a particle and the bounce integral) are called "nonadiabatic."

A. Adiabatic Processes

The most widely discussed mechanism for magnetospheric particle acceleration is due to adiabatic compression. As the particles are transported across magnetic flux shells conservation of the first adiabatic invariant (magnetic moment) implies a betatron acceleration ($E_\perp \propto B$). Conservation of the second invariant (bounce integral) implies a Fermi acceleration ($E_\parallel \propto \ell^{-2}$, with ℓ being the length of the field line between mirror points). For a dipole field $E_\perp \propto r^{-3}$ and $E_\parallel \propto r^{-2}$. Thus, as particles are transported inward their energy increases and the distribution tends toward a pancake pitch-angle distribution with $\langle E_\perp \rangle > \langle E_\parallel \rangle$. If pitch-angle scattering which violates the adiabatic invariants would occur without loss of particles into the atmosphere, the distribution function would remain isotropic and the energy would change as $E \propto r^{-8/3} \propto V^{-2/3}$ where V is the volume of a flux tube transported inward or outward. This would be the analogy to the adiabatic compression of an ideal monoatomic gas.

Even though radial transport occurs and this adiabatic acceleration takes place undoubtedly, it alone is not sufficient to explain the presence of MeV particles at say L = 20. For example, solar wind protons enter the magnetosphere with an energy of a few keV and would have an energy of only a few 100 keV at L = 20. Solar wind electrons would be even less energetic. Newly created ions at 100 R_J would have an energy of less than $17(m/m_p)$ keV and would not be energized to several tens of MeV by the time they reach L = 20. In addition, diffusion (or radial transport) is a slow process with a typical time scale of years.[1] The emission rate of energetic particles into interplanetary space requires that they are resupplied with a time scale of a few rotation periods (a few 10's of hrs).[16] Thus faster and more powerful mechanisms are required to produce energetic particles in the outer magnetosphere. Once they are produced they may gain additional energy by inward diffusion.

We note that only inward transport would give rise to an energy increase by the adiabatic mechanism. Thus the source for the particles must be in the outer magnetosphere. If the source were close to Jupiter (say in the Io torus) outward transport would adiabatically decrease the particle's energy.

B. Quasi-Adiabatic Processes

Nishida[17] and Sentman, Van Allen and Goertz[18] have proposed a "recirculation model" which overcomes the difficulty posed by the adiabatic deceleration of outward moving particles. They point out that in the inner magnetosphere (say at L < 10) the distribution

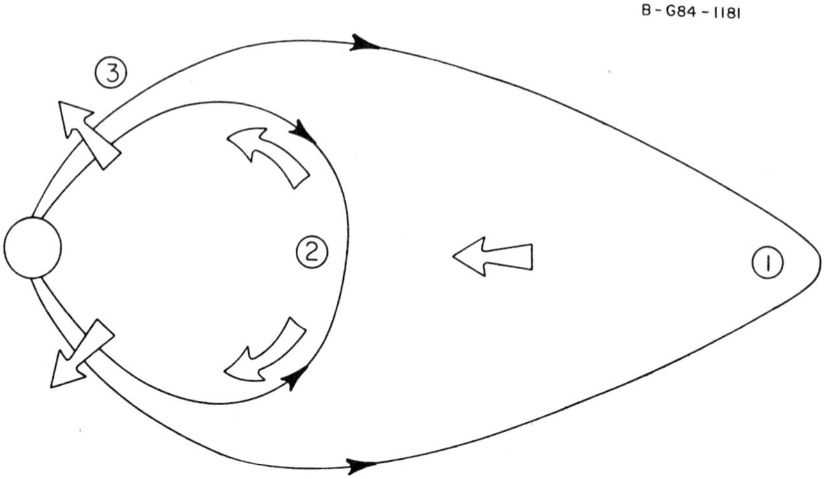

Fig. 3. A sketch of the "recirculation model". The arrows indicate the direction of the diffusive transport as described in the text.

function of inward diffusing particles (moving from point 1 to point 2 in Figure 3) will be anisotropic enough to generate electromagnetic electron and ion-cyclotron waves by an instability (see below). These waves will cause a pitch-angle scattering, thereby allowing some particles to mirror at low altitudes (point 3) where they are subject to meridional diffusion as a result of low frequency waves violating only the second adiabatic invariant. This diffusion transports particles toward large values of L because the source for these low altitude mirroring particles is at low L values. In this process the total energy of the particle is conserved. The net result is that the particles return to point 1 with field aligned energies comparable to the energy gained by adiabatic diffusion from point 1 to point 2. Pitch-angle scattering of this distribution will then produce a high energy nearly isotropic distribution at point 2 and the process can start over again. Particles can then, in principle, repeat the cycle many times and gain high energies. However, Sentman et al.[18] point out that the combination of several diffusive processes makes this recirculation a slow process.

The other quasi-adiabatic process was investigated by Goertz[19] and Borovsky et al.[20] They suggest that due to the day-night asymmetry of the magnetosphere, particles corotating from day to night and back will encounter a varying magnetic field strength. As the particles drift from noon to midnight they experience an adiabatic betatron and Fermi deceleration and the distribution will become more dumbbell-shaped (i.e., field aligned). Without pitch-angle scattering this energy loss would be reversible as the particles drift from midnight to noon. However, if non-adiabatic pitch-angle scattering (without change of energy) would occur some of the energy gained during the midnight to noon half of the orbit would be stored in the parallel component at noon and escape the deceleration during the noon to midnight half of the orbit. This mechanism is the analog of the magnetic pumping mechanism of Alfvén.[21] In the most advantageous case (pitch-angle scattering time = rotation period) the average energy can increase by a factor of 2 each cycle, i.e., the energization time would be equal to the rotation period (10 hrs). This mechanism may thus account for the energization of particles in the outer magnetosphere. However, some observations[16] require acceleration processes which occur on time scales much less than 10 hrs.

C. Non-Adiabatic Processes

Non-adiabatic processes violate all three adiabatic invariants. Three processes have been suggested to be operative in the Jovian magnetosphere: magnetic reconnection (or merging), parallel electric fields and plasma wave heating.

The magnetic field topology in the magnetodisc is favorable to the occurrence of sporadic and/or steady state reconnection. For steady state merging the average energy gain is equal to the magnetic energy density outside the current sheet divided by the density in the current sheet.[1] For $B = 10$ nT and $n = 0.1\ cm^{-3}$ one obtains an energy gain of $(B^2/\mu_0 n) \sim 5$ keV. Thus steady state merging cannot account for MeV particles. Sporadic (time-dependent) merging may produce higher energies as observed in the earth's magnetotail.[22] However, the theoretical understanding and modelling of particle acceleration in sporadic merging is not well understood.

As shown in Figure 1, we expect field aligned currents to be present in the Jovian magnetosphere, even though the magnitude and location of the field aligned current density and currents is not known. Whenever the field aligned current density exceeds a critical value (which is not well established theoretically) a parallel electric field occurs. The field aligned potential drop ($\phi_\parallel = -\int E_\parallel ds$) may be a significant fraction of the total EMF driving the current. The electric potential for corotation in the equatorial plane is $\phi = \Omega_J B_J R_J{}^2/L \simeq 376$ MV/L. If only a fraction of this would appear along the field lines a powerful linear accelerator could become active and cause the appearance of MeV particles. The potential drop between field lines in the equatorial plane would then be different from that in the ionosphere. In other words, the outer magnetosphere would slip relative to the ionosphere. Thus this model is sometimes called the "slippery clutch" model. This mechanism must not be confused with the partial corotation described above where the entire flux tube slips through the neutral atmosphere. The theory of the "slippery clutch" model has not been formulated in any detail.

Heating of particles by absorption of plasma waves is potentially very significant. For example, ion cyclotron waves, driven by field aligned currents, are believed to be responsible for the acceleration of 0^+ ions in the earth's magnetosphere.[23] However, the question of the ultimate power source is not answered by this model. Somehow the waves must be maintained at a sufficient level to provide efficient and fast heating. Recently, Barbosa et al.[24] have suggested that the low frequency MHD waves observed in the magnetodisc of Jupiter are sufficiently intense to accelerate injected (or locally produced) ions to MeV energies. Presumably the MHD waves are maintained by free energy in the low energy plasma.

THE IO-INTERACTION

As Io moves through the corotating plasma it distorts the electric and magnetic field in its vicinity. This distortion is due to the current driven through the ionosphere or body of Io by the motional electric field seen in the frame of reference moving with Io (~ 0.1 V/m). This current must, of course, be closed by field aligned currents which are toward Io on the side facing Jupiter and away from Io on the side facing away from Jupiter. Before the discovery of the Io torus, it was believed that the current closes in the ionosphere of Jupiter.[25] It is now generally believed that the current system is closed by polarization currents in the front of an Alfvén wave pulse wave pulse propagating along the magnetic flux tube.[26,27]

A simple picture of the interaction considers Io as a flat plate with the normal oriented parallel to the magnetic field (Figure 4). In that case the absorption of plasma and associated perturbations (sound waves) can be neglected. The plasma flow in the corotating frame satisfies

$$\frac{\partial^2 v}{\partial z^2} = \frac{1}{v_A^2(z)} \frac{\partial^2 v}{\partial t^2} \quad . \tag{6}$$

The boundary condition at z = 0 is

$$v(z = -\varepsilon) = v_< = v(z = +\varepsilon) = v_> \quad . \tag{7}$$

From Faraday's law and $\vec{E} + \vec{v} \times \vec{B} = 0$ we get

$$\frac{\partial v_>}{\partial z} - \frac{\partial v_<}{\partial z} = \frac{\mu_o}{B_o}\frac{\partial}{\partial t} I_{Io} = \frac{\mu_o}{B_o}\frac{\partial}{\partial t}\int_{-\varepsilon}^{\varepsilon} \Sigma_{Io} E_{Io}\, dz = 2\frac{\mu_o}{B_o}\Sigma_{Io}\frac{\partial}{\partial t} E_{Io} \quad . \tag{8}$$

For a uniform medium we have

$$v(z,t) = v(v_A t \mp z) \tag{9}$$

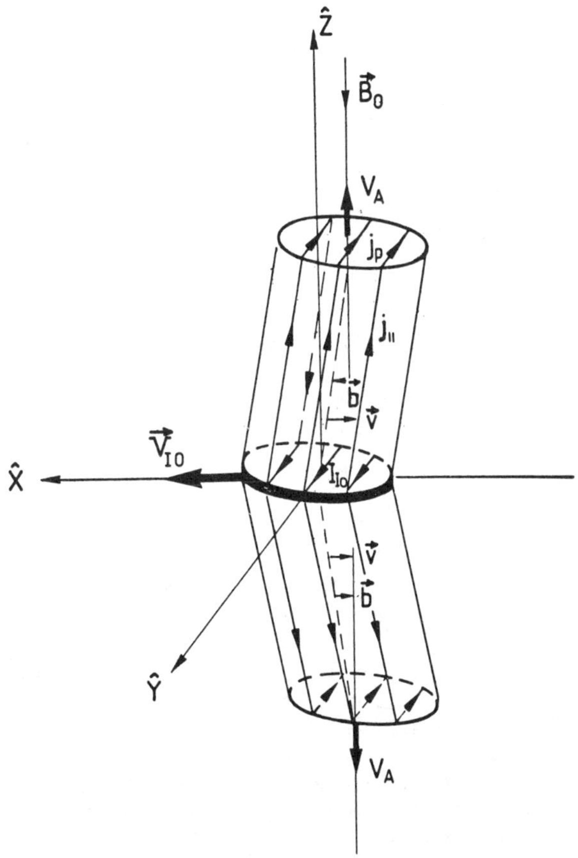

Fig. 4. A simple picture of the Io interaction. The field aligned currents are closed by a conductive current through Io and polarization currents in the leading fronts of the Alfven waves.

where the ∓ sign refers to waves propagating to positive and negative values of z, respectively. Writing

$$E_{Io} = B_o \, (v_{Io} - v(o)) \tag{10}$$

where $B_o = 0.02$ G is the background magnetic field at Io we see that Io launches two pulses propagating in opposite directions. The amplitude of each pulse is

$$A_v = v_{Io} \frac{\Sigma}{\Sigma + \Sigma_A} = k \, v_{Io} \tag{11}$$

where the Alfvén conductance Σ_A is

$$\Sigma_A = \frac{1}{\mu_o v_A} \quad . \tag{12}$$

The duration of each pulse (seen from a corotating frame) is

$$T = \frac{2R_{Io}}{v_{Io} - A_v} = \frac{2R_{Io}}{v_{Io}} \frac{1}{(1 - k)} = \frac{T_o}{(1 - k)} \quad . \tag{13}$$

The magnetic amplitude is

$$b(z,t) = \mp B_o \; v(z,t)/v_A \tag{14}$$

and the power flux

$$F = B_o^2 \; A_v^2/v_A \quad . \tag{15}$$

When v_A is not a constant the formulation becomes more complicated.[28]

For a thick atmosphere model of Io ($\int n \; dh > 10^{16} \; cm^{-2}$) the Io conductance has been estimated as 12 mhos.[1] The Alfvén conductance is of the order of 4 mhos.[28] Thus $\Sigma/\Sigma_A \approx 3$ and the velocity amplitude is 3/4 v_{Io}. The duration of the pulse is 260 seconds. This is considerably less than the time it takes the Alfvén wave to propagate to the Jovian ionosphere and back (> 700 seconds).[29]

Thus the information about the conductive properties of the Jovian ionosphere cannot return to Io in time to influence the interaction. Before the discovery of the torus the Alfvén travel time was estimated as 60 seconds. In that case several reflections would occur before Io moves out of the flux tube considered and the ionosphere would control the current to a certain degree as discussed by Hill et al.[1]

It has been shown by Goertz[27] and Goldstein and Goertz[30] that there exists a parallel electric field at the leading and trailing edge of the Alfvén wave pulse generated by Io (Figure 5). This electric field will accelerate ambient electrons which carry the field aligned current between the leading and trailing edge of the pulse. The currents are closed by polarization currents. The energy of the current carrying electrons has been estimated as several keV at high latitudes.[30] These electrons appear as beams in velocity space and may be subject to several electrostatic instabilities which lead to the emission of radio waves into conical sheets by mode coupling.[30] There is also a possibility of an enlarged loss cone of electrons coming up from the ionosphere on the side of the flux tube facing Jupiter because the parallel electric field would decelerate them. This form of free energy is

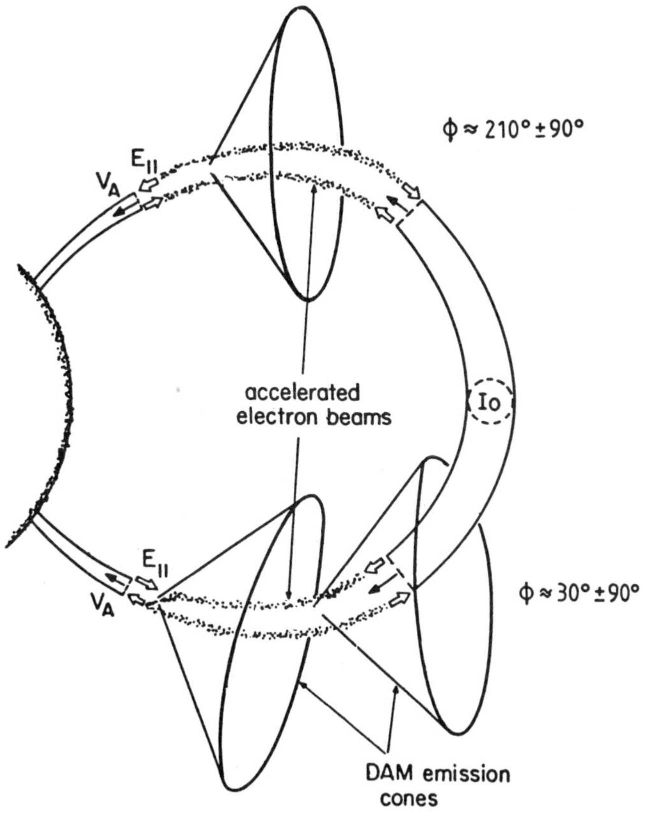

Fig. 5. A view of the Io-generated Alfven wave pulse. At the leading and trailing edges parallel electric fields accelerate electrons which carry field aligned currents. These currents are confined to the region between the two edges of the pulse and may lead to the emission of radio waves into conical sheets.

suggested to drive the earth's auroral kilometric radiation (AKR). For a more detailed discussion see Goertz.[31]

In any case the conical sheets would provide for a natural explanation of the DAM arcs (Goertz[31]) as discussed by Gurnett and Goertz.[32] The multiplicity of arcs has a simple explanation also. The Alfvén waves will be reflected at the Jovian ionosphere and propagate back to the opposite hemisphere where they are again reflected, etc. At any time there is a pattern of Alfvén waves attached to and carried around by Io as shown in Figure 6. Thus at any time many closely spaced DAM sources may be active. Each source traces out an arc-like structure in the frequency-time diagram (Figure 7) as the pattern rotates with Io. Of course, the

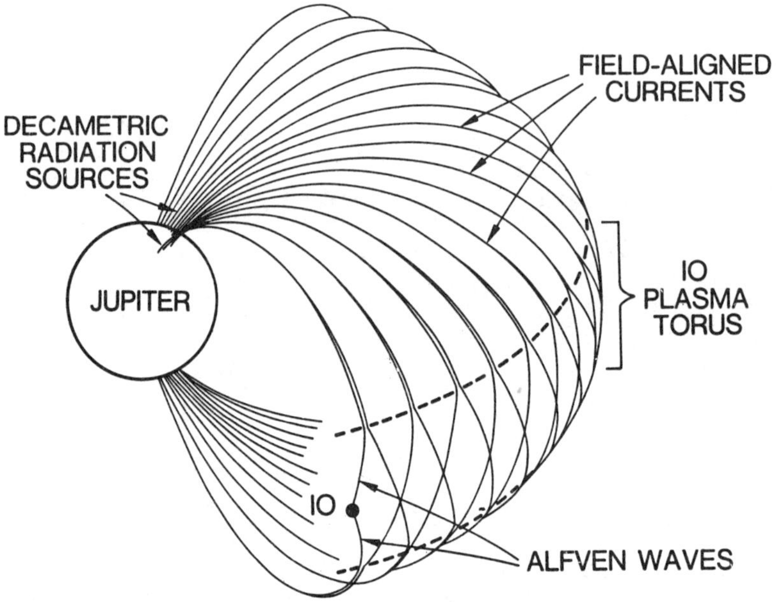

Fig. 6. The pattern of Alfven waves associated with Io. The traces indicate the position of one edge (say the leading edge) of the Alfvén wave.[32]

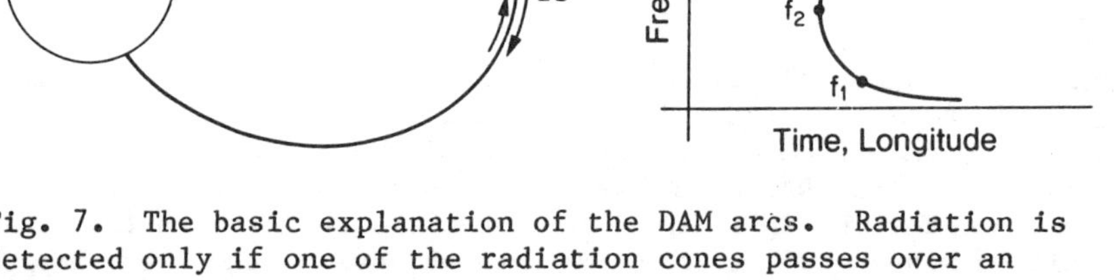

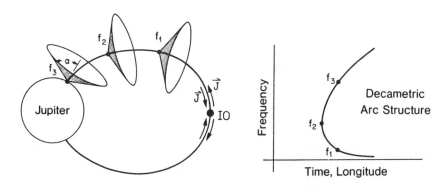

Fig. 7. The basic explanation of the DAM arcs. Radiation is detected only if one of the radiation cones passes over an observer. At any time there are, in general, two cones visible by an observer emitting at two different frequencies. These frequencies will vary as the observer rotates relative to the meridian along which the cones are emitted.

radio emission provides a damping for the Alfven waves in addition to the collisional damping in the ionosphere. Gurnett and Goertz[32] estimate that about 50 reflections are needed to damp the waves significantly (mainly by collisional damping). Thus at any time 50 DAM sources may be active and the whole Alfvén pattern may extend completely around Jupiter. However, this number depends on the ionospheric conductance which is poorly known. (For $\Sigma = 0.1$ mho only 10 reflections may occur.)

WAVE-PARTICLE INTERACTIONS

We have already mentioned the importance of wave-particle interactions several times. We now discuss the role of wave particle interactions for the maintenance of the auroral UV activity and the heating of the torus. Broadfoot et al.[33] have shown that the total radiated power from the polar regions which apparently map along field lines from the Io torus is 3×10^{12} W. The best estimates to account for this power requires (0.3 to 1.2) $\times 10^{14}$ W to be dissipated by energetic particle precipitation into the Jovian atmosphere.[34] The most liberal estimates for overall energy deposition of energetic electrons is less than 10^{13} W.[34] This falls short of the required input. Goertz[35] and Thorne[34] have thus suggested that energetic ions scattered into the atmosphere on field lines with $6 < L < 8$ are responsible for the UV emission. And, indeed, the flux of trapped ions with energies above 500 keV decreases significantly from $L = 8$ to $L = 6$. The observed deficit of ion-fluxes at $L = 6$ when compared with $L = 8$ above 500 keV corresponds to a loss of 3×10^{13} W. Extrapolating this to below 100 keV would be able to account for the 10^{14} W needed. No equivalent decrease of electron fluxes is seen. Thorne[34] also suggests that backscattered electrons (secondaries) are of fundamental importance for heating the Io plasma torus. Unfortunately, the ion-cyclotron waves which are most likely responsible for the scattering have not been measured. However, Thorne[34] has shown that the observed ion fluxes at $L = 8$ are large enough to cause rapid growth of ion cyclotron waves and strong pitch-angle diffusion. Gurnett and Goertz[36] on the other hand argue that energetic electrons produce electron cyclotron waves (whistlers) which mode convert into ion-cyclotron waves. They also show that pitch-angle diffusion by the anomalous gyroresonance between electron cyclotron waves and ions is quite strong and may be sufficient to explain the ion precipitation. In any case, the ultimate source of the UV power is the same as that responsible for the acceleration of ions in the outer magnetosphere.

SUMMARY

Various arguments suggest that the dissipation of Jupiter's rotational energy (6×10^{34}) is responsible for driving the diverse energetic phenomena such as auroral UV radiation, energetic particle acceleration and radio emissions. Rotational energy can be tapped by particle injection, magnetic pumping and field aligned electric fields (slippery clutch). We have pointed out that each mechanism requires field aligned currents coupling the outer magnetosphere to the Jovian ionosphere. Jupiter's magnetosphere is clearly less energetic than astrophysical magnetospheres, but more energetic than the earth's magnetosphere. Certain processes such as radial diffusion and wave particle interactions which occur near the earth also occur at Jupiter. Other processes which may be more important in astrophysical magnetosphere such as magnetic pumping, tapping of rotational energy occur at Jupiter, but are less important at the earth. Thus Jupiter is a kind of rosetta stone linking the well-studied magnetosphere of the earth with the inaccessible magnetospheres of astrophysical objects.

ACKNOWLEDGEMENTS

This work was supported by NASA grants NAGW-386, NSG-7632, NGL-16-001-043, and NSF grant ATM-811126.

REFERENCES

1. Hill, T. W., A. J. Dessler, and C. K. Goertz, Physics of the Jovian Magnetosphere, (ed. A. J. Dessler, Cambridge Univ. Press, 1983), p. 353.
2. Scarf, F. L., F. V. Coroniti, C. F. Kennel, and D. A. Gurnett, Vistas in Astron. 25, 263 (1982).
3. Hill, T. W., A. J. Dessler, and F. C. Michel, Geophys. Res. Lett. 1, 3 (1974).
4. Dessler, A. J., Icarus 44, 291 (1980).
5. Eviatar, A., and G. L. Siscoe, Geophys. Res. Lett. 7, 1085 (1980).
6. Hill, T. W., EOS 62, 25 (1981).
7. Hill, T. W., A. J. Dessler, and L. J. Maher, J. Geophys. Res. 86, 9020 (1981).
8. Hill, T. W., J. Geophys. Res. 84, 6554 (1979).
9. Belcher, J. W., Physics of the Jovian Magnetosphere, (ed. A. J. Dessler, Cambridge Univ. Press, 1983), p. 68.
10. Vasyliunas, V. M., Physics of the Jovian Magnetosphere, (ed. A. J. Dessler, Cambridge Univ. Press, 1983), p. 395.
11. Kennel, C. F., and F. V. Coroniti, Ann. Rev. Astron. Astrophys. 15, 389 (1977).
12. Hill, T. W., J. F. Carbary, and A. J. Dessler, Geophys. Res. Lett. 1, 333 (1974).
13. Krimigis, S. M., et al., Science 206, 977 (1979).
14. Siscoe, G. L., and D. Summers, J. Geophys. Res. 86, 8471 (1981).
15. Goertz, C. K., Space Sci. Rev. 23, 319 (1979).
16. Schardt, A. W., and C. K. Goertz, Physics of the Jovian Magnetosphere, (ed. A. J. Dessler, Cambridge Univ. Press, 1983), p. 157.
17. Nishida, A., J. Geophys. Res. 81, 1771 (1976).

18. Sentman, D. D., J. A. Van Allen, and C. K. Goertz, Geophys. Res. Lett. 2, 465 (1975).
19. Goertz, C. K., J. Geophys. Res. 81, 3145 (1978).
20. Borovsky, J. E., C. K. Goertz, and G. Joyce, J. Geophys. Res. 86, 3481 (1981).
21. Alfvén, H., Phys. Rev. 75, 1732 (1949).
22. Sarris, E. T., S. M. Krimigis, and T. P. Armstrong, J. Geophys. Res. 81, 2341 (1976).
23. Ashour-Abdalla, M., H. Okuda, and C. Z. Cheng, Geophys. Res. Lett. 8, 795 (1981).
24. Barbosa, D. D., A. Eviatar, and G. L. Siscoe, J. Geophys. Res. 89, 3789 (1984).
25. Goldreich, P., and D. Lynden-Bell, Astrophys. J. 156, 59 (1969).
26. Neubauer, F. M., J. Geophys. Res. 85, 1171 (1980).
27. Goertz, C. K., J. Geophys. Res. 85, 2949 (1980).
28. Goertz, C. K., Adv. Space Res. 3, 59 (1983).
29. Bagenal, F., J. Geophys. Res. 88, 3013 (1983).
30. Goldstein, M. L., and C. K. Goertz, Physics of the Jovian Magnetosphere, (ed. A. J. Dessler, Cambridge Univ. Press, 1983), p. 317.
31. Goertz, C. K., this volume.
32. Gurnett, D. A., and C. K. Goertz, J. Geophys. Res. 86, 717 (1981).
33. Broadfoot, A. L., et al., Science 204, 979 (1979).
34. Thorne, R. M., Physics of the Jovian Magnetosphere, (ed. A. J. Dessler, Cambridge Univ. Press, 1983), p. 454.
35. Goertz, C. K., Geophys. Res. Lett. 7, 365 (1980).
36. Gurnett, D. A., and C. K. Goertz, Geophys. Res. Lett. 10, 587 (1983).

PLANETARY RADIO WAVES

C. K. Goertz
Dept. of Physics and Astronomy, Univ. of Iowa, Iowa City, IA 52242

ABSTRACT

Three planets, the earth, Jupiter and Saturn are known to emit non-thermal radio waves which require coherent radiation processes. The characteristic features (frequency spectrum, polarization, occurrence probability, radiation pattern) are discussed. We distinguish between radiation which is externally controlled by the solar wind and internally controlled radiation which only originates from Jupiter. The efficiency of the externally controlled radiation is roughly the same at all three planets (5×10^{-6}) suggesting that similar processes are active there. We discuss briefly the maser radiation mechanism for the generation of the radio waves and general requirements for the mechanism which couples the power generator to the region where the radio waves are generated.

INTRODUCTION

The observation of escaping radio waves from a planet allows, in principle, the remote sensing of plasma conditions in the source region. However, the decoding of the information contained in the radio waves observed is usually not unique because several poorly understood processes are involved. Figure 1 shows a sketch of the flow of energy from the energy source to the observer. Somewhere a generator transforms energy stored in, for example, the solar wind or the motion of Jupiter's moon, Io, into low frequency electromagnetic energy (usually associated with magnetically aligned currents) which provides a coupling between the generator and the source. This energy is absorbed by the plasma in the source region. The resulting plasma distribution function contains free energy which can amplify thermal noise waves and lead to the rapid growth of these waves. In general the amplification rate depends on the wave frequency and polarization and the escaping waves will reflect this in their frequency spectrum and polarization. However, the waves must propagate through an anisotropic plasma before they reach the observer. This propagation can severely distort the characteristics of the waves. The characteristics of the observed radiation has been divided into its phenomenology and morphology by Smith.[1] The first term refers to observables and the second to the interpretive organization of data such as occurrence probability. The phenomenology is mainly determined by the radiation mechanism and propagation effects whereas the morphology is mainly controlled by the generator and coupling mechanisms.

The advantage of observing planetary radio waves lies in the ability to provide some parameters such as magnetic field strength, plasma densities and temperatures in this flow diagram by in-situ

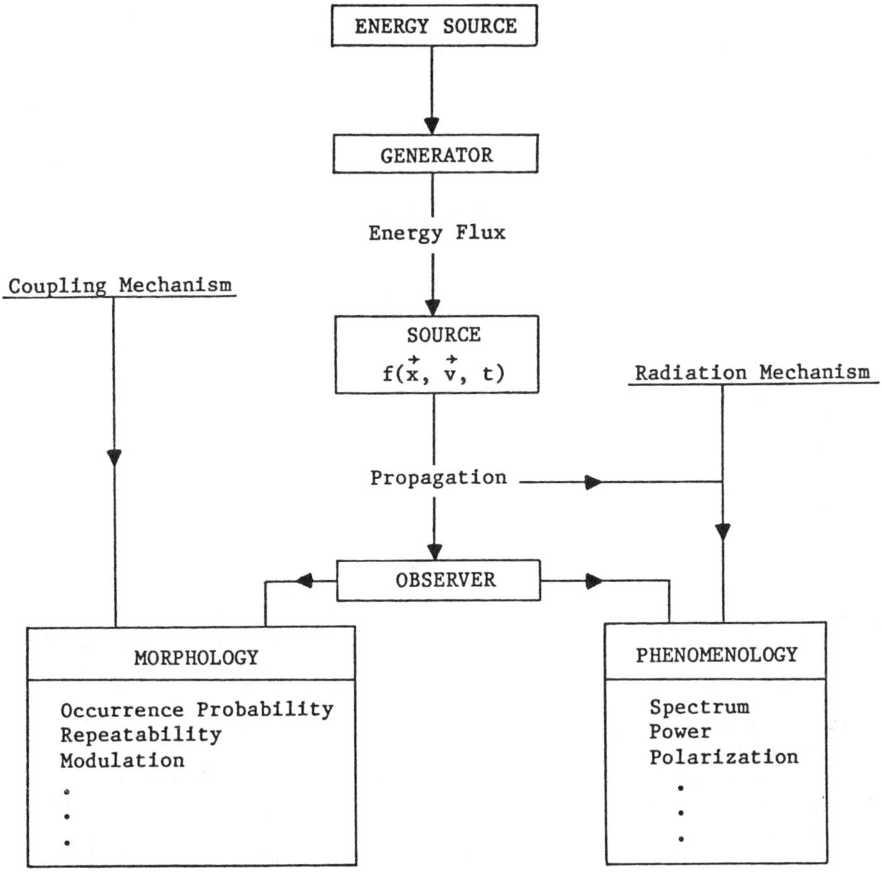

Fig. 1. The flow of energy from the energy source to the observer.

observations. Nevertheless, our understanding of the complete process sketched in Figure 1 is still rudimentary. Perhaps the humility gained by years of unsuccessful attempts to explain the major parts of the emission process is the most valuable result of studying planetary radio waves for the astrophysicists.

We note that radio waves are defined as electromagnetic waves which can escape from the source and propagate through free space to an observer. As we shall see, most planetary radio wave components require a coherent radiation process and are not produced by incoherent synchrotron radiation as most radio waves from "known" astrophysical sources. In fact, the only planet which is known to emit synchrotron radiation at a significant level is Jupiter. In other words, the explanation of the radiation requires the full use of plasma physics and not just single particle theory.

It is obvious that a complete treatment of this complex subject is beyond the scope of this paper. Fortunately, reviews of the observational characteristics[2] and current theories of radio wave generation[3] are easily available.

For the uninitiated, the jargon of planetary radio physics can be confusing. The following abbreviations are commonly used.

AKR	Auroral Kilometric Radiation (Earth)
DIM	Decimetric Radiation (Jupiter)
DAM	Decametric Radiation (Jupiter)
HOM	Hectometric Radiation (Jupiter)
bKOM	Broadband Kilometric Radiation (Jupiter)
nKOM	Narrowband Kilometric Radiation (Jupiter)
SKR	Saturn's Kilometric Radiation (Saturn)
SED	Saturnian Electrostatic Discharge Radiation (Saturn)

The names refer, in broad sense, to the wavelength range over which the radiation is most intense except for SED which occurs over an extremely broad frequency range.

OBSERVED PROPERTIES

In Figure 2 (taken from Kaiser and Desch[4]), three dynamic spectrograms extending over 24 hours are shown. The intensity of the radiation received by the Voyager Planetary Radio Astronomy (PRA) instrument is indicated by the darkness.[5] It is at once obvious that in each case the intensity is modulated by the rotation of each planet.

Earth's AKR extends from below 100 kHz to above 500 kHz. The radiation is most intense when the northern magnetic dipole points towards the spacecraft.

The radiation from Jupiter is the most complex one extending from less than 20 kHz to 40 MHz. The DIM radiation between 100 MHz and several GHz is not shown. That component is due to synchrotron radiation by MeV electrons trapped in the inner regions of the Jovian magnetosphere. It is reasonably well understood.[2] The lower frequency components shown in Figure 1 display a very complex morphology which has been extensively reviewed by Carr et al.[2] The component above 5 MHz has been observed for almost 30 years from the ground. No radiation above 40 MHz has ever been detected (except, of course, DIM at much higher frequencies). This DAM radiation extends to below 1 MHz as the Voyager observations show and consists of storms of emission lasting up to several hours. The occurrence of storms repeats every 9.6 hours (the rotation period of Jupiter) and is controlled by the moon, Io. A remarkable feature of these spectrograms are the arc-like traces. There are vertex-early arcs and vertex-late arcs. One also distinguishes between the great arcs at higher frequencies and lesser arcs at lower frequencies. The curvature of the lesser arcs is stronger.

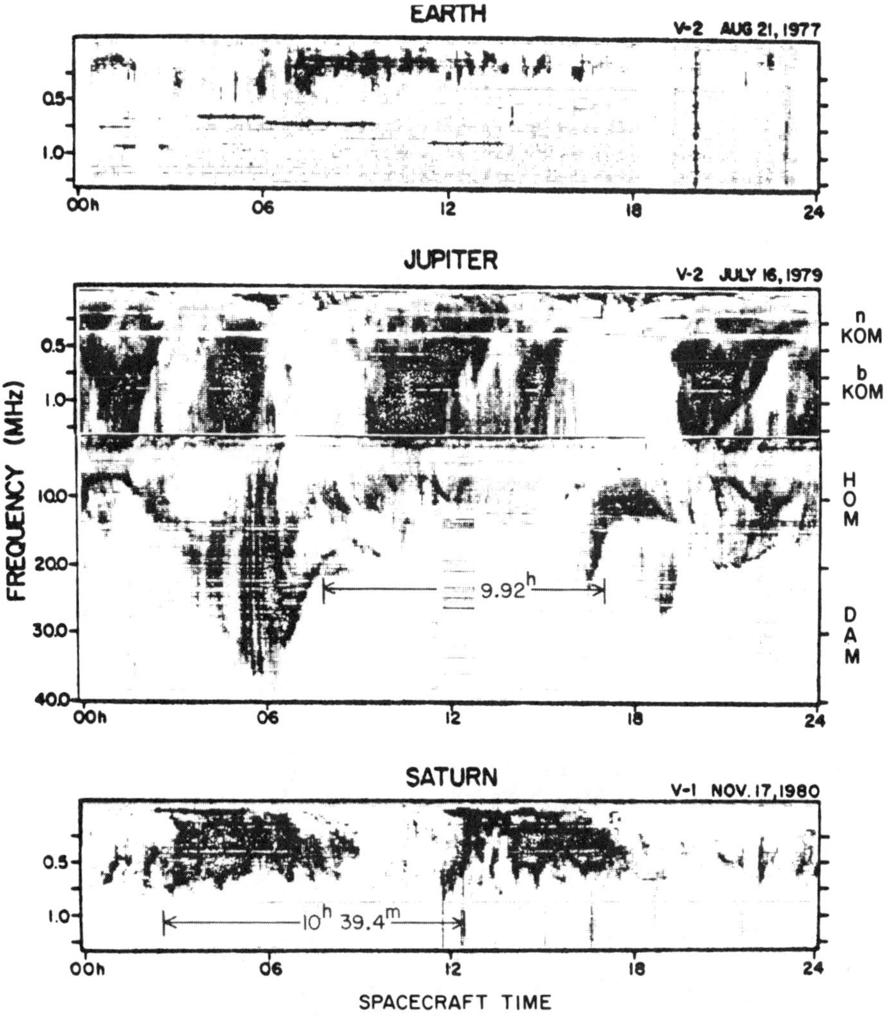

Fig. 2. Voyager observations at about 100 planetary radii from the respective planets. The observations are in the form of a dynamic spectrum with increasing darkness proportional to increasing intensity. The different components are indicated. This figure is adopted from Kaiser and Desch.[4]

For a more detailed discussion of these structures, we refer the reader to the excellent review article by Carr et al.[2] Below about 100 kHz Voyager discovered two distinct kilometer wavelength emissions. The nKOM emission occurs near 100 kHz over a narrow range of frequency (20 to 40 kHz). Its repetition period is 3 to 5% slower than that of DAM. The bKOM extends from below 100 kHz to above 1 MHz. Like the nKOM its intensity peaks at 100 kHz. Its repetition period is equal to that of DAM. The probability of receiving radiation peaks when the northern magnetic pole tips toward the observer. Between 300 kHz and 3 MHz, Voyager discovered a separate component called HOM which may or may not be an extension of the DAM component to lower frequencies.

The Saturnian SKR extends from below 100 kHz to 800 kHz. The repetition period is 10h 39.4m which is the inferred rotation period of Saturn's magnetic field. The probability of receiving SKR peaks when the Saturnian longitude 110° points toward the sun. Thus the source of SKR is not like a lighthouse which rotates a continuous beam but switched on and off.

The SED component extends over a very broad frequency range (from less than 100 kHz to above 60 MHz) and consists of very short broadband bursts of radiation. It is believed to be due to lightning in the equatorial atmosphere of Saturn. We will not discuss it further because it is not very intense and is only observable close to Saturn. In addition to these components, there is also an escaping continuum radiation at all three planets.[28] This radiation is also weak and cannot be detected at distances larger than several times the radius of the planetary magnetosphere. It is believed to be due to coupling of upper hybrid electrostatic waves to electromagnetic waves at density gradients.[28]

Carr et al.[2] point out that the local time and latitude coverage of the Jovian and Saturnian radiation is severely limited when compared to the earth's case where nearly the whole 4π sphere of the sky has been covered by satellite observations. It is quite conceivable that there are more components to the Jovian and Saturnian radiation which have escaped detection.

In Figure 3 we compare the average spectra of the three planetary radiations all normalized to a distance of 30 planetary radii from the center of the planet. Jupiter is the strongest radio source with Saturn and earth comparable in strength. The upper cutoff frequencies (40 MHz at Jupiter, roughly 1200 kHz at Saturn and roughly 700 kHz at the earth) vary roughly as the equatorial magnetic field strength of each planet (4 G at Jupiter, 0.4 G at Saturn, and 0.3 G at the earth) although the correlation is not linear.

The local time variation of the received power is well known for AKR. It is most powerful when observed from above the dusk to midnight sector. When observed from above the noon sector AKR power is down by two orders of magnitude. Neither the nKOM nor the DAM component seem to vary in power with local time of observation. However, bKOM is most intense when viewed from above the noon sector. SKR is also strongest when observed from this vantage point. However, in light of the limited local time coverage, these conclusions must be regarded as tentative.

It is interesting to note that some components whose power is local-time dependent are controlled by variations of the solar wind parameters, i.e., they are externally controlled. For example,

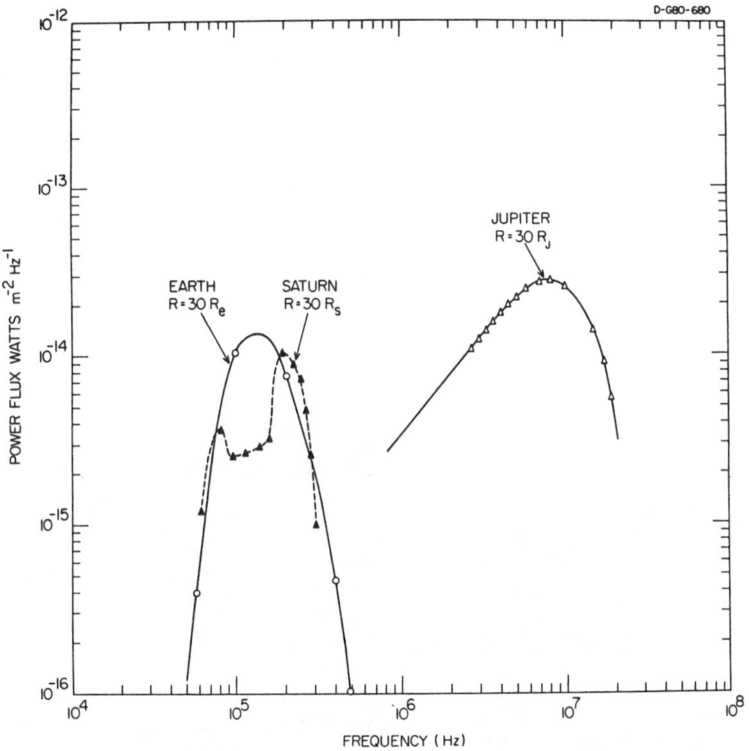

Fig. 3. The average spectra of the three known planetary radio waves.

Gallagher and D'Angelo[6] have shown that the 3-day averaged AKR intensity correlates best with the solar wind speed. Systematic investigations with higher time resolution have not been published, but may reveal better correlations with other solar wind parameters. Gurnett[7] has shown that AKR correlates well with the occurrence of discrete auroral arcs (Figure 4) and Voots et al.[8] correlate AKR with the auroral electrojet (AE) index. Auroral activity, however, is controlled by the direction of the interplanetary magnetic field[9] and the best AKR predictor (on short time scales) may have to involve the same parameter.

The case of solar wind control over the Jovian radiation is much less clear cut. The DAM component is so strongly controlled by Io (see below) that long-term modulations by the solar wind are difficult to detect, especially when the monitoring of DAM by one station is not continuous. In addition, the time delay between solar wind fluctuations observed at the orbit of the earth (1 AU) and those at Jupiter (5 AU) is not accurately known. These difficulties have been eliminated by the Voyager observations which have monitored the radiation continuously for several years. The projection of measured solar wind parameters at the spacecraft location to Jupiter is much more reliable. The Io control of the

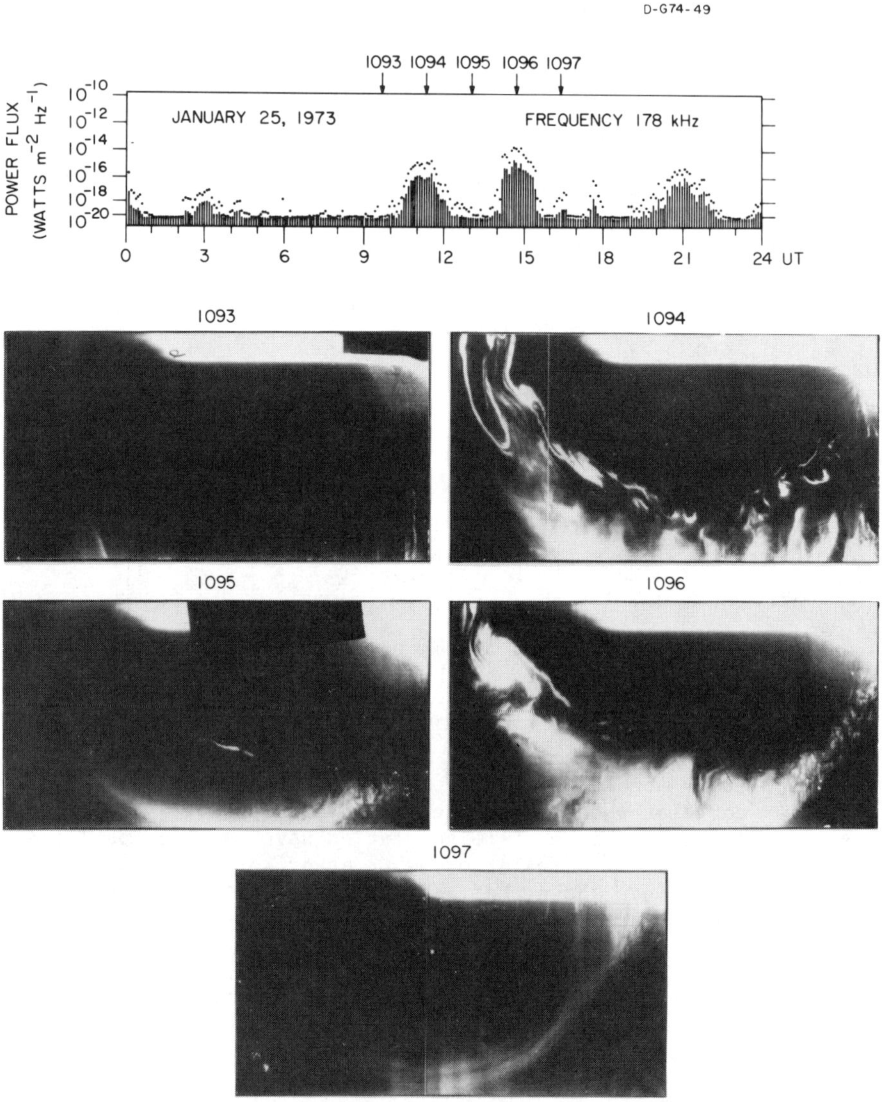

Fig. 4. The correlation with AKR intensity (top panel) with auroral activity (lower panel pictures). AKR is strong during periods of enhanaced auroral activity. This figure is from Gurnett.[7]

DAM is still difficult to take out, but the HOM component is not controlled by Io. Barrow and Desch[10] have found long-term intensity modulations of HOM which could be due to a solar wind control. The correlation with solar wind density is significant, that with solar wind speed is not. However, HOM is not only controlled by the solar wind because many changes in HOM intensity are unaccompanied by solar wind density changes and vice versa.

Saturn's SKR emission, on the other hand is strongly modulated by the solar wind ram pressure or density.[11] The most spectacular evidence for this control comes from a study of Desch[12] which shows that during the times Saturn was immersed in Jupiter's magnetotail SKR disappeared completely. During these episodes the plasma density at the spacecraft location (outside Saturn's magnetosphere) dropped by several orders of magnitude.[13,14]

Using these results Desch and Kaiser[15] have recently discovered a remarkable result. They have calculated the ratio of radio power in the solar wind controlled components (AKR, HOM, SKR) to solar wind power incident on the three planet's magnetosphere. This ratio is the overall efficiency of transforming the solar wind power into radiation power. All three efficiencies are very similar, namely 5×10^{-6}, even though the respective powers vary by more than a factor of 100. This remarkable and significant result suggests that the coupling and emission processes are very similar in all three cases.

The polarization of the received radiation is an important clue to the radiation mechanism. AKR is predominantly left-hand (LH) polarized when viewed from the northern hemisphere. This corresponds to a clockwise rotation of the electric field vector when the wave is approaching the observer. It has been suggested by Kaiser and Desch[4] that AKR observed from the south would be right-hand (RH) polarized. SKR's polarization is reversed from that of AKR[16] and does not seem to depend on the local time of the observer. Since the orientation of Saturn's magnetic dipole is opposite to that of the earth in both cases the polarization indicates that the received radiation is in the extraordinary (X) mode. However, this does not prove that it is generated in this mode because the polarization may switch as the wave propagates through the anisotropic plasma and mode coupling between the ordinary (O) mode and X-mode may occur before the radiation is received. The highest frequency part of DAM is strongly RH polarized. This radiation is believed to originate in the northern hemisphere of Jupiter where the magnetic field is pointing toward the observer. Thus this radiation also appears to be in the X-mode. The bKOM associated with the northern hemisphere is LH when viewed from the dayside and RH when viewed from the nightside. The nKOM is LH when the observer is north of the Jovian current disk and RH when the observer is in the southern lobe of the Jovian magnetotail. The polarization seems to be a function of the observer's local time, also.[4] Intermediate frequency (1 - 10 MHz) radiation has a mixed polarization. Table 1 summarizes the observed properties of planetary radio waves.

Table I Summary of planetary radio wave observations

	EARTH AKR	JUPITER bKOM	nKOM	DAM	SATURN SKR
Frequency Range	50-700 kHz	20-1000 kHz	50-175 kHz	0.5-40 MHz	3-1200 kHz
Average Power	30 MW	800 MW	100 MW	30 GW	1 GW
Polarization	LH	LH	LH	RH	RH
Solar Wind Control	Yes	?	No	No	Yes

IO-CONTROL OF DAM

One of the most fascinating aspects of DAM is the apparent control the innermost Galilean satellite, Io, has over it. Bigg[17] showed that the probability of receiving DAM at the earth is not only a function of the longitude facing the observer, the so-called central meridian longitude (CML), but also of the angle between Io's position and its superior geocentric conjunction (γ_I). The center panel of Figure 5 is a typical occurrence probability plot. Increasing probability is indicated by increasing darkness. The regions of enhanced probability are called sources. We see that the probability is generally larger in the CML range from 200° to 270°. This is the A source. The region from 105° to 180° is called the B source and the region centered on CML = 340° is the C source. It is also clear that the B source is most active when Io is on the dawn side of Jupiter ($\gamma_I \sim 100°$) whereas the A and C sources are most active when Io is near dusk ($\gamma_I \sim 240°$). Thus one distinguishes between, for example, Io-A ($\gamma_I \sim 240°$) and non Io-A sources. Voyager observations indicate that the radiation is always observed in these CML source ranges, however, at a variable intensity. For less sensitive earth-based receivers the probability may appear to be low due to a detection threshold effect. It has been suggested by Gurnett and Goertz[18] that the A, B, and C sources are always active. However, when Io is at either $\gamma_I \sim 100°$ or $\gamma_I \sim 240°$ the source intensity is enhanced. This would suggest that the distinction between Io and non-Io sources is somewhat artificial. It has been proposed by several authors (see e.g., Goertz[19]) that Io triggers the radiation in the Jovian ionosphere about 15° ahead of it, i.e., at a longitude 15° less than the sub-Io longitude. The B source is then triggered at a longitude of

205° whereas the A source is triggered at 140° and the C source at 245°. To aid the reader, we have drawn lines of constant sub-Io longitude in Figure 5. The triggering is symmetric about the longitude of 200°. This is the longitude to which the northern tip of the magnetic dipole is tilted.[20] It is also interesting to note that this is the dividing CML value between vertex-early and vertex-late arcs.

The most widely accepted interpretation of these findings assumes that the radiation is emitted into a hollow cone as shown in Figure 6. The radiation is observed when the observer (usually in the solar ecliptic plane) is situated along the line of intersection between the cone and the ecliptic plane. A single source can be observed from two different longitudes. The variation of the source longitudes with the declination of Jupiter is consistent with this interpretation.[21] Goertz[19] has pointed out that the B source really consists of two slightly separated sources B_1 and B_2. Thus Io triggers two sources which are more widely separated when Io is in the afternoon sector (A and C) than when it is in the morning sector (B_1 and B_2 sources).

RADIATION MECHANISMS

The theoretical problem of planetary radio waves can be summarized as follows: By what sporadic mechanism is the plasma distribution function $f(\vec{x}, \vec{v}, t)$ in the source region modified so that its free energy can be converted efficiently into escaping electromagnetic waves. Unfortunately, except for the case of AKR, we have no in-situ measurements of the distribution function. Several different mechanisms, each requiring a different source distribution function, predict similar observables. Thus the various theories are all viable even though some are more plausible.

The fact that the cut-off frequencies scale with the planetary magnetic fields and that the frequency spectra are repeatable over many planetary rotations has suggested to many that the radiation is emitted near the local electron gyrofrequency. Polarization characteristics indicate that AKR, SKR, and DAM are in the X-mode. The high intensities require coherent processes.

The various theories can be grouped into four categories. They either deal with "direct" or "indirect", "linear" or "non-linear" processes. Direct refers to particle-generated waves whereas indirect processes involve wave-wave interactions (mode conversion). Linear theories use only the linearized Vlasov equations whereas non-linear theories use the full instrumentarium of modern non-linear plasma physics. For a review of the theories see Goldstein and Goertz[3] and Grabbe[22] Table 2 summarizes the four types of radiation mechanism and gives the references to the first papers. In the following we will only describe the most widely discussed linear direct maser mechanism of Wu and Lee.[23]

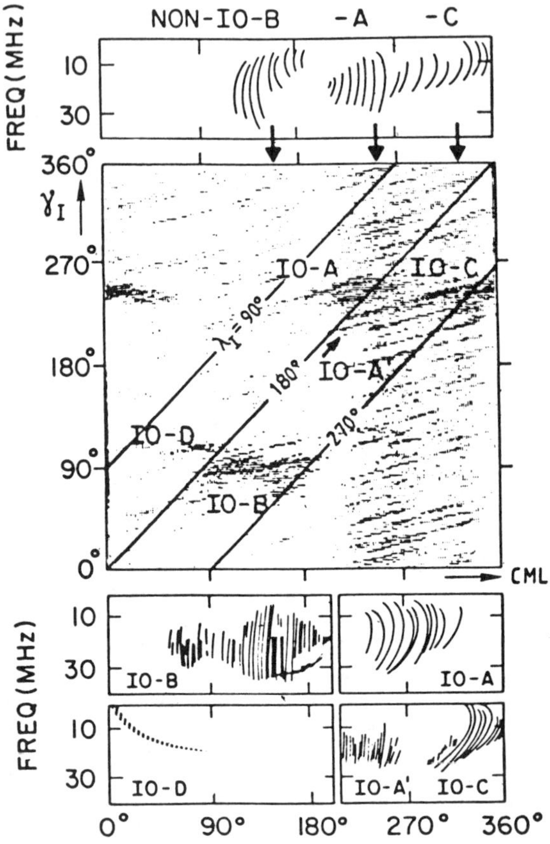

Fig. 5. Occurrence probability of the Jovian DAM radiation. At the top the spectral characteristics (e.g., arcs) of each source are displayed.

In a plasma with free energy stored in thermal energy, ordered motion, anisotropies, potential mechanical energy, electrostatic energy, etc, a multitude of instabilities can occur when the free energy exceeds a critical level. A small perturbation of the electromagnetic fields grows exponentially in time ($\sim \exp(\gamma t)$). When the growth rate, γ, is positive one speaks of coherent or stimulated emission. Several authors have derived equations for the growth rate. Here we use the general expression given by Melrose[24] for a homogeneous, magnetized, gyrotropic plasma

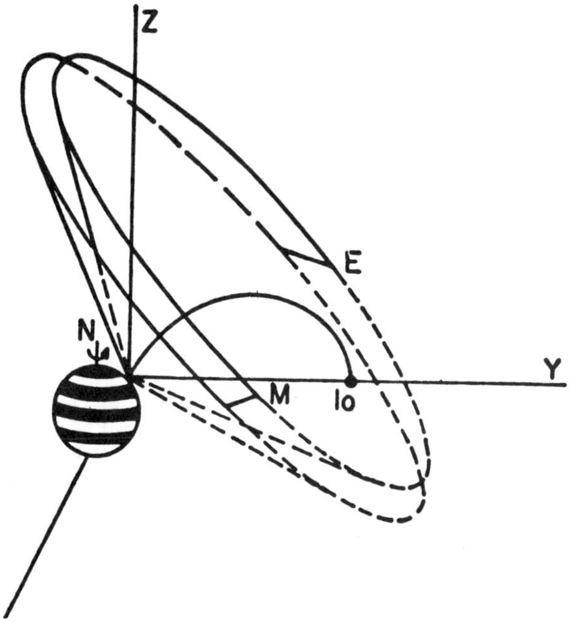

Fig. 6. The hollow cone radiation model. The cone intersects the ecliptic at M and E.

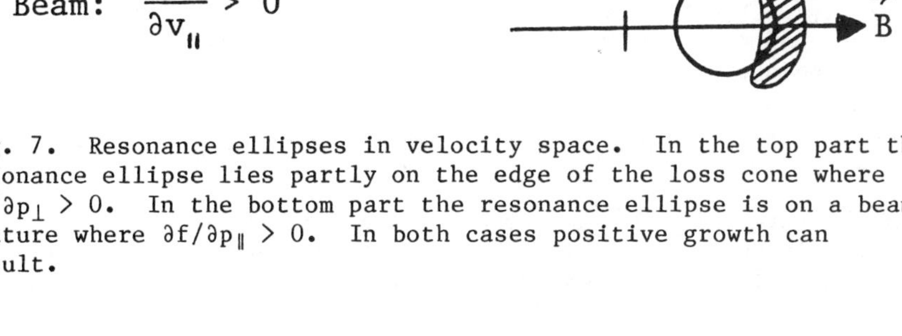

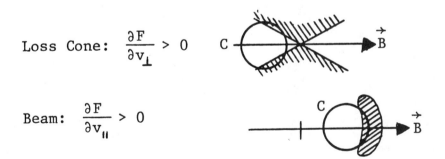

Fig. 7. Resonance ellipses in velocity space. In the top part the resonance ellipse lies partly on the edge of the loss cone where $\partial f/\partial p_\perp > 0$. In the bottom part the resonance ellipse is on a beam feature where $\partial f/\partial p_\parallel > 0$. In both cases positive growth can result.

$$\frac{\gamma^{\pm}}{\omega}(k) = \sum_{\nu} \frac{2\pi^2 q^2 c^2}{\omega^2 |k_{\parallel}|} k_{\parallel} n^{\pm} \int dp_{\parallel} \int dp_{\perp} \, p_{\perp}^2 \, \delta(p_{\parallel} - p_{\parallel o}) \times \frac{\theta_{\nu}^{\pm} G_{\nu}^{\pm}}{(m^2 c^4 + p^2 c^2)^{1/2}} \tag{1}$$

where the ± sign refers to contributions from the ions (+) and electrons (-). The various terms in equation (1) are defined as follows.

$$\theta_{\nu}^{\pm} = |J_{\nu \pm 1}(z) E_R + J_{\nu \mp 1} E_L + (2^{1/2}) \frac{v_{\parallel} k_{\perp}}{v_{\perp} |k_{\perp}|} J_{\nu}(z) E_{\parallel}|^2 / W(k_{\perp} k_{\parallel}) \tag{2}$$

is a measure of the wave polarization indicated by the subscripts on the electric fields. R and L stand for right-handed or left-handed electromagnetic modes and the subscript $\parallel$ refers to the longitudinally polarized waves. The wave energy density is

$$W = (|E(\bar{k})|^2 \omega \, \partial\Lambda / \partial\omega / 4\pi\lambda_{ss}) \tag{3}$$

and the function θ is independent of the wave amplitude. The trace of the conductivity tensor Λ and the dispersion tensor λ_{ss} are defined by Melrose.[25] The argument of the Bessel function is

$$z = k_{\perp} v_{\perp} / \Omega \quad . \tag{4}$$

The wave vector $\bar{k}$ has a component $k_{\parallel}$ parallel to the background magnetic field and $\bar{k}_{\perp}$ perpendicular to it.

The sign of the growth rate is determined by the distribution function through

$$G_{\nu}^{\pm} = k_{\parallel} p_{\perp} \frac{\partial f}{\partial p_{\parallel}} + (\omega \gamma_r m^{\pm} - k_{\parallel} p_{\parallel}) \frac{\partial f}{\partial p_{\perp}} \tag{5}$$

where γ_r is the relativistic Lorentz factor ($\gamma_r = (1 - v^2/c^2)^{-1/2}$). The resonance condition is expressed by the δ function in (1) with the resonance momentum

$$p_{\parallel o} = \gamma_r^{\pm} m^{\pm} \beta_{\parallel o} C = [1 - \nu\omega_c^{\pm}/\gamma_r \omega](\omega/k_{\parallel}) m^{\pm} \gamma_r^{\pm} \quad . \tag{6}$$

The interpretation of the resonance condition is straight-forward. A particle is resonant with the wave if the Doppler-shifted wave frequency (i.e., the frequency seen in the particle's frame) equals ν times its relativistic gyrofrequency.

$$\omega - k_\parallel v_{\parallel o} = \nu\, \omega_c^\pm / \gamma_r^\pm \quad . \tag{7}$$

The integer ν indicates the change of perpendicular energy upon emission of a photon. For ν = 0 the change is zero. We see from (2) that for $k_\perp = 0$ (z = 0) ν = 0 can only lead to emission of electrostatic waves. This resonance is called Landau resonance. Positive values of ν correspond to a decrease of $p_\perp$. This resonance is called the normal gyroresonance. Only the RH mode can be generated (ν = +1). Negative values of ν correspond to an increase of $p_\perp$ and are called anomalous gyroresonances. Only the LH mode can be generated (ν = -1). Using the resonance condition (6) and (5) we can write

$$G_\nu^\pm = k_\parallel p_\perp \left.\frac{\partial f}{\partial p_\parallel}\right|_{p_\parallel = p_{\parallel o}} + \nu\, \omega_c^\pm m^\pm \left.\frac{\partial f}{\partial p_\perp}\right|_{p_\parallel = p_{\parallel o}} \quad . \tag{8}$$

For growth we need G > 0 somewhere along the resonance contour in velocity space. Wu and Lee[23] point out that the relativistic gyroresonance condition ((6) or (7)) describes an ellipse in velocity space as shown in Figure 7. (For $\gamma_r = 1$ the contour would be a straight line parallel to the $v_\perp$ axis.) This relativistic feature is most important for ω close to ω_c and $k_\parallel \ll k_\perp$, i.e., perpendicular propagation. Two possibilities for gyroresonance growth exist. Either the distribution has a loss-cone feature (top of Figure 4) or a beam-like feature. In both cases, the integral over G can be positive. Landau resonance requires a beam-like feature ($\partial f/\partial p_\parallel > 0$) for growth. Parallel propagating waves can also be amplified by gyroresonance. However, it is believed that the radio waves are generated at high latitudes. Parallel propagating waves would then not be beamed towards an observer located in the ecliptic. Thus amplification of parallel propagating waves ($k_\parallel \gg k_\perp$) is not considered to be a viable radiation mechanism. Amplification of the nearly perpendicular propagating fast X-mode ($\omega > \omega_{ce}$) is the most frequently suggested radiation mechanism. This agrees with the observed polarization of AKR, SKR, and DAM. The emission geometry would be a thin conical sheet which would account nicely for the sources of DAM. The fast X-mode can escape freely along a path of decreasing magnetic field strength. This would also be true of the O-mode (which has the opposite polarization). The slow X-mode ($\omega < \omega_{ce}$) could not freely escape because it is absorbed when its frequency becomes equal to the local electron gyrofrequency for parallel propagation or the local upper hybrid frequency for perpendicular propagation.

The existence of this stop band for the X-mode has lead Oya[25] to postulate a generation mechanism which produces the slow X-mode by coupling between beam excited upper-hybrid resonance electrostatic waves and plasma inhomogeneities. The slow X-mode then couples to the escaping O-mode at a point where its frequency equals the local plasma frequency. This requires very high plasma frequencies to account for the high frequency parts of DAM, i.e., large plasma densities in the source region (> $10^7 cm^{-3}$) which are difficult to accept. In addition, the received radiation seems to be in the X-mode and not in the O-mode. However, this is not a firm conclusion for DAM because it is not certain that the source is in the northern hemisphere. If it were in the southern hemisphere, the observed radiation would have to be in the O-mode. Oya's mechanism is an example of the indirect linear mechanisms.

A critical question for all linear mechanisms is the magnitude of the growth rate. Since the source size is limited the waves are amplified only for a finite time. It is at present not clear whether the linear growth rates calculated on the basis of observed distribution functions in the earth's auroral regions where AKR is generated are sufficient. For DAM and SKR this is a completely open question because the distribution functions are not known in the source regions.

Non-linear mechanisms may be required. For a discussion of these we refer the reader to Goldstein and Goertz.[3] Usually they involve the generation of electrostatic waves by electron beams that in turn produce radio waves by three-wave coupling mechanisms. Since several potentially inefficient processes are involved, the overall efficiency may also be a problem.

Table II Summary of generation mechanisms

	Linear Direct	Linear Indirect	Non-Linear Direct	Non-Linear Indirect
Process	Maser	Mode coupling between slow x → 0	EIC + beam → fast x	Beam → ES Waves → Fast x + 0
Free Energy	Loss Cone (beam)	Beam	Beam	Beam
References	23	25	22	29

COUPLING MECHANISMS

The low frequency modulation of planetary radio waves indicates that they are either coupled to the solar wind (AKR, SKR, HOM) or the moon Io (DAM) or the Io plasma torus (nKOM). Theoretical considerations of generation mechanisms indicate that electron beams or loss cone features in the distribution function are required to exist in the source region. These two possibilities may not be different because the back scattered albedo populations of a downgoing beam would have a loss cone distribution. In any case the generator must provide a slowly varying energy flux into the source region. This can either be in the form of particles precipitating into the loss cone or in the form of a low frequency MHD wave. Particle precipitation would be mainly into regions of low magnetic surface field, i.e., the southern hemisphere at Jupiter. DAM, however, is believed to come from the high field regions in the north. Precipitation by pitch angle scattering would not produce a beam and decrease the sharpness of a loss cone distribution, i.e., reduce the free energy in the distribution function. Thus it is not believed to provide an adequate explanation. For a variety of reasons the most widely discussed coupling scenario involves field aligned Birkeland currents and field aligned electric fields which accelerate electrons and enlarge the atmospheric loss cone. The mechanisms by which field aligned electric fields can be maintained are subject to intense research efforts in space physics. Some of those which are believed to be active in the earth's auroral zone where AKR is generated are discussed by Kennel[26] and Baker.[9] A particular coupling model for the Io effect is discussed by Goertz.[27]

SUMMARY

Considerations of the generation mechanisms reveal that almost any anisotropic distribution of energetic charged particles can be unstable against the emission of the escaping fast X-mode. The most promising direct maser mechanism emphasizes a loss cone distribution. Indirect mechanisms, usually require a beam distribution. It is unlikely that we will ever know much more about the distribution functions in the source region of SKR, DAM and KOM. The similarities between these emissions and AKR suggests that an understanding of AKR will help to explain the other planetary radio waves. Obviously, such a view cannot be defended rigorously. Also, even for AKR there is no generally accepted theory, although the maser mechanism seems to be most promising. A firm understanding of the maser (or any other coherent radiation) mechanism may be very helpful to astrophysicists when considering emission from RSCVN stars.[30]

Finally, the question of how much planetary radio waves may teach us about astrophysical processes is a relevant one for this conference. I believe that planetary radio waves are fundamentally different from most astrophysical radio waves because they, in general, require coherent collective plasma processes. Since there are no strong indications of second harmonic emissions, the energy of the emitting electrons must be relatively low. The much lower spectral resolution of astrophysical observations does not allow a definitive test for the presence of coherent radiation from astrophysical sources although it is difficult to believe that it should not be significant. A systematic search for coherent astrophysical radio sources should be undertaken. Even though the energy of the radiating electrons is small in planetary radio sources the emitted power is large. To maintain these emissions, generators need to produce this power. It seems that the final energy source is either the energy (kinetic, thermal and electromagnetic) stored in the solar wind or in the moon, Io. The generator mechanism produces field aligned Birkeland currents which may dissipate their energy in regions where ($\overline{E} \cdot \overline{j} > 0$), i.e., where field aligned electric fields exist. Thus the question of how and where field aligned currents are produced in astrophysical radio sources is an important one. However, field aligned currents alone do not necessarily produce distribution functions with free energy. Parallel electric fields are also essential and ideal MHD ($\overline{E} + \overline{v} \times \overline{B} = 0$) is not sufficient to describe the physics of planetary radio source. It would be very surprising if it would suffice in the more complex astrophysical radio sources. Finally, the distribution functions in the source region cannot be described by simple parameters such as density and temperature. Higher order moments of the distribution function are most important and assumptions such as local thermal equilibrium are quite irrelevant in planetary radio sources.

ACKNOWLEDGEMENTS

This work was supported by NASA grants NAGW-386, NAGW-364 and NGL-16-001-043.

REFERENCES

1. Smith, R. A., Jupiter, (ed. T. Gehrels, Univ. Arizona Press, 1979), p. 1146.
2. Carr, T. D., M. D. Desch, and J. K. Alexander, Physics of the Jovian Magnetosphere, (ed. A. Dessler, Cambridge Univ. Press, 1983), p. 226.
3. Goldstein, M. L., and C. K. Goertz, Physics of the Jovian Magnetosphere, (ed. A. Dessler, Cambridge Univ. Press, 1983), p. 317.
4. Kaiser, M. L., and M. D. Desch, NASA Technical Mem. 83863 (1984).
5. Warwick, J. W., J. B. Pearce, R. G. Pelcher, and A. C. Riddle, Space Sci. Rev. 21, 309 (1977).
6. Gallagher, D. L. and N. D'Angelo, Geophys. Res. Lett. 8, 1087 (1981).
7. Gurnett, D. A., J. Geophys. Res. 79, 4227 (1974).
8. Voots, G. R., D. A. Gurnett, and S.-I. Akasofu, J. Geophys. Res. 81, 2259 (1977).
9. Baker, D., this volume.
10. Barrow, C. H., and M. D. Desch, Geophys. Res. Lett., in press, (1984).
11. Desch, M. D., and H. D. Rucker, J. Geophys. Res. 88, 8999 (1983).
12. Desch, M. D., J. Geophys. Res. 87, 4549 (1982).
13. Lepping, R. P., L. F. Burlaga, M. D. Desch, and L. Klein, Geophys. Res. Lett. 9, 885 (1982).

14. Kurth, W. S., J. D. Sullivan, D. A. Gurnett, F. L. Scarf, H. S. Bridge, and E. C. Sittler, J. Geophys. Res. 87, 10,373 (1982).
15. Desch, M. D., and M. L. Kaiser, Nature, submitted (1984).
16. Warwick, J. W., et al., Science 212, 239 (1981).
17. Bigg, E. K., Nature 203, 1008 (1964).
18. Gurnett, D. A., and C. K. Goertz, J. Geophys. Res. 86, 717 (1981).
19. Goertz, C. K., Adv. Space Res. 3, 59 (1983).
20. Acuña, M. H., K. W. Behannon, and J. E. P. Connerney, Physics of the Jovian Magnetosphere (ed. A. Dessler, Cambridge Univ. Press, 1983), p. 1.
21. Goertz, C. K., Nature 229, 151 (1971).
22. Grabbe, C. L., Rev. Geophys. Space Sci. 19, 627 (1981).
23. Wu, C. S., and L. C. Lee, Astrophys. J. 230, 621 (1979).
24. Melrose, D. B., Astrophys. Space Sci. 2, 171 (1968).
25. Oya, H., Planet. Space Sci. 22, 687 (1974).
26. Kennel, C. F., this volume.
27. Goertz, C. K., this volume.
28. Kurth, W. S., D. A. Gurnett, and R. R. Anderson, J. Geophys. Res. 86, 5519 (1981).
29. Ben-Ari, M., Ph.D. Thesis, Tel-Aviv University, Tel-Aviv (1980).
30. Dulk, G. A., Astrophys. J. 273, 249 (1983).

AN OSCILLATORY INSTABILITY OF INTERSTELLAR MEDIUM RADIATIVE SHOCK WAVES

James N. Imamura
Los Alamos National Laboratory, Los Alamos, NM 87545

ABSTRACT

Observations of the radiative shock waves produced during the late stages of supernova remnant evolution cannot be understood in the context of steady state shock models. As a result, several more complicated scenarios have been suggested. For example, it has been proposed that several shocks are producing the emission or that one shock, which is in the process of making the transition between the adiabatic and the radiative phases of its evolution, produces the emission. In this paper, we suggest another explanation. We propose that supernova remnant shock waves are subject to an oscillatory instability. By an oscillatory instability, we mean one where the postshock cooling region periodically varies in size on a time scale determined by the postshock plasma cooling time. An oscillatory instability may be able to produce the types of behavior exhibited by supernova remnant radiative shocks in a natural way.

INTRODUCTION

A radiative shock wave is a shock in which the cooling time scale of the plasma downstream of the shock transition is less than the characteristic flow time of the postshock plasma. The thickness of the transition region is on the order of the particle mean free path, λ_p, while the thickness of the cooling region, λ_{cool}, is determined by the postshock velocity and cooling time of the plasma. Such a shock is composed of three regions: (1) a precursor in which the inflowing plasma can be heated, ionized, or dissociated by an energy flux from behind the shock; (2) a transition region in which the bulk kinetic energy of the incoming plasma is converted into thermal motions; and (3) a more extended region in which the shock-heated plasma radiates away its internal energy relaxing to its final state. In general, $\lambda_p \ll \lambda_{cool}$ and the ionization, dissociation, or heating length scale of the precursor. Because of this, most models of radiative shock waves take the transition region to be a discontinuity and only model the cooling region and precursor in detail.

Radiative shock waves arise in a wide variety of interesting astrophysical situations. For instance: (1) they can occur in X-ray binary systems[1,5]. In these cases, plasma flows from the "normal" star in the binary system to its companion, a compact object (e.g., a white dwarf or neutron star). If the accreting plasma approaches the compact object radially or approximately radially, it forms a stand-off shock and then cools as it settles onto the surface of the object. (2) Radiative shocks can occur during the late stages of supernova remnant evolution[3,10].

Immediately after the initial supernova outburst,[1] the ejecta from the outburst expand with a velocity of ~ 10^4 km s^{-1}; sweeping up the surrounding interstellar medium. When the amount of swept up mass is small, the ejecta move at constant velocity. This "free" expansion continues until the ejecta have interacted with an amount of mass approximately equal to their own, after which time they are strongly decelerated and assume a structure which is quite nicely modeled as a Sedov blast wave. The blast wave phase lasts until the postshock cooling time scale becomes less than the postshock flow time scale. This occurs around the time that the shock velocity has dropped to two hundred kilometers per second. The cooling leads to the formation of a dense shell near the outer edge of the remnant which, driven by the hot plasma in the interior of the remnant, acts like a piston driving a radiative shock wave into the interstellar medium. (3) Radiative shock waves occur during the interaction of stellar winds and the interstellar medium. If the shock velocities are sufficiently low, the plasma cooling time scale will be shorter than the flow time scale and radiative shocks will be produced. Examples of this are the Herbig-Haro phenomenon[14] and the late stages of the evolution of stellar wind blown bubbles[16].

In this work, we concentrate our remarks to the radiative shocks produced by supernova remnants, but our results also apply to the shocks produced by the interaction of stellar winds and the interstellar medium.

The structure and appearance of steady interstellar medium radiative shock waves have been the subjects of several extensive theoretical efforts in the last decade[10]. However, when the results of the calculations are compared to observations, the agreement is not always good. For example, the observations of the old supernova remnants, Vela and the Cygnus Loop, cannot be fit by unique steady state shock models. The observations require that time-dependent phenomena be present or that there be several shocks with a range of shock velocities present[12]. The resolution of this problem is not readily apparent. Because of this, it is of interest to consider effects which could lead to such scenarios. Instabilities of the postshock cooling regions of radiative shock waves are good candidates for such processes.

The stability properties of the cooling regions of radiative shock waves are incompletely understood. Two basic types of instabilities have been considered. There are thermal instabilities[2,9], that is, instabilities which occur at approximately constant pressure. Such instabilities tend to lead to "clumping" of the cooling plasma through the amplification of preshock density fluctuations. They mainly affect the observable features of radiative shock waves; they do not affect the shock dynamics. There are also "oscillatory" types of instabilities in which cooling and dynamical processes both play major roles[4-8]. In oscillatory instabilities, the size of the cooling region varies in size with a characteristic oscillation period on the order of the postshock plasma cooling time scale. Whether the shock is unstable to oscillatory motions is determined by the form of the plasma cooling function. We consider oscillatory instabilities in this paper. There have not been any detailed calculations of time-dependent interstellar medium radiative shock waves and thus, there are no direct demonstrations that they are unstable to oscillatory motions. However, studies of the stability properties of radiative shock waves using power law cooling functions proportional to $\rho^2 T^\alpha$, have suggested that interstellar medium shocks may be unstable[4].

The rest of this paper is organized as follows. In Section II, the qualitative nature of the instability is discussed and the results of the linear and nonlinear stability analyses are presented, and in Section III, interstellar medium shock waves are discussed in the context of oscillatory instabilities.

SHOCK STABILITY PROPERTIES

A. Qualitative Picture

The oscillatory instability of radiative shock waves can be understood by considering planar shock waves. Consider the situation where plasma flows from the $x = +\infty$ direction along the x-axis towards a stationary wall situated in the y-z plane at $x = 0$. The plasma has a velocity $-v_{in}$, density ρ_{in}, and pressure P_{in} (=0). This highly supersonic plasma forms a strong shock at a distance x_s from the wall and then cools as it settles onto the wall. The characteristic time scale of the shocked plasma is the postshock plasma cooling time, τ_{cool}, defined as the ratio of the internal energy of the postshock plasma to the emissivity of the postshock plasma. The distance x_s is of the order of the product of the absolute value of the postshock plasma velocity and τ_{cool}. Whether a perturbation of the position of the shock front (and therefore of the shock velocity) grows or damps depends upon the form of the plasma cooling function. To see this, consider a power law cooling function of the form $\Lambda_3 = \Lambda_0 \rho^\beta T^\alpha$, where Λ is the plasma emissivity in units of ergs cm^{-3} s^{-1}, Λ_0 is a constant, T is the temperature, and α and β are constants. Consider the fluid in a frame in which the shock velocity would be zero if it were in equilibrium, but allow for a small nonzero shock velocity, v_s. The cooling time scale and postshock velocity are then found to be

$$\tau_{cool} = \frac{\text{Internal Energy of the Plasma}}{\text{Plasma Emissivity}}$$

$$= \tau_o \frac{(v_{in}+v_s)^{2(1-\alpha)}}{\rho_{in}^{\beta-1}} \qquad (1)$$

and

$$v_{post} = -\frac{v_{in} - 3v_s}{4} \qquad (2)$$

Here $\tau_o = (9/8)\,(16k/3\mu m_o)^{\alpha}\,4^{-\beta}\,\Lambda_o^{-1}$, μ is the mean molecular weight, m_o is the atomic mass unit, and v_{post} is the postshock velocity. Defining $\lambda_{cool} = 0.25|v_{post}|\,\tau_{cool}$ and assuming that $|v_s/v_{in}| << 1$, we have

$$\lambda_{cool} = \tau_o\,\frac{v_{in}^{3-2\alpha}}{4\rho_{in}^{\beta-1}}\,[1-(1+2\alpha)\frac{v_s}{v_{in}}]$$

$$= \lambda_s + \delta\lambda. \qquad (3)$$

Here λ_s is the steady shock thickness, $\delta\lambda$ is the perturbation of the shock thickness. Equation (3) shows that if $\alpha \gtrsim -1/2$, λ_{cool} is smaller than its equilibrium value if $v_s > 0$ (i.e., if the shock front is perturbed away from the wall), and is larger than its equilibrium value if $v_s < 0$ (i.e., if the shock front is perturbed towards the wall). Thus, one expects that the cooling region will be stable against small perturbations to the shock front position if $\alpha > -1/2$. However, because this estimate ignores the detailed structure of the cooling region, it does not yield the quantitatively correct value of the critical temperature exponent α or the dependence of the stability properties on the exponent of the density dependence β. It does indicate, however, the qualitative effect of varying the temperature exponent α on the stability of radiative shock waves, that is, for large values of α, radiative shock waves are expected to be stable against oscillatory motions.

B. Detailed Calculations

Linear and nonlinear studies of radiative shock waves with cooling functions $\Lambda = \Lambda_o\rho^{\beta}T^{\alpha}$ have been carried out. Values of β = 1 and 2, and values of α ranging from -2 to 2 have been considered. All calculations assumed that the electron and ion temperatures were equal, that viscosity and electron thermal conduction were negligible, and that geometrical effects were small. For interstellar medium shocks, these assumptions are usually justified, however, for radiative shocks produced by accretion onto compact objects, these assumptions break down in several situations[6,15]. The details of our linear calculations can be found in Refs. 4 and 5, and the details of our nonlinear calculations can be found in Refs. 5 and 6. In general, there is very good quantitative agreement between the linear and nonlinear analyses (however, see Ref. 6 for a discussion of the slight differences in the nonlinear calculations). Because of this, only the results of the linear analyses for the onset of instability in terms of the α value and the oscillation frequencies are presented.

Radiative shock waves are found to be capable of oscillating in several distinct modes, which are called, in order of increasing oscillation frequency, the fundamental (F), the first overtone (1O), the second overtone (2O), and so on. For $\beta = 2$ cooling functions, i.e., cooling functions where the loss processes are due to particle collisions, such as bremsstrahlung, collisionally excited line radiation, etc., the F, 1O, and 2O modes are stable for $\alpha \gtrsim 0.4$, 0.8, and 0.8, respectively. For $\beta = 1$ cooling functions, i.e., cooling functions where the loss processes are due to single particle processes such as Compton or cyclotron cooling, the F, 1O, and 2O modes are stable for $\alpha \gtrsim 0.05$, 0.14, and 0.04, respectively. For both types of cooling functions, the oscillation frequencies are $\sim$ 0.3-0.4 (v_{in}/x_s) for the F mode, $\sim$ 0.6-1.0 (v_{in}/x_s) for the 1O mode, and $\sim$ 1.3-1.5 (v_{in}/x_s) for the 2O mode. Note that (v_{in}/x_s) is $\propto 1/\tau_{cool}$ showing that the oscillation periods are on the order of the postshock plasma cooling time scale.

In general: (1) the larger the value of α, i.e., the stronger the temperature dependence of the cooling function, the more stable the shocks are likely to be; and (2) the smaller the value of β, i.e., the weaker the density dependence of the cooling function, the more stable the shocks are likely to be.

Examples of the behavior of radiative shock waves in the nonlinear regime are presented in Figure 1. The the shock luminosity, L_s, as a function of time, is presented for $\beta = 2$ and $\alpha = 1$, 1/2, and 1/3 models. The linear analysis predicts that the F mode will be stable for $\alpha \gtrsim 0.4$, and that the 1O and 2O modes will be stable for $\alpha \gtrsim 0.8$. The nonlinear analysis is in good agreement with this prediction.

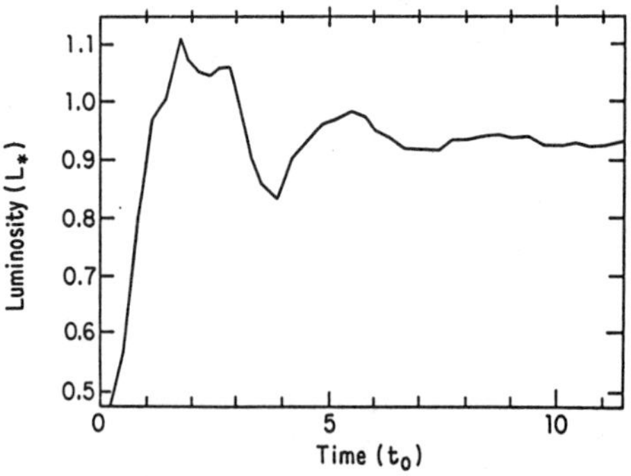

Figure 1a: the shock luminosity as a function of time for a power law cooling function of the form $\Lambda \propto \rho^2 T^1$. The luminosities are in units of the average luminosity and the times are in units of $2\pi(x_s/v_{in})$.

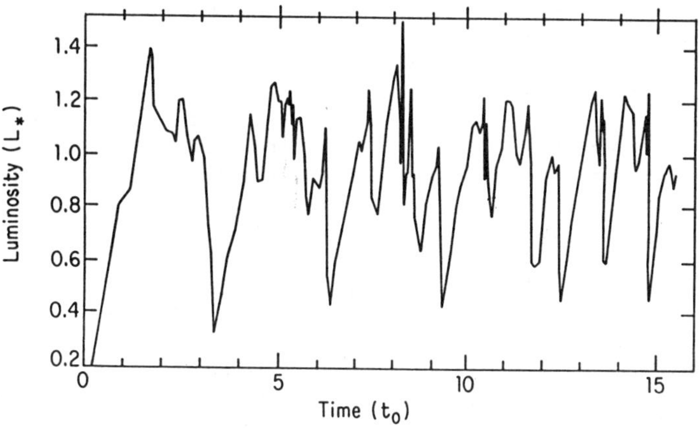

Figure lb: Same as Figure la, except that the cooling function has the form $\Lambda \propto \rho^2 T^{1/2}$.

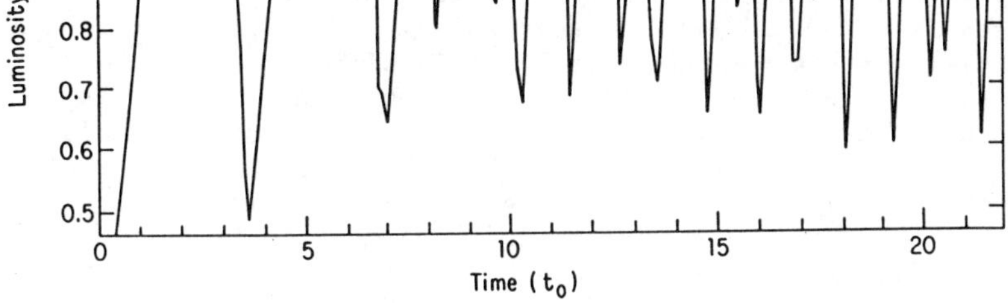

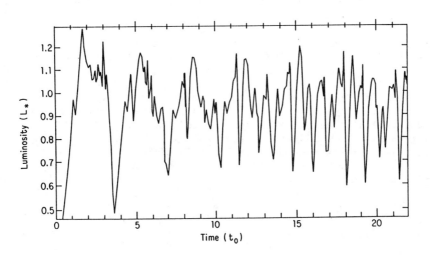

Figure lc: Same as Figure la, except that the cooling function has the form $\Lambda \propto \rho^2 T^{1/3}$.

INTERSTELLAR MEDIUM RADIATIVE SHOCK WAVES

The oscillatory instability may occur in interstellar medium radiative shock waves even though there isn't a fixed surface onto which the plasma flows. In the interstellar medium, radiative shocks are driven by a cold, dense shell of plasma whose motion can be assumed to be constant as the cooling time of the plasma is short compared to the time over which the layer evolves. This is the assumption normally made in the study of interstellar medium shock waves and is equivalent to the statement that the column density of the swept up matter is large compared to the column density through the cooling region. The application of the oscillatory stability results to interstellar medium shocks requires that the perturbed velocity go to zero at the swept up layer. However, it is not *a priori* obvious that the pressure perturbation also goes to zero at the swept up layer. The effect of requiring the pressure perturbation to go to zero at the dense layer, rather than the velocity perturbation, was checked by a linear analysis. The change in the boundary condition had no effect on the onset of instability or the oscillation frequencies.

The cooling losses of interstellar plasma are primarily due to particle collisions and thus the $\beta = 2$ results are the relevant ones. In general, however, interstellar medium cooling functions are not a single power law in T and so the results of the stability analyses just presented are not strictly applicable[11]. However, because most of the radiated energy comes from plasma at temperatures close to that found near the shock wave, only requiring that the cooling function be a power law for temperatures around the shock temperature may be sufficient. Using the equilibrium interstellar medium cooling function[13], the emissivity can be approximated as $\Lambda \propto \rho^2 T^{0.55}$ for temperatures in the range 10^5 to 4×10^7 K. Thus, for these shock temperatures, we expect that interstellar medium radiative shock waves will be unstable. However, because nonequilibrium ionization effects substantially alter the cooling function[11], this result is only suggestive. Detailed time-dependent numerical hydrodynamic calculations are needed to verify this claim.

Comparisons of the spectral observations of supernova remnants with shock wave emission models generally yield shock velocities in the range 90-140 km s^{-1} [12]. The oscillatory instability may occur in this regime, but the time scale is too long to observe time variations. The cooling time scale for the interstellar medium case is on the order of hundreds of years. Thus, the instability should thus manifest itself as deviations of the observed spectra from those predicted by the steady shock models. There is a great deal of evidence for such effects, particularly when UV data are combined with optical data[12]. Other evidence is the high [O III]/Hβ line ratio observed in parts of many supernova remnants. Raymond *et al.*[12] argue for nonsteady emission and consider three possible explanations. The first is the thermal instability of McCray *et al.*[9]. They consider this an unlikely explanation because the steady models in the 2×10^4 to

2×10^5 K range reproduce the observed line intensities and because the spectrum appears uniform over the IUE aperture. The other two explanations are that the shock wave is just in the process of making the transition from an adiabatic to a radiative shock wave, either where it comes into contact with a dense cloud or a homogeneous medium. We believe that this special situation is unlikely because the evidence for nonsteady behavior is so widespread.

Thus, we are led to propose that the oscillatory instability is the cause of the apparent nonsteady emission[4]. If interstellar medium shocks oscillate in low order modes, it is possible that their structures in different temperature ranges will reflect different shock velocities and thus lead to spectra which appear to contain contributions from steady state shocks of several shock velocities. More detailed calculations are needed to verify this claim.

This work was performed under the auspices of the United States Department of Energy.

REFERENCES

1. Aizu, K., Progr. Theoret. Phys., 49, 1184 (1973).
2. Avedisova, V. S., Sov. Astron.-AJ, 18, 283 (1974)
3. Chevalier, R. A., Ann. Rev. As. Astr., 15, 175 (1977).
4. Chevalier, R. A. and Imamura, J. N., Ap. J., 261, 543 (1982).
5. Imamura, J. N., Ap. J. (1985), in press.
6. Imamura, J. N., Wolff, M. T. M., and Durisen, R. H., Ap. J., 276, 667 (1984).
7. Langer, S., Chanmugam, G., and Shaviv, G., Ap. J. (Letters), 245, L23 (1981).
8. _____., Ap. J., 258, 289 (1982).
9. McCray, R., Stein, R. F., and Kafatos, M., Ap. J., 196, 565 (1975).
10. McKee, C., and Hollenbach, D. J., Ann. Rev. As. Astr., 18, 219 (1980).
11. Raymond, J. C., Ap. J. Suppl., 39, 1 (1979).
12. Raymond, J. C., Black, J. H., Dupree, A. K., Hartmann, L., and Wolff, R. S., Ap. J., 238, 881 (1980).
13. Raymond, J. C., Cox, D. P., and Smith, B. W., Ap. J., 204, 290 (1976).
14. Schwartz, R. D., Ann. Rev. As. Astr., 21, 209 (1983).
15. Shapiro, S. L., and Salpeter, E. E., Ap. J., 198, 671 (1975).
16. Weaver, R., McCray, R., Castor, J., Shapiro, P., and Moore, R., Ap. J., 218, 377 (1977).

THE WINDS IN CATACLYSMIC VARIABLE STARS

France A. Córdova and Edwin F. Ladd*
Los Alamos National Laboratory, M.S. D436, Los Alamos, NM 87545

Keith O. Mason
Mullard Space Science Laboratory, University College London, U.K.

ABSTRACT

Ultraviolet spectrophotometry of two dwarf novae, CN Ori and RX And, at various phases of their outburst cycles, confirms that the far UV flux increases dramatically about 1 - 2 days after the optical outburst begins. At this time the UV spectral line profiles indicate the presence of a high velocity wind. The detectability of the wind depends more on the steepness of the spectrum, and thus on the flux in the extreme ultraviolet, than on the value of the far UV luminosity. The UV continuum during outburst consists of (at least) two components, the most luminous of which is located behind the wind and is completely absorbed by the wind at the line frequencies. Several pieces of evidence suggest that the UV emission lines that are observed in many cataclysmic variables during quiescence originate in a location different from that of the wind in the binary, and are affected little by the outburst.

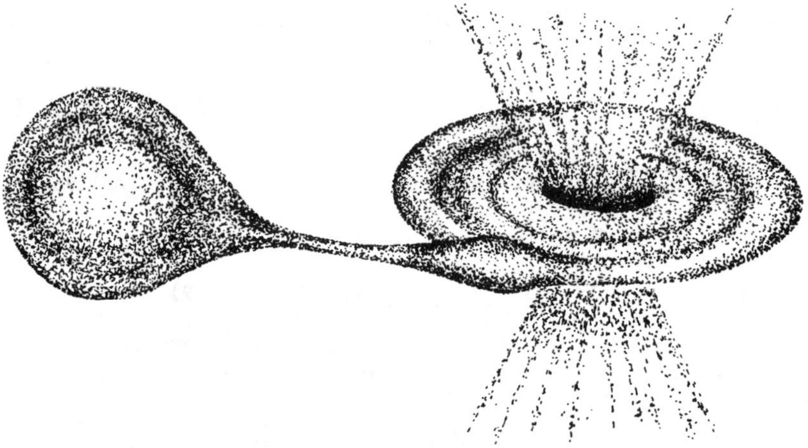

Fig. 1 Schematic of a cataclysmic variable, depicting a degenerate dwarf accreting matter, via a disk, from a Roche-lobe filling, low-mass companion star. The observations suggest that an extended wind comes from the disk center.

* also, Hopkins Observatory, Williams College

INTRODUCTION

Ultraviolet spectroscopy has shown that high velocity winds emanate from many cataclysmic variable (CV) stars[1]. CVs are binary systems in which a low-mass red star overflows its Roche lobe and transfers material via an accretion disk onto a degenerate dwarf (see illustration, Figure 1). The winds are seen only in CVs with high luminosity such as the dwarf novae during outburst and the novalike stars. In systems in which the disk is viewed face-on, broad, shortward-shifted absorption lines with terminal velocities of 3000 - 5000 km s^{-1} are observed. The terminal velocity is similar to the escape velocity from the surface of a white dwarf, suggesting that the wind emanates from near this star. In many CVs some of the line profiles also have emission components, hence the common appellation "P Cygni profile" after the well-known, luminous, mass-losing star that has roughly similar line profiles. When the disk is seen edge-on, as in the eclipsing novalike systems UX UMa[2,3] and RW Tri[4], the line radiation from the wind appears to be entirely in emission. The emission line profiles in RW Tri and UX UMa are asymmetric, peaking at wavelengths longer than the rest wavelength of the line. This may result from partial absorption on the blueward side of the line. An interpretation of these profiles is that they are formed in an accelerating flow that is not projected against a UV continuum source[4]. The UV emission lines are not eclipsed by the red star to the same degree as the continuum; therefore the emission line forming region must be large compared to the UV continuum emitting region. These properties are consistent with a wind which is accelerated from the region of the inner disk or white dwarf and which extends substantially above the disk.

Estimates for the mass-loss rate due to the wind are in the range 10^{-11} to 10^{-10} $M_\odot$ yr^{-1}, or about 10^{-3} to 10^{-2} of the deduced mass accretion rate onto the white dwarf[1]. The only viable mechanism yet proposed for accelerating the winds in CVs to the high observed velocities is radiation pressure in the lines, analogous to the mechanism driving the winds of OB stars. This view is supported by the fact that the momentum rate of the radiation from the inner disk, $L_{tot}/c = 10^{24}$ g cm s^{-2}, is of the same order as the momentum rate of the wind.[5] It is difficult, however, to determine either momentum rate to within a factor of ten: the spectrum, and hence the radiant energy, is as yet unknown; and the mass loss rate is extremely uncertain because of lack of knowledge of the ionization structure, geometry, and homogeneity of the wind. The material in the wind is thought to be photoionized rather than collisionally ionized[3,4], although the nature of the photoionizing spectrum is unclear.[6]

Because cataclysmic variables as a class undergo large changes in brightness, they offer an unique opportunity to study how changes in the radiation field affect the wind. This is particularly true of the dwarf novae, which have outbursts of a few magnitudes amplitude lasting several days and recurring on timescales of a few weeks. The outburst in a dwarf nova is believed to be caused by an increase in the rate of mass accretion through the disk.

We and other groups have been engaged in programs to measure the ultraviolet spectra of many dwarf novae at various stages of their outbursts. All the observations presented here have been made using a low resolution (~6 Å) spectrometer with a wide (10" x 20") aperture on the International Ultraviolet Explorer (IUE) satellite. We discuss UV data that were taken by us, together with data from the IUE archives. In this paper we present some preliminary, new results of the investigation into the behavior of the line spectrum in two CVs as a function of outburst epoch. A more complete analysis, including studies of several additional CV systems, is in preparation by the authors.

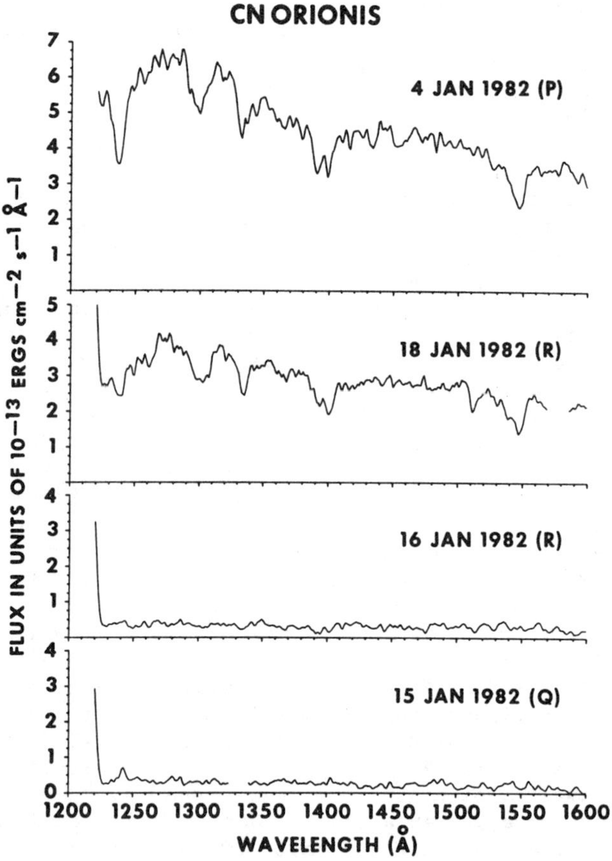

Fig. 2a Selected ultraviolet line spectra for various outburst states of of CN Orionis from 1200 Å to 1600 Å. The legends indicate the dates of the IUE observations. The letters in parentheses following the dates indicate the visual outburst state: quiescence (Q), rise (R), decline (D), and peak (P).

THE IUE SPECTRA: Observations, Deductions

1. Spectral Lines

A comparison of the spectra of dwarf novae in quiescence reveals that they fall into two types: those having emission lines of considerable equivalent width (EW as high as 80 Å), and those having weak or no detectable emission lines (EW <8 Å). Examples of the two types are shown in the bottom panels of Figures 2 a and b, where we plot the spectra of CN Ori and RX And in the far UV wavelength range 1200 Å - 1600 Å. The ultraviolet spectrum of RX And in quiescence exhibits a number of emission lines, the most prominent being C IV 1550 Å, Si IV 1400 Å, and N V 1240 Å. Of these lines, only N V is possibly detected in the quiescent-state spectrum of CN Ori.

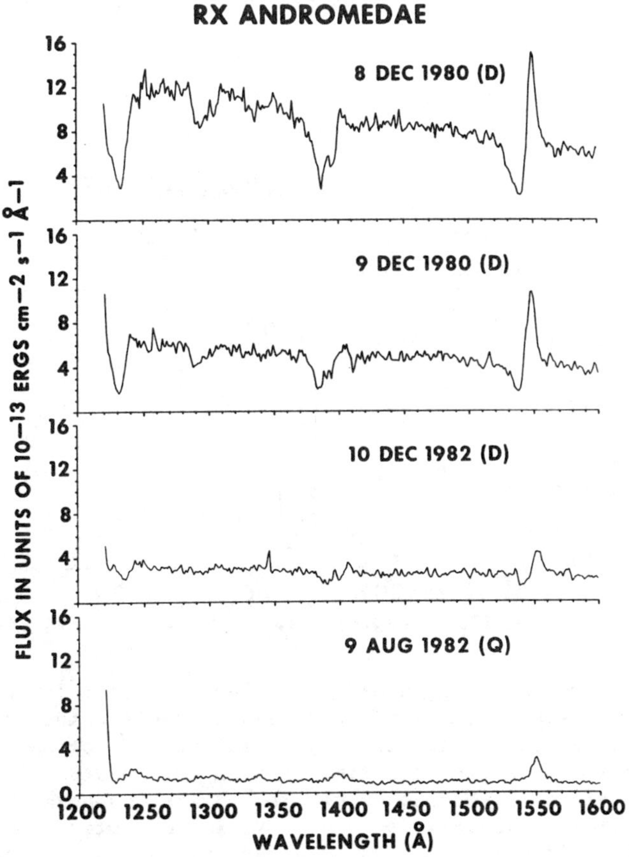

Fig. 2b Same as Fig. 2a, but for RX Andromedae. Note presence of emission lines in quiescence (bottom panel) and outburst.

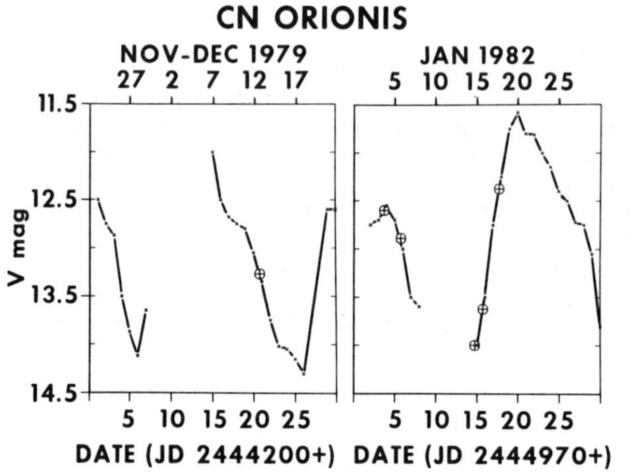

Fig. 3 Visual light curves of CN Ori (above) and RX And (below). Encircled crosses indicate times of IUE observations.

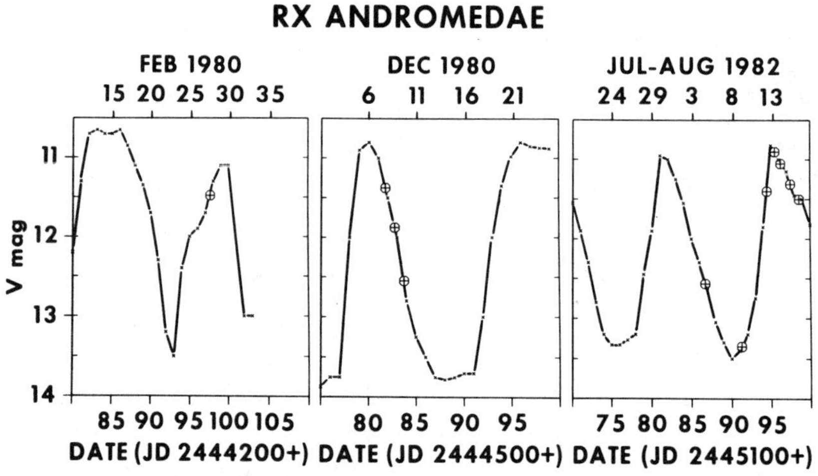

The remaining panels in Figure 2 show representative spectra of these stars at various stages in their outburst cycles. The optical state of the stars at the epoch of each spectrum is illustrated in Figure 3, where the time of each IUE spectrum is marked on plots of the visual light curves of the stars composed from data of the American Association of Variable Star Observers. Figure 2 illustrates that in the brighter states both stars develop broad absorption lines which are shifted shortward of the rest wavelength. RX And's C IV and Si IV line profiles have distinct emission components, but none of CN Ori's lines have such a feature.

Various parameters of the spectral lines are listed in Table 1[1]. These include the equivalent width (EW) of the emission and absorption components, the (interpolated) continuum level at the rest wavelength of the spectral line, the "blue" edge velocity (v_{blue}) of the absorption component when it is present, and the "red" edge velocity (v_{red}) of the emission component; v_{blue} represents an approximate estimate of the terminal velocity under the assumption that the line broadening is caused by a wind.

Table 1, in combination with the information about the outburst state from Figure 3, reveals how the flux and EW of the line components vary as a function of outburst phase and UV and optical continuum brightnesss levels. The EW of the absorption components correlate positively with the continuum brightness level; i.e., the amount of absorbed flux grows faster than the continuum. The flux in the emission component of RX And's lines does not vary by more than a factor of three during the outburst cycle. Although, on average, this flux is somewhat higher when the continuum is higher, the EW of the emission component is inversely correlated with UV continuum brightness.

Inspection of Figure 2 shows that the flux levels at the bottom of the absorption lines in RX And stay nearly the same throughout the outburst (when observations are compared for the <u>same</u> outburst). This is also true of another dwarf nova, AB Draconis, for which we have taken IUE spectra throughout an outburst. Indeed, the spectral slope defined by the flux levels at the bottom of the lines is the same as that of the general continuum during the quiescent state. In contrast, the general continuum slope during outburst becomes much steeper (i.e. bluer) than it is in quiescence. This suggests that there may be two sources of continuum flux during the outburst: a component that is similar in strength, or slightly stronger than, the quiescent state component, and an additional, steeper component that arises during the outburst and contributes most of the far UV outburst light. The wind must be optically thick in the UV resonance lines and be in front of the second continuum component in order to absorb all the radiation of this component at the wavelengths of the lines.

The following evidence suggests that the emission and absorption components originate in different places:

(a) The ratios of the fluxes and equivalent widths of the lines of different elements are not the same for the emission and absorption components, suggesting that these components are formed under different physical conditions.

---

[1] The errors on the quantities given in Table 1 are roughly 10% in the Equivalent widths, Line fluxes, and Continuum fluxes; 30% in v_{blue} for the outburst spectra and 50% in v_{blue} for the quiescent spectra; and 50% in all measurements of v_{red}. A full discussion of the error analysis is given in a paper in preparation.

Table 1a: Spectral Line Data for CN Orionis

Line: NV (1240 Å)

U.T. Date of SWP Observations	Emission⁺ EW	Emission⁺ Flux	Absorption⁺ EW	Absorption⁺ Flux	Continuum Flux⁺⁺ at Line Center	v_{blue} (km s^{-1})	v_{red} (km s^{-1})
1979 Dec 13.38	--	--	3.9	8.3	2.1	5000	--
1982 Jan 4.36	--	--	3.5	20	5.6	2500	--
Jan 6.40	--	--	2.5	7.4	3.0	3600	--
1982 Jan 15.18	--	--	--	--	0.35	--	--
Jan 16.18	--	--	--	--	0.39	--	--
Jan 18.22	--	--	4.8	17	3.4	4300	--

Line: Si IV (1400 Å)

U.T. Date of SWP Observations	Emission⁺ EW	Emission⁺ Flux	Absorption⁺ EW	Absorption⁺ Flux	Continuum Flux⁺⁺ at Line Center	v_{blue} (km s^{-1})	v_{red} (km s^{-1})
1979 Dec 13.38	--	--	4.5	7.6	1.7	4600	--
1982 Jan 4.36	--	--	4.1	19	4.6	2800	--
Jan 6.40	--	--	6.1	15	2.5	4200	--
1982 Jan 15.18	--	--	--	--	0.27	--	--
Jan 16.18	--	--	6.7	2.0	0.30	2700	--
Jan 18.22	--	--	3.9	11	2.8	3000	--

Line: C IV (1550 Å)

U.T. Date of SWP Observations	Emission⁺ EW	Emission⁺ Flux	Absorption⁺ EW	Absorption⁺ Flux	Continuum Flux⁺⁺ at Line Center	v_{blue} (km s^{-1})	v_{red} (km s^{-1})
1979 Dec 13.38	--	--	4.4	6.6	1.5	4400	--
1982 Jan 4.36	--	--	4.2	6.0	3.6	2500	--
Jan 6.40	--	--	3.6	6.8	1.9	4100	--
1982 Jan 15.18	--	--	--	--	0.17	--	--
Jan 16.18	--	--	--	--	0.32	--	--
Jan 18.22	--	--	4.6	11	2.3	2900	--

⁺Equivalent width (EW) in Angstroms; spectral line flux in units of 10^{-13} erg cm^{-2} s^{-1}.

⁺⁺Continuum flux in units of 10^{-13} erg cm^{-2} s^{-1} $Å^{-1}$.

Table 1b: Spectral Line Data for RX Andromedae

Line: NV (1240 Å)

U.T. Date of SWP Observations	Emission+		Absorption+		Continuum Flux++ at Line Center	v_{blue} (km s^{-1})	v_{red} (km s^{-1})
	EW	Flux	EW	Flux			
1980 Feb 28.16	9.1	17	--	--	1.8	--	1700
1980 Dec 8.33	0.4	4.2	11	130	12	4700	3500
Dec 9.40	2.7	14	5.7	30	5.3	4100	3300
Dec 10.31	2.2	7	1.3	3.7	3.0	4100	3400
1982 Aug 5.38	9.2	18	--	--	2.0	--	2800
1982 Aug 9.69	7.7	10	--	--	1.4	--	3700
Aug 13.04	--	--	11	180	17	4000	--
Aug 13.96	--	--	11	240	22	3300	--
Aug 14.78	--	--	12	240	20	4600	--
Aug 16.04	--	--	6.6	63	9.6	3800	--
Aug 16.82	--	--	10	105	10	5000	--

Line: Si IV (1400 Å)

U.T. Date of SWP Observations	Emission+		Absorption+		Continuum Flux++ at Line Center	v_{blue} (km s^{-1})	v_{red} (km s^{-1})
	EW	Flux	EW	Flux			
1980 Feb 28.16	9.8	1.3	--	--	1.4	--	1200
1980 Dec 8.33	0.5	4.2	8.9	78	8.8	5400	2100
Dec 9.40	1.3	6.2	7.1	34	4.8	5300	2100
Dec 10.31	1.1	2.9	5.0	13	2.7	3900	2400
1982 Aug 5.38	3.4	6.7	--	--	1.9	--	1600
1982 Aug 9.69	6.7	7.2	--	--	1.1	--	1400
Aug 13.04	--	--	9.3	110	12	4700	--
Aug 13.96	--	--	8.1	120	15	4600	--
Aug 14.78	--	--	10	140	14	4800	--
Aug 16.04	--	--	9.6	98	10	4600	--
Aug 16.82	0.7	5.6	8.4	66	7.7	4700	--

Table 1b: RX And, continued

Line: C IV (1550 Å)

U.T. Date of SWP Observations	Emission[+] EW	Emission[+] Flux	Absorption[+] EW	Absorption[+] Flux	Continuum Flux[++] at Line Center	v_{blue} (km s^{-1})	v_{red} (km s^{-1})
1980 Feb 28.16	43	45	--	--	1.0	--	2600
1980 Dec 8.33	8.6	58	8.8	63	6.7	5400	2100
Dec 9.40	12	62	7.7	35	4.3	5100	2000
Dec 10.31	6.8	17	3.2	7.8	2.5	2600	3200
1982 Aug 5.38	19	29	--	--	1.5	--	2600
1982 Aug 9.69	24	22	--	--	0.9	--	2600
Aug 13.04	4.8	49	8.9	95	10	5000	1800
Aug 13.96	3.4	42	11	150	13	4300	2400
Aug 14.78	4.0	45	12	140	12	5400	1700
Aug 16.04	3.2	26	10	90	8.4	4500	2300
Aug 16.82	9.1	57	8.5	55	6.3	4700	3100

(b) Theoretical mass-loss models don't fit the absorption and emission components simultaneously. This could be because the wind is not spherical, or because an additional emission component arises elsewhere than in the wind, or both.

(c) The data presented here together with similar published data on other dwarf novae[7] reveal that the stars which show prominent emission lines during quiescence also display emission line components during outburst, while those stars having no emission lines during quiescence exhibit no emission line components during outburst. Yet the velocity-shifted absorption components of the lines seen in CN Ori behave like those in RX And, AB Dra and other stars which have emission components. Thus it appears that the wind produces the absorption lines, and the line emission is extra.

Emission lines appear to be absent in those stars that have steep UV spectra in quiescence ($\alpha > 2.0$; see next section), suggesting that the presence of these lines may correlate with the slope of the continuum. A steep spectrum could indicate a relatively small disk, or a hotter, more luminous white dwarf, or both. The size of the disk will be influenced by the orbital period, mass ratio, and mass accretion rate, while the luminosity of the central star is governed by its mass and the mass accretion rate. Simultaneous data over a wide energy range (i.e. soft X-ray to near infrared) would help determine the relative contributions of the central star, accretion disk, and companion star to the total flux, and constrain the factors on which the presence of emission lines during quiescence depends.

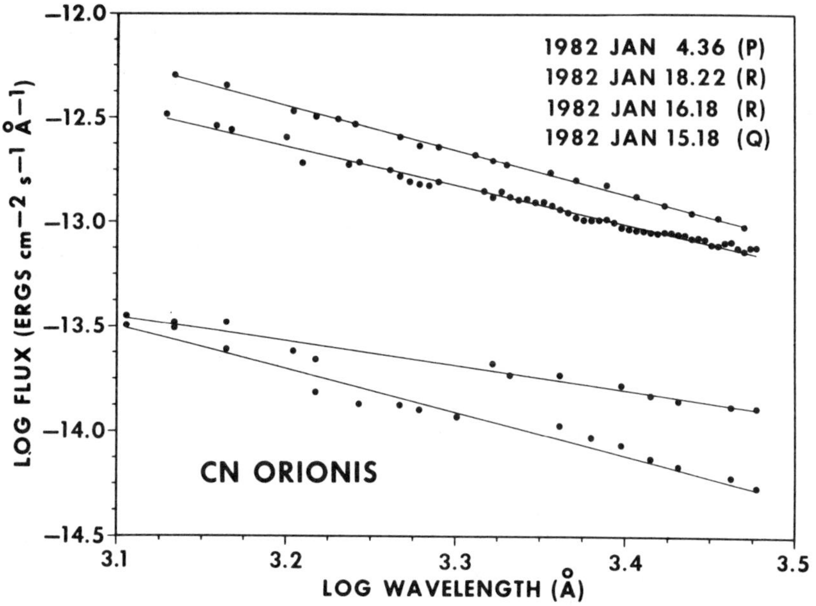

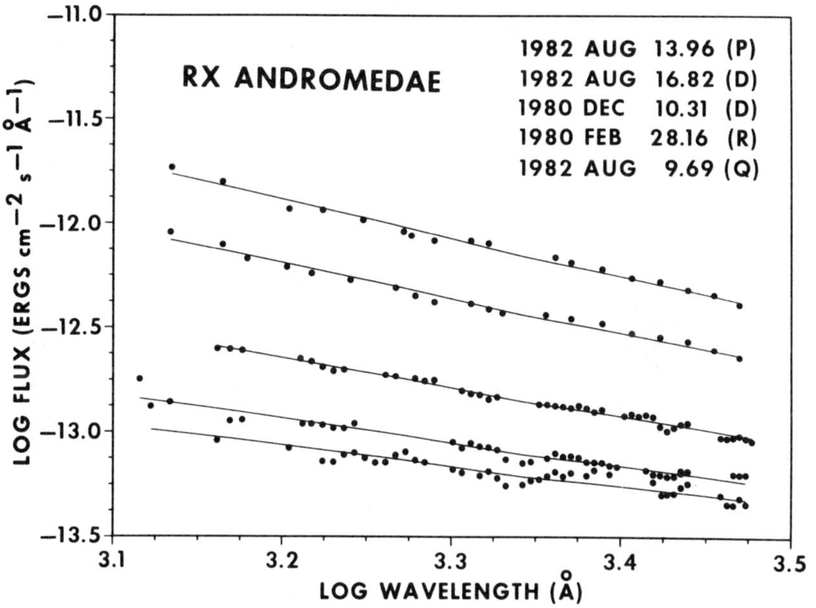

Fig. 4 The ultraviolet continuum spectra in various outburst states with model power law fits: CN Ori (above), RX And (below). The legends are ordered in descending flux level. The spectral slopes are given in Table 2. Other notations as in Fig. 2.

2. Continuum Slope

To determine the continuum slope of the UV spectra, we have integrated the data in bins between 25 Å and 100 Å wide, avoiding spectral lines. We have fit each spectrum with a power law (i.e., $F_\lambda \propto \lambda^{-\alpha}$) modified by reddening (which is measured to be E(B-V) = 0.0 for CN Ori and 0.02 for RX And). The results of these fits, together with the value of the continuum at C IV and the V mag and outburst state of the star, appear in Table 2. Some examples of the continuum spectral fits are shown in Figure 4.

For RX And the spectral slope becomes systematically steeper as the luminosity increases. The power-law slope, α, varies from 1.0 in quiescence to 1.9 near the peak of the outburst. In the case of CN Ori the slope of the continuum during quiescence is indistinguishable from that at the peak of the outburst, in both cases being about 2.0. A spectrum of CN Ori taken during the rising phase of the outburst, however, has a much flatter distribution, with a slope of 1.0. The flattening of the spectrum during the optical rise to outburst has been reported for another dwarf nova, VW Hyi, and is interpreted as indicating that the outburst starts in the outer (i.e. cooler) parts of the accretion disk.[7]

Table 2: Spectral Slope Compared with Other Parameters

RX Andromedae

U.T. Date	Power law Slope, α ($F_\lambda \propto \lambda^{-\alpha}$)	Continuum Flux at 1550 Å (ergs cm^{-2} s^{-1} $Å^{-1}$)	V mag+	Outburst State	Wind?
1980 Feb 28.16	1.1	1.0	11.5	R	no
1980 Dec 8.33	1.8	6.7	11.4	D	yes
Dec 9.40	1.8	4.3	11.9	D	yes
Dec 10.31	1.4	2.5	12.6	D	yes
1982 Aug 5.38	1.2	1.5	12.6	D	no
1982 Aug 9.69	1.0	0.9	13.3	Q	no
Aug 13.04	1.6	10.0	11.4	R	yes
Aug 13.96	1.9	13.0	10.9	P	yes
Aug 14.78	1.8	12.0	11.1	D	yes
Aug 16.04	1.7	8.4	11.3	D	yes
Aug 16.82	1.7	6.3	11.5	D	yes

CN Orionis

U.T. Date	Power law Slope	Continuum Flux	V mag	Outburst State	Wind?
1979 Dec 13.38	2.0	1.5	13.25	D	yes
1982 Jan 4.36	2.1	3.6	12.6	P	yes
Jan 6.40	2.0	1.9	12.85	D	yes
1982 Jan 15.18	2.1	0.2	14.0	Q	no
Jan 16.18	1.2	0.3	13.65	R	no
Jan 18.22	1.8	2.3	12.35	R	yes

+ V = 0 corresponds to 3.92 x 10^{-9} erg cm^{-2} s^{-1} $Å^{-1}$ at 5400 Å.

TRIGGERING THE WIND

The visual magnitude of the star does not, in general, provide a reliable estimate of the luminosity at higher energies. For example, the maximum UV flux we have detected from CN Ori occured on 1982 Jan 4, when the star was undergoing an optical outburst and had a V mag of 12.6. Two weeks later, CN Ori was detected at a somewhat brighter V mag of 12.4 on the rise to its next outburst. The far UV flux at this time, however, was only one-half the level during the Jan 4 observation. A second illustration of this is provided by RX And. The visual magnitude during the 1980 Feb. 28 observation (on the rise to outburst) was two magnitudes higher than during the 1982 Aug. 9 observation (quiescence), yet the far UV fluxes were nearly the same. In fact, most of the discrepancies in comparing the UV and visual light curves occur on the rise to outburst.

The inference from the data presented here is that dwarf novae brighten at optical wavelengths before they brighten in the far UV, and this is supported by the observed flattening of the spectral slope during the rise to outburst (e.g. CN Ori's 1982 Jan. 16 spectrum). There is a delay following the spectral flattening, after which the far UV flux increases by at least an order of magnitude. A delay of 1 - 2 days is indicated from the available data (ref. 7 and this paper). In some stars the spectrum is then much steeper than during quiescence (e.g. RX And), while in other stars (e.g. CN Ori), it is the same as during quiescence. During the decline both UV and V fall together (cf. Table 2).

For the dwarf novae, the shortward-shifted absorption lines, which indicate the presence of a wind, appear only <u>after</u> the optical outburst is well underway, i.e., about the time that the far UV luminosity increases substantially. The development of the wind-like profiles in the spectral lines is thus probably associated with the steepening of the spectrum that follows the initial flattening during the rise. The detection of these spectral line features, however, is not a smooth function of the local UV continuum flux. For instance, spectra of RX And taken on 1980 Dec. 10 and 1982 Aug. 5, both far down on the decline from maximum outburst light, have far-UV continuum levels that differ by less than a factor of two, yet the former spectrum shows marked shortward-shifted absorption, whereas the latter spectrum exhibits no evidence for any absorption. The <u>slope</u> of the UV continuum, however, is different, $\alpha = 1.4$ for the former spectrum, and $\alpha = 1.2$ for the latter. In fact, the wind is only observed in RX And when $\alpha > 1.4$. This suggests that the wind is detected when either or both the EUV and soft X-ray flux are enhanced.

In the previous Section we suggested that there might be two separate contributors to the far UV flux. One is observed during quiescence and changes very little during the outburst. The second contributor is responsible for most of the increase in the far UV luminosity during the outburst. The spectral lines demonstrate that the wind is apparently in front of the latter UV source; when this contributor turns on near the peak of the optical outburst, all of the source's continuum flux is absorbed at the resonant line frequencies, and accelerates the wind material. When this high-energy source diminishes, the wind is no longer accelerated due to lack of radiation pressure in the spectral lines, and we can no longer detect its presence.

This work is supported by the US Dept. of Energy and the UK Science and Engineering Research Council. The authors are grateful to the AAVSO for supplying their visual data, and to Dr. P. Szkody for supplying IUE tapes with the 1980 December data on RX And.

REFERENCES

1. Côrdova, F.A., and Mason, K.O., Future of Ultraviolet Astronomy Based on Six Years of IUE Research (NASA Conf. Publ. 2349, 1984) p. 377.
2. Holm, A.V., Panek, R.J., and Schiffer III, F.H., Astrophys. J., 252, L35 (1982).
3. King, A.R., Frank, J., Jameson, R.F., and Sherrington, M.R., Mon. Not. Astron. Soc., 203, 677 (1983).
4. Côrdova, F.A., and Mason, K.O., Astrophys. J., 290, 671 (1985).
5. Côrdova, F.A., and Mason, K.O., Astrophys. J., 260, 716 (1982).
6. Kallman, T.R., Astrophys. J., 272, 238 (1983).
7. Hassall, B.J.M., Pringle, J.E., Schwarzenberg-Czerny, A., Wade, R.A., and Whelan, J.A.J., Mon. Not. Astron. Soc., 203, 865 (1983).

NEW PARADIGMS FOR BLACK HOLE ACCRETION FROM HIGH RESOLUTION SUPERCOMPUTER EXPERIMENTS

John F. Hawley
Theoretical Astrophysics, Caltech, Pasadena, Ca. 91125

Larry L. Smarr
Astronomy Department, University of Illinois, Urbana, Il. 61801

ABSTRACT

From a series of high resolution supercomputer calculations, new paradigms emerge for black hole accretion with angular momentum. We have discovered that the centrifugal barrier, or funnel wall, plays a key role in shock heating the inflowing supersonic fluid, producing a stationary thick disk with bipolar outflow. The infall dynamics are calculated in detail and followed until a steady state is reached. We delineate a number of distinct regimes of steady state flow in the two parameter solution space defined by the fluid's angular momentum and the disk thickness. The essential physics is highlighted with selected numerical examples. The same code can calculate general relativistic magnetohydrodynamical flows near black holes.

INTRODUCTION

Observations of quasars, active galactic nuclei (AGN), and astrophysical jets have been increasing in number and scope for two decades. The fundamental power source is, in all likelihood, gravitational, with a supermassive black hole providing the potential well (see, e.g., Blandford, this volume). The enormous range in luminosity (10^{39} to 10^{48} ergs s^{-1}) probably reflects a wide variance in accretion rates during the lifetime of a "standard" sized black hole[1] of mass $\simeq 10^8 M_\odot$. In addition to this long timescale accretion rate variation, there is growing observational evidence that the accretion process is variable on timescales[2] as short as hours, comparable to the inner orbital periods around such supermassive black holes. Many of the AGN have bulk bipolar flows emerging from them. In certain cases, this outflow forms narrow, highly collimated, supersonic jets.

To date, most theoretical models of black hole disk accretion have been analytic stationary approximations. These yield insight into the time averaged properties of accretion disks, but the highly variable nature of AGN indicates dynamic flows may be necessary to explain the observations.

We have addressed this issue by using a time dependent, fully general relativistic two-dimensional hydrodynamics code which is documented[3,4] in papers I and II. The results in this paper come from a series of high resolution numerical experiments run with this code on the CRAY-1 supercomputer at the Max Planck Institute for Astrophysics in Garching, West Germany. These experiments begin the investigation of realistic dynamic fluid behavior in the gravitational field of a black hole. This paper summarizes some of these new, generic phenomena we have discovered in the near hole region.

SOLVING THE ACCRETION EQUATIONS

In our present study we do not intend to provide detailed astrophysical models of central engines. The physics required to do this would include many complicated effects such as magnetohydrodynamics (MHD), viscosity, radiation hydrodynamics, and relativistic plasma physics. This conference marks an important step toward sorting out the essential physics. We expect that the development of a detailed theory should take at least as long as was required for understanding stellar evolution, where progress was aided by the simplifications of spherical symmetry and a relatively detailed knowledge of nuclear physics. The field of black hole accretion physics is still too immature to hope to obtain such complete models.

We can, as a first step, investigate highly resolved axisymmetric hydrodynamics in near hole regions with the current generation of supercomputers. As additional physics is added in later calculations, fluid dynamical effects will remain, and dominate in some regimes. Our near term goal, therefore, has been to write and use a code which can map out the phenomenology of black hole hydrodynamics solution space. We use time explicit, Eulerian finite differencing to solve the equations of general relativistic ideal hydrodynamics. The particular investigation is of the behavior of fluid with angular momentum in the fixed Kerr black hole metric. In general our results involve phenomena which could not previously be examined, including multiple shock structures, vortical motions and the nonlinear development of instabilities. In addition, the development of carefully tested numerical techniques for hydrodynamics is the requisite first step towards solving the more general and complicated problems involving the MHD equations, viscosity, and radiation transport.

The first attempt to model the behavior of an ideal fluid in a Kerr black hole metric was made by James R. Wilson in 1971 when he studied the axisymmetric infall of cold gas with nonzero angular momentum[5]. This two-dimensional code used cylindrical coordinates and was the predecessor to the present code. During the middle 1970's, Wilson[6-8] added the general relativistic MHD equations to the code. In 1980, this code, including the MHD terms, was rewritten into spherical polar coordinates by Smarr and Wilson.

We have since taken the 1980 hydrodynamics (dropping the MHD terms) code and developed the current production version. Extensive effort has gone into calibration against analytically determined solutions and the comparison of various finite difference techniques. This development involved working out numerous analytic solutions in terms of the code variables and then carrying out hundreds of computer runs to transform the original 1980 prototype version into an accurate astrophysical research tool. This work is detailed in references 3, 4, 9, and 10.

We briefly remark on the resources necessary to carry out this work. Although much of the calibration work was accomplished on a VAX 11/780, the high resolution runs each require several hours of CRAY-1 time. The total project requirements to date have been some 75 Cray hours (roughly two full VAX years). The great advantage of using a supercomputer is illustrated by the fact that the VAX 11/780 runs in Paper II were done with a grid resolution of 40x30 and took 8 hours of compute time for a full evolution. In contrast, the CRAY-1 (some 200 times faster) could run with a grid resolution of 160x160, taking only two hours to reach the same evolutionary time. A comparison of the resolution of VAX and CRAY runs are displayed in reference 9.

Finally, modern computer imaging is essential to be able to probe the gargantuan amount of numerical output from a supercomputer run. The black and white contour line drawings that we are limited to in this paper are completely inadequate to sort out the physics of the resulting flows. Full color images (at least 512x512x8 bits) and image processing techniques are necessary to represent the information contained in the solution at one time level. The amount of number crunching to produce these images from the raw numerical data in a short time also requires a supercomputer. Finally, the dynamics can be understood only with the use of highly time-resolved color image movies of the sort we have shown at this conference. Our work with these essential tools has only been possible because of the fine imaging facilities at the University of Illinois (VIP System) and at the Max Planck Institute for Astrophysics.

We mention the above points to emphasize that the technology and manpower requirements for performing these calculations, let alone solving more complicated equations, such as MHD, are approaching the requirements (and dollar investments) of observational astronomy. The style of this research in other areas of science actually has a long history[11]. We would like to emphasize that in all phases of our project (the numerical techniques, the unravelling of the hydro physics, the code management, and the color imaging), we benefited greatly from the shared experience of our colleagues James Wilson, Karl-Heinz Winkler, and Michael Norman.

Although it is outside the scope of this paper to describe the expanded MHD version of this code, the reader should be aware that Wilson has used the cylindrically symmetric MHD code to extensively investigate black hole magnetospheres[6-8,12,13]. In those papers one will find the generalization of our equations to MHD and a number of rather complex solutions. No new solutions have been calculated with the code since those papers. It remains a future task of some graduate student to carefully calibrate and improve the MHD portion of the code to achieve the standards of rigor which characterizes our non-MHD code described herein. A first step in this direction was an analytic analysis of the relativistic MHD equations[14] which included a comparison with the elegant work of Thorne and Macdonald[15]. For a recent review of idealized analytic black hole magnetospheres, see papers in this volume by Blandford and by Lovelace.

The hydrodynamic calculations are performed in a fixed black hole gravitational field; for our accretion problems this field is represented by the Kerr metric $g^{\mu\nu}$ expressed in Boyer-Lindquist coordinates[16]. The Kerr metric represents a rotating black hole. The work reported herein uses the simpler special case of a nonrotating Schwarzschild black hole. At large distances the Schwarzschild coordinates become flatspace spherical coordinates. The metric defines the lapse function $\alpha = (-g^{tt})^{-1/2}$, the shift vector $\beta_i = g_{ti}$ and the three-metric $\gamma_{ij} = g_{ij}$. The dynamic matter is described by the ideal fluid stress-energy tensor $T^{\mu\nu} = (\rho + \rho\varepsilon + P)U^{\mu}U^{\nu} + Pg^{\mu\nu}$ where ρ is the baryon density, P is the pressure, ε is the specific internal energy, and U^{ν} is the 4-velocity of the matter. In our work, we assume that the mass in the gas is much less than the mass of the hole. Therefore we can ignore the gravity of the gas. The reader is referred to several introductions to relativistic hydrodynamics[16,17] for background on notation.

Before beginning our discussion we remind the reader of the equations of general relativistic hydrodynamics[3] as they are applied in our code. First there is the equation of baryon conservation

$$\partial_t D + \frac{1}{\sqrt{\gamma}} \partial_i (\sqrt{\gamma}\, D\, V^i) = 0, \tag{1}$$

where $D = W\rho$, U^{α} is the 4-velocity, $V^i = U^i/U^t$ is the transport velocity, γ is the determinant of the 3-metric, and $W = (1 - V_i V^i)^{-1/2}$ is the redshift factor equal to $U^t\alpha$ where α is the lapse function. Momentum conservation is written

$$\partial_t S_j + \frac{1}{\sqrt{\gamma}} \partial_i (S_j V^i \sqrt{\gamma}) + \alpha\partial_j P + \frac{1}{2} (\partial_j g^{\mu\nu}) \frac{S_\mu S_\nu}{S^t} = 0, \tag{2}$$

where $S_\nu = \rho h W U_\nu$, with relativistic enthalpy $h = 1 + \varepsilon + P/\rho$. Finally, we write the equation of energy conservation

$$\partial_t E + \frac{1}{\sqrt{\gamma}} \partial_i (\sqrt{\gamma}\, V^i E) + P\partial_t W + \frac{P}{\sqrt{\gamma}} \partial_i (\sqrt{\gamma}\, V^i W) = 0, \tag{3}$$

where $E = \rho\varepsilon W$. In these equations $j = r, \theta, \phi$; $i = r, \theta$; and μ and $\nu = r, \theta, \phi, t$. The equation of state that we currently use is that for an ideal gas, $P = \rho\varepsilon(\Gamma - 1) = \frac{E}{W}(\Gamma - 1)$. In this case, the enthalpy reduces to $h = 1 + \Gamma\varepsilon$. For a complete discussion and derivation of these equations, including their Newtonian limit, see Paper I. Our methods for solving these equations, and the calibration tests of our code, are detailed in Paper II.

In this work, we use geometric units in which $G = c = M$ (the black hole mass) = 1. To recover physical c.g.s. units, multiply the geometric value by the appropriate values of G, c, M. For example, the unit distance is

$$r = 1.5 \times 10^{5}\ (M/M_{\odot})\ \mathrm{cm}, \qquad (4)$$

and the unit time,

$$t = 5 \times 10^{-6}\ (M/M_{\odot})\ \mathrm{s}, \qquad (5)$$

where $M_{\odot}$ is the mass of the sun. The temperature of the fluid is given by the specific internal energy, ε, in units of c^2. If we assume the fluid's unit mass to be that of a proton, the temperature is

$$T = \varepsilon m_{p} c^{2}/k = 10^{13} \varepsilon\ \mathrm{K}. \qquad (6)$$

Therefore, as is standard convention, we measure distance and time in units of M, and specific internal energy as dimensionless units of $c^2 = 1$. For a complete discussion of geometric units, see reference 16. For scaling, keep in mind that the two most popular astrophysical black hole masses are $10\ M_{\odot}$ (Cygnus X-1) and $10^{8}\ M_{\odot}$ (AGN central engine).

FUNNEL HYDRODYNAMICS

There are three forces to consider in a purely hydrodynamic accretion flow: gravitational, centrifugal, and thermal pressure gradient. In a cold Keplerian thin disk[18], gravitational and centrifugal accelerations alone balance in the equatorial plane. As pressure becomes important, the vertical thickness of the disk increases, forming the so-called thick disk (see schematic diagram in Fig. 1). In the stationary thick disk[19], all three forces are everywhere in balance. Our work focuses on the dynamic properties of both thin and thick disks where the forces are <u>not</u> in balance.

The central problem in black hole accretion models is how to convert the gravitational potential energy of infalling matter into energy flowing away from the hole (bulk material flow or electromagnetic radiation). Such a process allows the black hole to be the central engine of an AGN. Much attention has been focused on whether this outflow can be highly collimated into a narrow jet in the near hole region and shot out along the orbital axis. A key feature of the thick disk is the <u>funnel</u> (see Fig. 1), a region along the orbital axis kept empty of fluid by centrifugal acceleration. The existence of these funnels has been argued to be an integral part of the jet formation process (see refs. below).

To define the "funnel wall", we must consider the total "effective potential" around the black hole which represents the combined effects of gravity and rotation. This potential is parameterized by the fluid's specific angular momentum

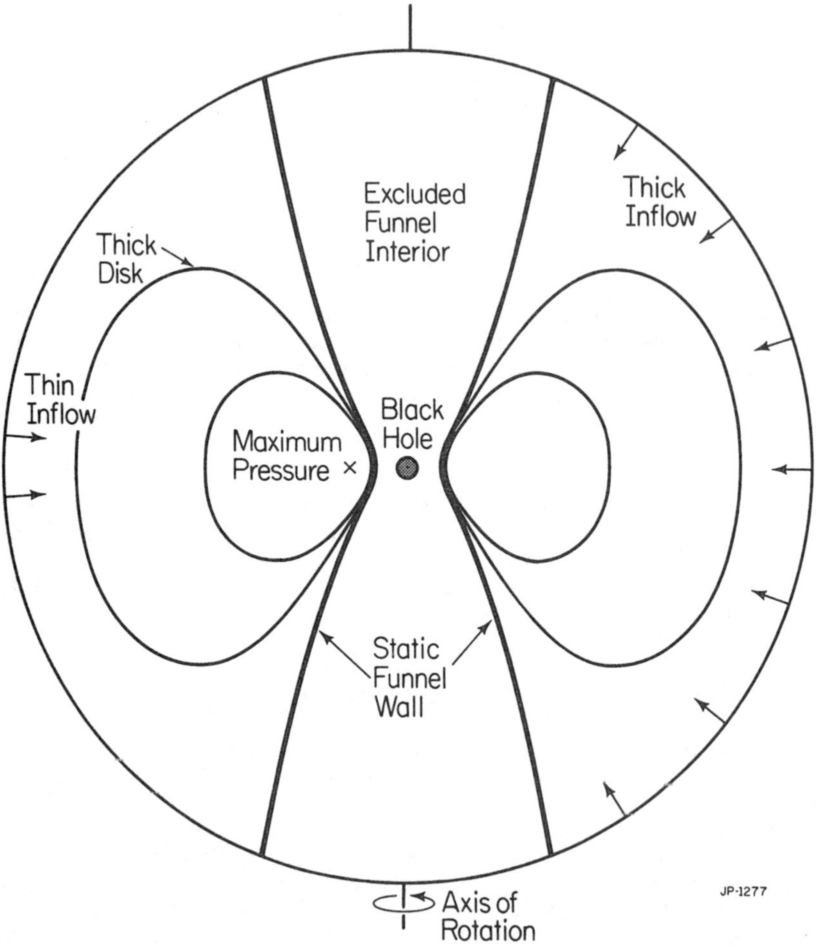

Fig. 1. Schematic diagram of our black hole accretion runs. The outer boundary is r = 100 M. Two major classes of outer boundary conditions are the thick and thin inflows.

$\ell = - U_\phi / U_t$. The effective potential determines the orbit of a test particle with a given value of angular momentum ℓ (see Figure 2). In general relativity, unlike Newtonian gravity, qualitatively different behavior in a test particle's orbit occurs for different ranges of ℓ. There are two critical values of ℓ which divide the possible orbits into three types[19,20]. We will confine the present discussion to the nonrotating Schwarzschild black hole; generalizations to the Kerr black hole are in the literature.

The first critical angular momentum is the "marginally stable" value $\ell_{ms} = 2\sqrt{3}$ M $\simeq$ 3.46 M, which is the angular momentum

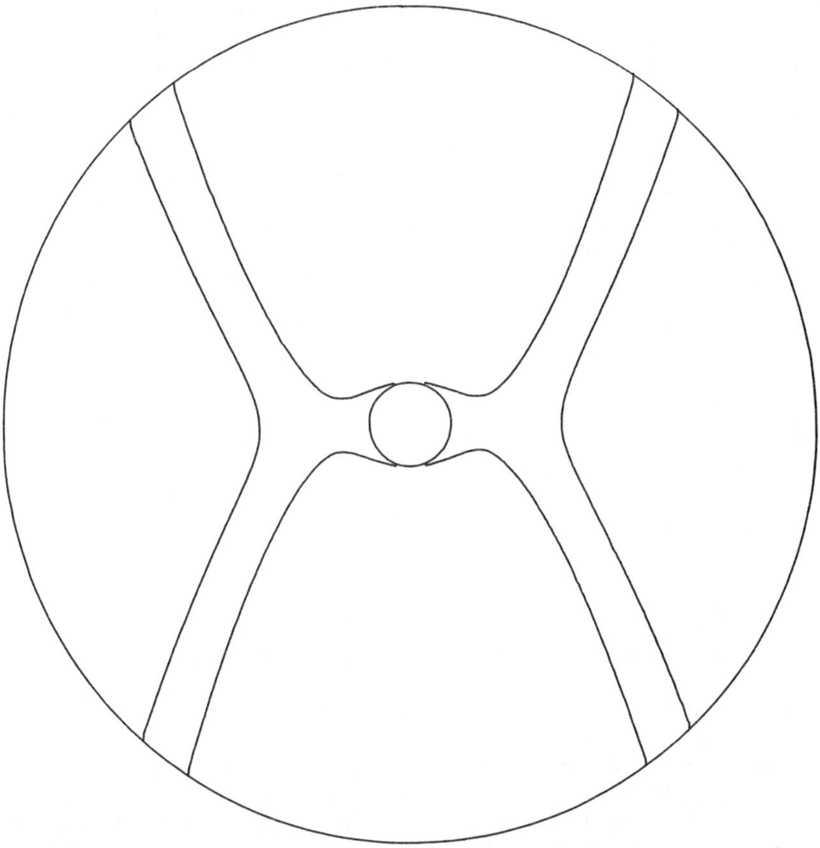

Fig. 2. The effective potential as a function of r along the equator for a Schwarzschild (nonrotating) black hole. From bottom to top the curves are for U_ϕ = 3.0 M, U_ϕ = $2\sqrt{3}$ M (marginally stable), U_ϕ = 3.8 M, U_ϕ = 4.0 M (marginally bound), and U_ϕ = 4.5 M. Note the important general relativistic effect: the potential does not monotonically increase as r approaches the horizon.

and energy of the closest stable circular orbit around a black hole. This orbit is located at r = 6 M with energy U_t = −0.943. Note that the specific binding energy per unit particle mass of a particle orbit is $U_t + 1$, in units of c^2. This shows that the marginally stable circular orbit has a binding energy of 0.057 c^2. The other critical value of ℓ is the "marginally bound" value, ℓ = 4.0 M, which is the smallest amount of angular momentum which will keep a particle, falling from rest at infinity (marginally bound orbit with U_t = −1.0), from falling into the hole.

These critical values of ℓ arise[16] from a purely relativistic effect---the "pit in the potential." The relativistic centrifugal effective potential is the generalization of the Newtonian effective potential, $- M/r + \ell^2/(2r^2)$ (see reference 16, Box 25.6). The relativistic potential has a maximum in the black hole geometry, whereas the corresponding Newtonian potential for a point mass goes to infinity as $r \to 0$. If $\ell < \ell_{mb}$ then this potential maximum drops below the energy ($U_t = -1.0$) where it can provide a turning point for a marginally bound orbit. If $\ell < \ell_{ms}$ the potential curve ceases to have a minimum and no circular orbit is possible (Fig. 2). Neither of these effects can occur in Newtonian gravity, thus providing new modes of accretion specific to black holes.

This one-dimensional analysis has been extended off the equatorial plane to determine the two-dimensional locus of turning points for a marginally bound test particle with a given angular momentum[19-22]. This locus is termed the "funnel wall"; by definition, the test particle is excluded from the interior of this wall. Two qualitatively different topologies of funnel walls result (see Fig. 3). For $\ell > \ell_{mb}$, the funnel wall crosses the equator and no inflow into the hole from any angle can occur. As ℓ decreases below ℓ_{mb}, the wall opens around the equator allowing some fluid to flow into the hole. Thus ℓ provides a crucial parameter describing fluid behavior.

This description of the funnel wall derives from a consideration of the dynamics of test particles, i.e., pressureless, noninteracting gas. One of the goals of our investigation is to determine whether this paradigm of the test particle funnel applies to the more realistic case of gas with pressure. How does the funnel affect such flows and what are its implications for jet production and disk formation?

Lynden-Bell[23] described the funnels in the context of jets, noting that their geometry was ideal for the production of a bidirectional flow inside the evacuated funnel interior. One speculative model, explored by many investigators[24-26], requires the coupling of electromagnetic radiation effects to the hydrodynamics. This model envisions that the intense electromagnetic radiation in the funnels of a thick disk accelerates relativistic plasma out along the orbital axis, i.e., the interior of the funnel. The accretion disk is radiation supported with a surface luminosity on the order of 10^{46} $(M/10^8 M_\odot)$ ergs s^{-1}. Given a sufficiently steep funnel wall, provided by the appropriate ad hoc angular momentum distribution in the thick disk, a narrow, relativistic, and highly collimated jet results. Unfortunately, it seems that radiation drag in the funnel prevents jet acceleration to high velocities[27]. In any case, many radio jets are observed to emerge from galaxies with nuclear luminosity much less than 10^{46} ergs s^{-1}, thus ruling out such radiation pressure supported thick disks in these sources[28]. Both of these points raise serious questions about the viability of this particular black hole jet production mechanism.

Rather than adding additional physics, such as radiation in the funnel, our investigation has centered on whether purely hydrodynamical processes can cause outflow of heated material from gas initially inflowing toward the hole. Because the centrifugal

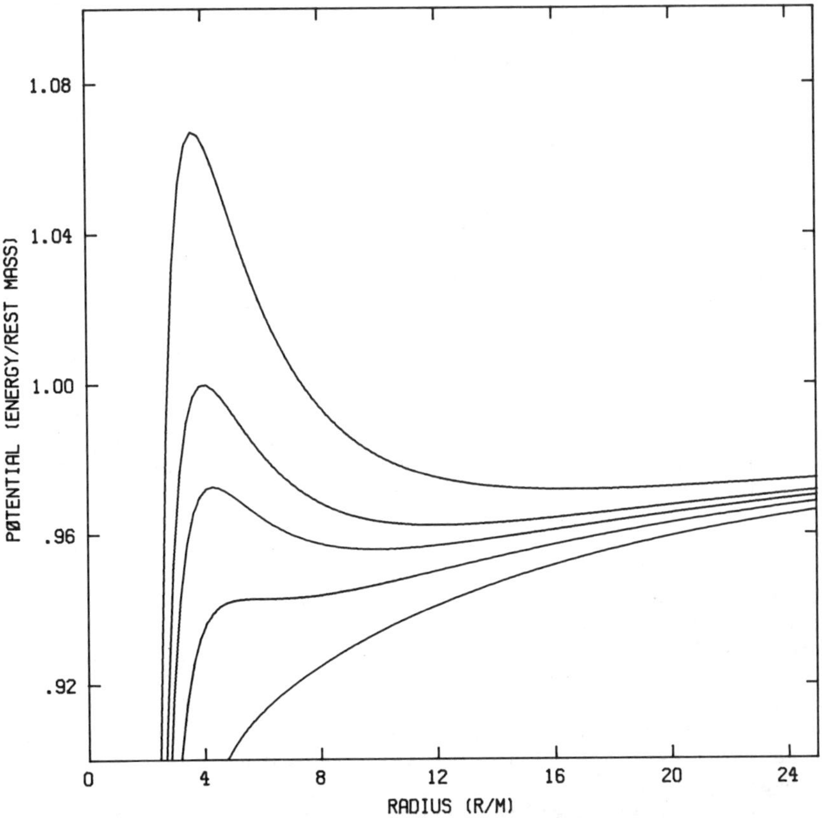

Fig. 3. The funnel walls for ℓ = 3.8 M and ℓ = 4.5 M are depicted. The ℓ = 4.5 funnel wall crosses the equator at r = 7 M.

forces in a hydrodynamic flow will keep the interior of the funnel empty, it has been previously assumed that hydrodynamic infall could not produce jets. The funnel wall was believed to divide the infalling fluid from the vacuum funnel interior. Even if fluid were heated near the hole, it was thought that confinement from the funnel wall and ram pressure from the infalling fluid would prevent the hot fluid's escape.

However, we have discovered that the solution to the full hydrodynamic equations provides a way out of this seeming difficulty. The funnel wall behaves as if it were an actual physical wall and creates a standoff shock. The shocked fluid then flows down towards the equator, hugging the interior of the standoff shock, leaving an evacuated region between the shocked fluid and the funnel wall. The surface of the infalling fluid therefore does not occur at the funnel wall. This creates an escape route out along which hot fluid may jet from the near hole region. Thus, from these hydrodynamic solutions come new black hole flow paradigms which supplant those based upon test particle notions. These new scenarios are described below.

In our hydrodynamic models, as in the test particle case, we expect qualitatively different behavior depending on whether ℓ is greater or less than ℓ_{mb}. Below we describe several representative cases which highlight the essential physics.

DISK DYNAMICS

Consider first the fate of cold fluid with angular momentum beginning to fall onto a naked hole. This gas may come from cold intragalactic gas surrounding a central black hole. In the standard picture, one assumes that the specific angular momentum is very high, $\ell >> \ell_{mb}$. As it falls towards the hole, it settles down into orbits lying in the equatorial plane at large distance ($r > 100$ M) from the hole. This gas forms a Keplerian thin disk, with a gradual mass flow inward determined by viscous torques. The heating which accompanies these torques is assumed to be promptly radiated away. These assumptions mean that the disk is a steady state fluid.

The total energy radiated over this secular infall timescale is equal to the binding energy of the last stable circular orbit. Because the energy release is gradual, the instantaneous luminosity is lowered. (For a review of thin disks, see Lamb, this volume). If the viscous heating proceeds too rapidly to be promptly radiated, the thin disk is assumed to "puff up" and become a pressure supported thick disk. The interior properties of the thick disk are then determined by viscous and radiation processes.

These standard models represent one extreme of black hole accretion, namely a flow which is dominated by viscosity and initially has high ℓ. Our models study the other extreme: the free dynamical infall of low $\ell \simeq \ell_{mb}$ gas. When this gas is introduced at large radii ($r \simeq 100$ M) it cannot support itself in a Keplerian orbit, as gas in the standard thin disk model can. Such an inflow thus represents the generalization of the Bondi spherically symmetric accretion problem (see refs. in Paper I) to the case of low angular momentum accretion. The gas must spiral in dynamically until it reaches the orbit appropriate to its angular momentum. We pose the question: can this cold gas then settle into some form of thin or thick disk while obeying the laws of hydrodynamics? If so, what are the generic properties of such disks?

A second goal is to study the possible flow patterns at the inner regions of disks. Pressure supported thick disks are believed to have nearly constant specific angular momentum along their inner edges, an angular momentum low enough to allow for inflow into the hole over the potential cusp (i.e., through a hole in the funnel wall). This inflow scenario was first suggested by B. Paczyński (unpublished) and has been discussed in detail[19]. Our investigation will shed light on the hydrodynamic properties of such nonviscous inflows. Furthermore, in the standard analytic thin disk model[29], one assumes no further heating as the gas dynamically spirals into the hole from the last stable circular orbit at r = 6 M. We investigate this assumption using the equations of hydrodynamics.

The infalling gas might already be hot ($\varepsilon \simeq$ binding energy) at r = 100 M. We have investigated some of these models as well. However, for the purposes of comparing the flow behaviors of different accretion scenarios, we will assume the gas is initially cold and restrict the discussion in this paper to only two parameters: 1) the initial geometric thickness of the inflow, and 2) the value of ℓ. Our models further assume that all the gas has the same values of (ℓ, U_t) as it enters the grid at r = 100 M. Other inflow boundary conditions can and have been studied with our code[10].

We will describe four archetypical models which we believe delineate qualitatively different features in this two parameter space. In order, we will examine: 1) thick inflow with $\ell > \ell_{mb}$, 2) thin inflow with $\ell > \ell_{mb}$, 3) thin inflow with $\ell_{ms} < \ell < \ell_{mb}$, and 4) thick inflow with $\ell_{ms} < \ell < \ell_{mb}$. Thick and thin inflow with $\ell < \ell_{ms}$ is discussed briefly. Table I summarizes our mapping out of the solution phenomenology of this parameter space.

Table I Flow Regimes For Black Hole Accretion

Angular Momentum	Thin Inflow	Thick Inflow
$\ell < \ell_{ms}$	All Gas Flows Into Hole On Dynamical Timescale	
$\ell_{ms} < \ell < \ell_{mb}$	Entire Flow Into Hole Shock Heating Near Horizon Nozzle Flow Onto Hole	Thick Disk Forms Some Funnel Outflow
$\ell > \ell_{mb}$	Two Temperature Disk Forms Hot Wind Outflow No Flow Into Hole	Thick Disk Forms Large Funnel Outflow No Flow Into Hole

THICK DISK FORMATION AND HOLLOW JETS

First consider an accretion problem with a geometrically thick inflow with angular momentum $\ell > \ell_{mb}$. The cold ($\varepsilon = 10^{-5}$, h $\simeq$ 1.0) uniform density fluid is continually sent onto the grid at the outer boundary (r = 100 M) at every point up to the funnel wall defined by its angular momentum (ℓ = 4.5 M > ℓ_{mb}) and energy (U_t = -1.0). Inflow occurs with a radial freefall velocity. The outer boundary condition for this model has been modified by allowing the inflow velocity to vary to permit outflow at the boundary if the hydrodynamical solution requires it. If no

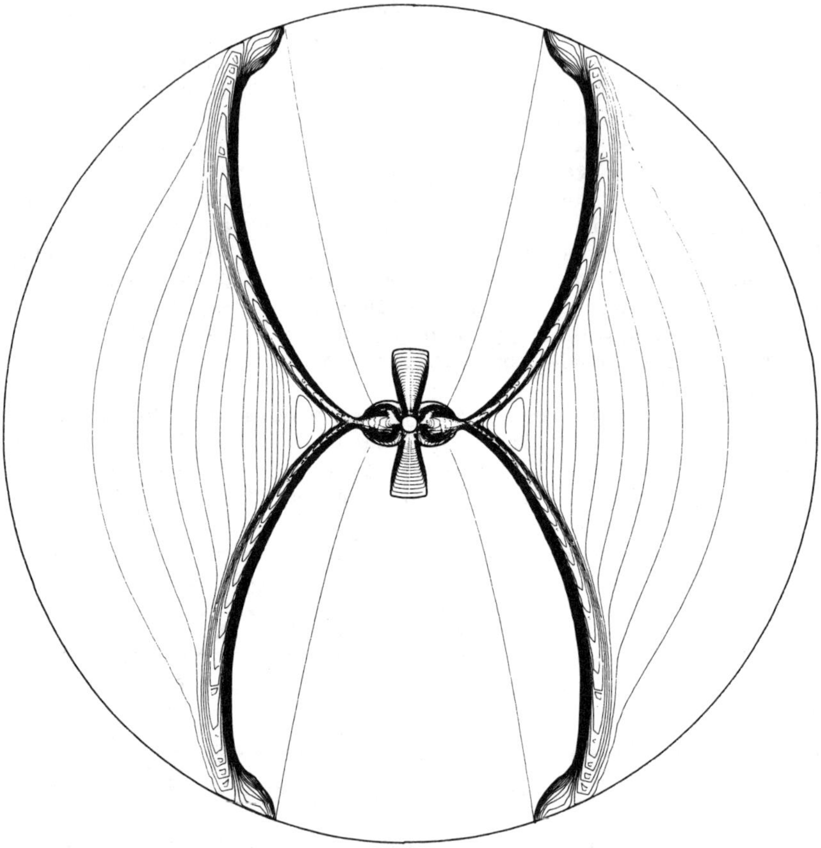

Fig. 4. The ℓ = 4.5 M thick inflow run at t = 500 M, roughly one freefall time from the outer boundary of r = 100 M. The funnel wall is shown with a single dark line. The heavy convergence of dark lines is the rapid falloff of density at the edge of the infalling fluid. Just outside this is the standoff shock. The last remnants of the initial Bondi accretion fluid is draining into the north and south poles of the black hole. The calculation resolution is 160 zones in θ from 0 to π, and 160 zones in r, staggered logarithmically, from 2 M to 100 M.

outflow occurs, steady inflow continues throughout the calculation. The initial condition is created by placing a Bondi, zero angular momentum solution (spherically symmetric) on the grid and matching the density and pressure to those of the incoming fluid at the outer boundary. This fluid will smoothly fall into the hole ahead of the high-ℓ gas coming from the boundary, lessening the transients which would be associated with fluid pouring onto a vacuum grid.

The funnel wall appropriate to this hU_t and ℓ is the outer line in Figure 3. It is clear that all the fluid will hit a

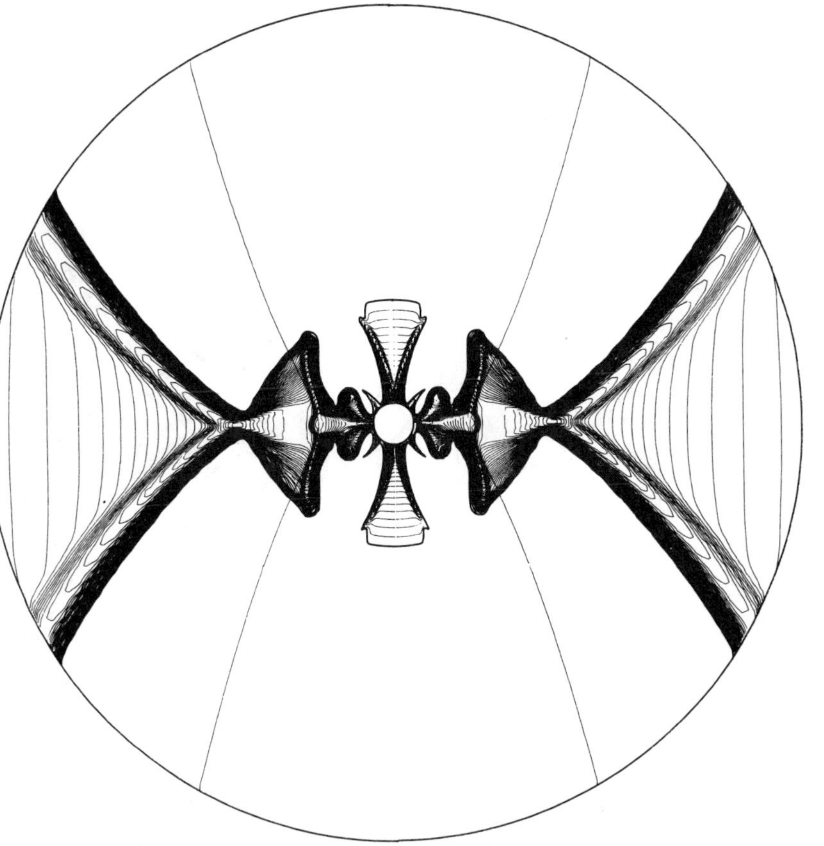

Fig. 5a. Same run as Figure 4 but at t = 570 M. This is a closeup with the outer edge at r = 30 M.

wall, a hydrodynamic fluid shocks. It is perhaps misleading to say that the fluid shocks at the funnel wall; it is the centrifugal deceleration (increasing towards the wall) which causes the shock. Nevertheless, the idea of a shock at a wall is a useful working analogy.

To understand the dynamic nature of the funnel region in this hydrodynamic accretion flow, it is useful to look in some detail of fluid density. The contour spacing is such that the fluid density doubles with every four contour levels. The first of these figures is Figure 4, at one freefall time from the outer boundary (t = 500 M). It shows that even though the fluid is first sent onto the grid radially at every point from the equator up to the static funnel wall, the flow does not remain radial. As the high angular momentum supersonic flow begins to feel the centrifugal deceleration, funnel wall standoff shocks form. These standoff shocks turn the flow sharply towards the equator; the

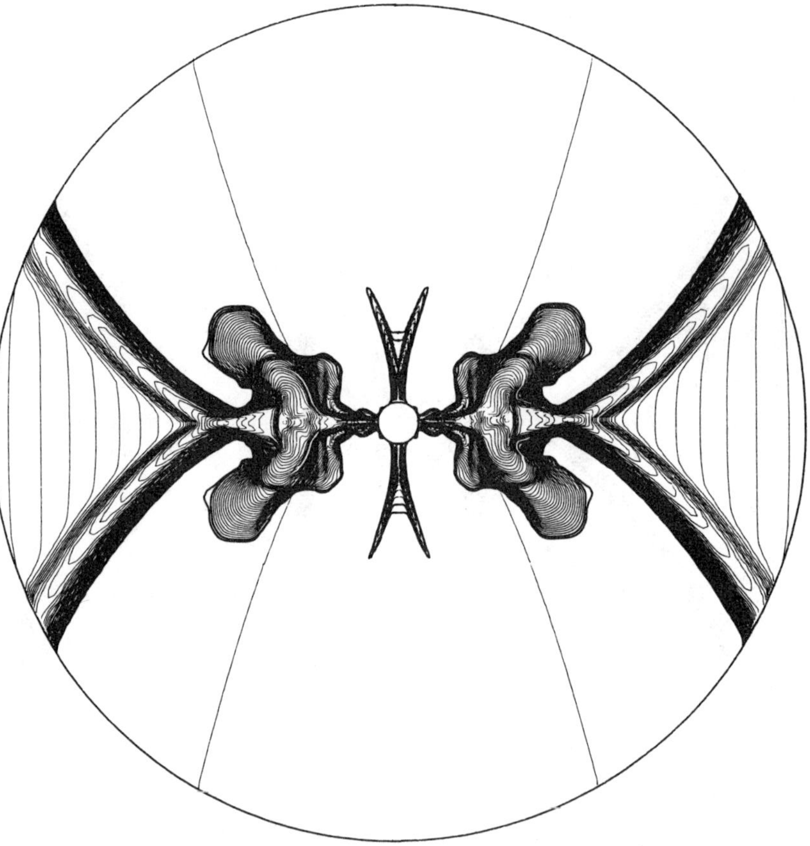

Fig. 5b. Same run as Figure 4 but at t = 620 M. This is a closeup with the outer edge at r = 30 M.

shocked fluid then flows down along the inside of the shock surface. In Figure 4, the heavy black curve lines represent the surface of the inflow while the closely spaced contour lines slightly outside of that surface represent the shock. Note that there is a large region of vacuum between the surface of the fluid and the static funnel wall (thin black line). This vacuum is centrifugally supported by the rotation of the fluid. This region will later act as the "escape route" for the shock heated gas. Note that as a leftover from our Bondi initial conditions, the last remnants of the zero angular momentum fluid inside the funnel wall can be seen falling into the north and south poles of the black hole.

The high-ℓ shocked fluid flows toward a point on the equator where the funnel wall standoff shocks cross. This crossing point occurs at $r \simeq 16$ M which is very close to the circular orbit appropriate to this specific angular momentum. After all the

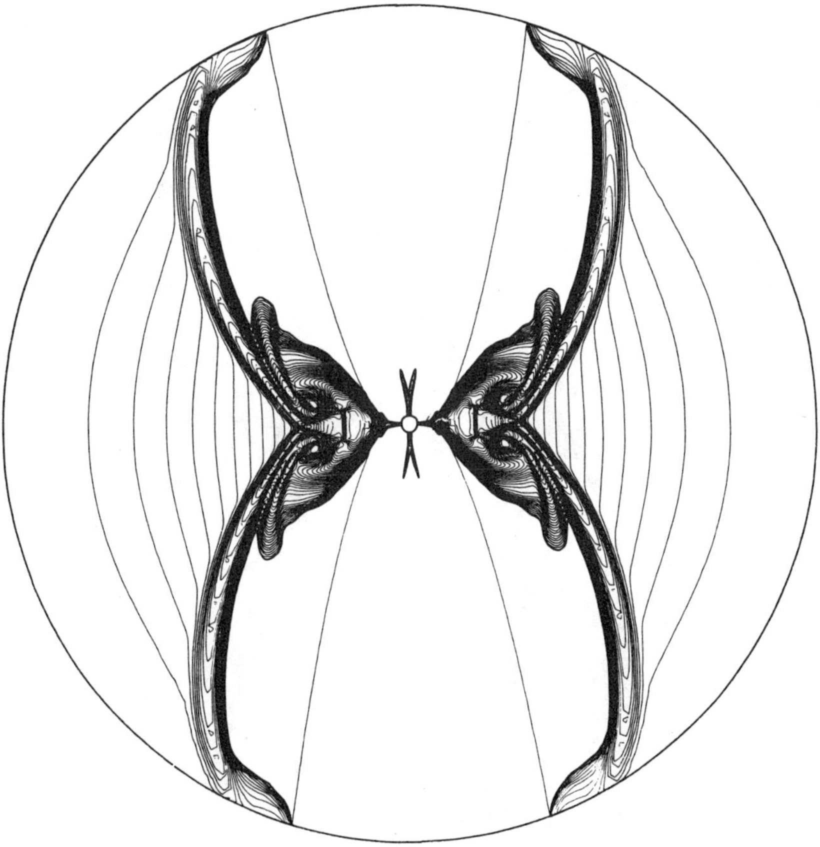

Fig. 6. Same run as Figure 4 but at t = 775 M. Outer edge is at r = 100 M.

heated fluid (now with $\varepsilon \simeq 0.01$) converges into the equatorial plane, it expands as it accelerates toward the hole.

At the end of the expanding equatorial stream, the fluid is centrifugally decelerated at the centrifugal barrier. In Figure 4 (t = 500 M), the enormous impulsive pressure gradient forces have temporarilly overwhelmed the centrifugal deceleration and are forcing some fluid through the funnel wall and into the hole. Two closely spaced stages of the rebound which follow are shown in closeup in Figure 5 (at times: a] t = 570 M and b] t = 620 M). In Figure 5a, one can see a strong shock forming near the funnel wall, as well as another pair of shocks forming even closer to the hole. We find that despite the close proximity of the flow to the horizon itself, these shocks nonetheless propagate upstream away from the hole. Although a small amount of material is forced through the funnel wall and into the hole by the high pressure gradients, most of the shocked gas rebounds backward from the

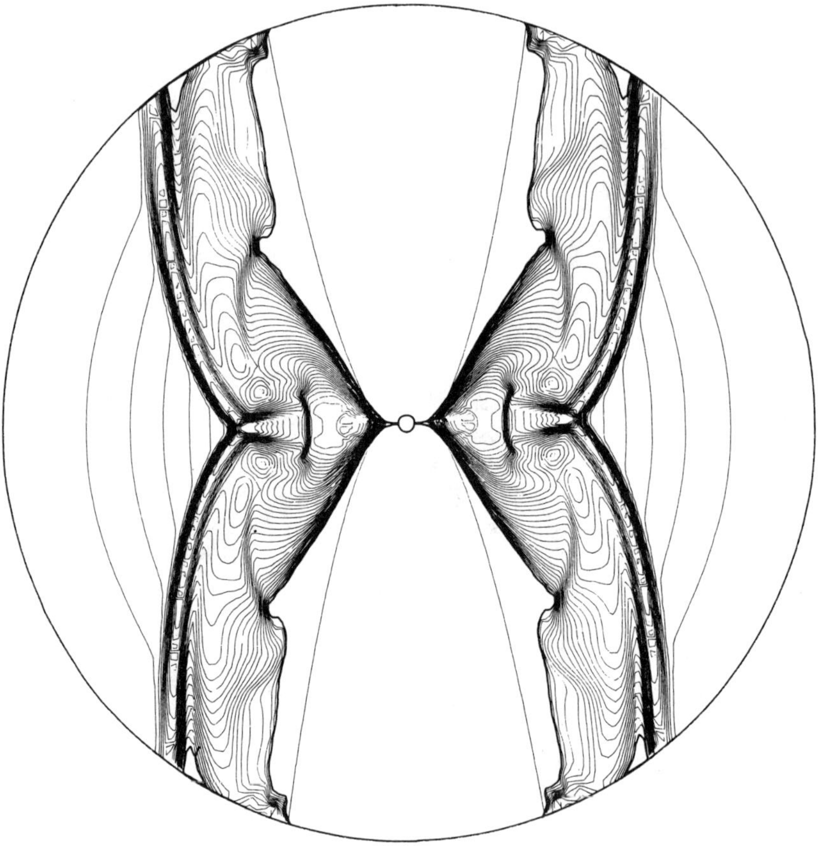

Fig. 7. Same run as Figure 4 but at t = 1300 M. Outer edge is at r = 100 M.

hole. As the shock formed off the funnel wall moves backwards, its central region becomes a "Mach ring", i.e., an accretion shock front perpendicular to the equator. Hereafter, we refer to this backward moving shock as the "accretion shock." Another example of such near hole shocks has previously been discussed[9] and illustrated in detail in Figures 3-5 of that paper.

The black hole accretion shock acts much like an accretion shock in a neutron star or white dwarf accretion flow which brings the supersonic flow to subsonic velocities. There is, of course, no solid surface for a black hole such as exists on a neutron star or white dwarf; however, our calculations demonstrate that the centrifugal barrier can, for all practical purposes, act like a solid surface off of which the fluid can shock. This is the most important aspect of black hole inflows with angular momentum which our computer runs have uncovered. Even a relatively low value of ℓ can create an effective surface for the black hole.

The hot, subsonic post-shock fluid then begins (Fig. 5b, t = 720 M) to splash back from the centrifugal barrier into the vacuum region evacuated by the previously formed funnel wall standoff shocks. By time t = 775 M (Fig. 6) this backflow has run into the infalling fluid and the "backflow standoff shock" forms to divert the flow away from the equator and up along the interior of the surface of the inflowing fluid. This processed fluid is now freely flowing up along the channel between the funnel standoff shocks and the static funnel wall. Much later at t = 1300 M (Fig. 7; note the last three figures, 5a, 5b, and 6 were separated by only 50 M in time), the full flow pattern has been established.

The thick inflow shock structures are schematically illustrated in Figure 8. To quantify the Figure, we follow the thermal history of an infalling gas cell through this shock

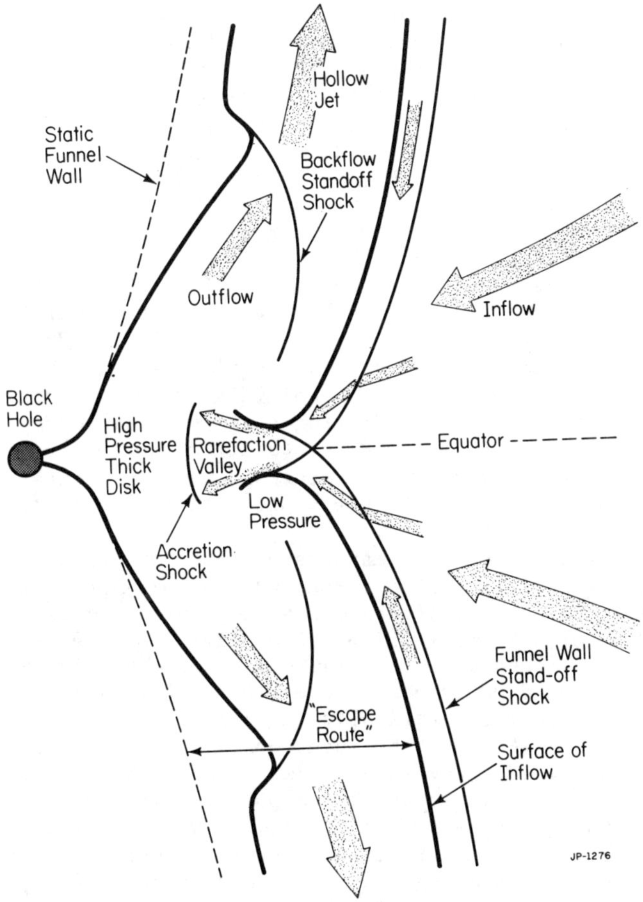

Fig. 8: Schematic diagram of shock structures and flow features seen in Fig. 7. These late time structures seem generic and stable.

structure. It begins at the outer boundary with $\varepsilon = 10^{-5}$ and $U_t = -1.0$. On the fall from the outer boundary, adiabatic heating increases the temperature of the fluid by 50% prior to the first shock heating. The standoff funnel shocks raise the temperature to an average of 0.007 (the temperature and shock strength decrease as the height above the equator increases). The standoff shocks do not bring the gas to rest, rather it accelerates through the rarefaction valley until it hits the accretion shock. There its radial motion is stopped, forming a thick pressure supported disk.

The reason that the accretion shock was able to stop the gas can be understood in terms of the energy of the circular orbit and the shock strength. In order for the shock to bring the fluid to rest, it must convert all the kinetic energy to internal energy, i.e., it must lower the fluid's energy $-U_t$ from 1.00 to the value associated with the circular orbit. Since $hU_t = (1 + 4\varepsilon/3)\, U_t$ is conserved in the shock, this means that the temperature ε must be raised accordingly. For fluid with a specific angular momentum of $\ell = 4.5$ M, the circular orbit is located at $r = 15.3$ M, with an energy $-U_t$ of 0.970. This requires that ε to rise to 0.023. We find in our model that the pressure maximum of the disk is actually located at $r = 15$ M with an $\varepsilon = 0.02$, and an energy $-U_t = 0.970$, in very close accord with our analytic expectations. It should be noted that these results are consistent with a numerical error in the energy of about 5%.

The high pressure of the hot disk drives shocked fluid into the lower pressure "escape route" between the standoff funnel shocks and the static funnel wall (see Figs. 5b, 6, 7). It is once again shocked by the "backflow standoff shock" which diverts the outflow and prevents it from colliding with the infalling fluid. A portion of the outflowing fluid finally leaves the outer grid with $\varepsilon = 0.01$ and $-U_t = 0.987$ (still slightly bound). Many of the shock features and flow patterns shown in Figure 8 also occur in the propagation of supersonic jets in the Newtonian nongravitational regime[30-35].

This $\ell = 4.5$ M model demonstrates that hot, pressure supported thick disks can be formed as a natural part of the hydrodynamic structures of accretion flows with angular momentum $\ell > \ell_{mb}$. However, the disk which is formed is not the stationary structure envisioned by the analytic disk theory. The disk is but a density enhanced pattern in a complicated inflow and outflow. Further, when dynamic infall plays a role in disk structure or formation, the disk cannot be isentropic. To bring fluid to rest requires a backwards traveling shock, a shock which will decrease in strength as it climbs out of the potential. Thus, the jump in entropy of the gas forming the thick disk will decrease with time.

We further conclude that while this accretion scenario is effective in shock heating fluid and driving an outflow, it does not produce a collimated outflow. Nothing has reduced the specific angular momentum in the fluid and the U_ϕ term in the Euler equation remains important. The flow seeks to move away from the orbital axis to the extent that it is permitted. In this thick flow, this means up along the funnel wall. Thus, these hydrodynamic processes form hollow conical jets. The same angular momentum which formed a funnel in the first place guarantees that the outflow generated from this same fluid will be driven along the exterior of the funnel wall, not up the funnel interior.

It is also apparent from this run that purely hydrodynamic process do not produce highly energetic outflows. The hollow jets do not have a total energy significantly in excess of the energy of the infalling fluid. Nonetheless, this model does show one new way in which outflow of hot gas from near the hole can occur. But clearly, an acceleration mechanism other than hydrodynamic shocks is required to produce the observed highly energetic jets. Further, if there is to be near hole collimation, then the angular momentum which makes these jets hollow must be contended with whenever the disk supplies the jetted material.

An interesting facet of these hollow jets is that they effectively line the funnel walls with hot material. The funnel is, by definition, an evacuated column extending out away from the hole. The funnel can therefore serve as a channel through which radiation emitted by the hot outflow escapes to infinity. The hydrodynamic process has produced hot gas and placed it in such a position as to allow it to effectively radiate into the vacuum funnel interior. Whether this new picture of a black hole accretion flow can overcome the objections for radiation driven jets, metioned above, remains to be seen.

TWO TEMPERATURE DISK WITH HOT WIND

We now consider the identical startup parameters, except that the width of the inflow is narrow and the fluid enters the outer boundary as a thin ring. The fluid freefalls until it encounters the centrifugal deceleration near the funnel wall. An accretion shock forms which heats the fluid to a temperature of $\varepsilon \simeq 0.002$, but not high enough to bring the fluid to rest. The hot, shocked fluid can no longer be confined to the equatorial plane; it flows outward, above and below the thin disk in a hot wind.

An intermediate stage in the process is depicted in Figure 9 at t = 1650 M after the start of the infall (again the freefall time is about 500 M). As before, the contours represent the logarithm of density. The inflowing fluid is the wedge along the equator. The accretion shock is nestled into the funnel wall and the flow emerging from this shock surface moves up and back around the inflow. Note that there are no standoff funnel wall shocks because the fluid encounters the funnel wall only in the equatorial plane. As more and more fluid is processed through the accretion shock, the thickness of the disk increases and a vertical pressure gradient develops. A fully dynamic two temperature disk forms with heat provided by the shock. Such a disk does not resemble the traditional equilibrium thick disk[36]. It is instead a two temperature structure with cool (pre-shock) gas inflowing along the equator and a surrounding wind of hot (post-shock) gas flowing out.

In contrast to the thick inflow, the resulting outflow looks less like a hollow jet and more like a hot wind. Because the heated fluid is not confined by an inflow along the funnel standoff shocks as in the previous case, it is free to flow almost radially outwards, surrounding the inflow. Eventually, it provides sufficient pressure above and below the inflow, to confine the inflow in an equatorial stream. The lack of geometric confinement and insufficient accretion shock strength causes a standing shock to form rather than the slowly outward traveling accretion shock.

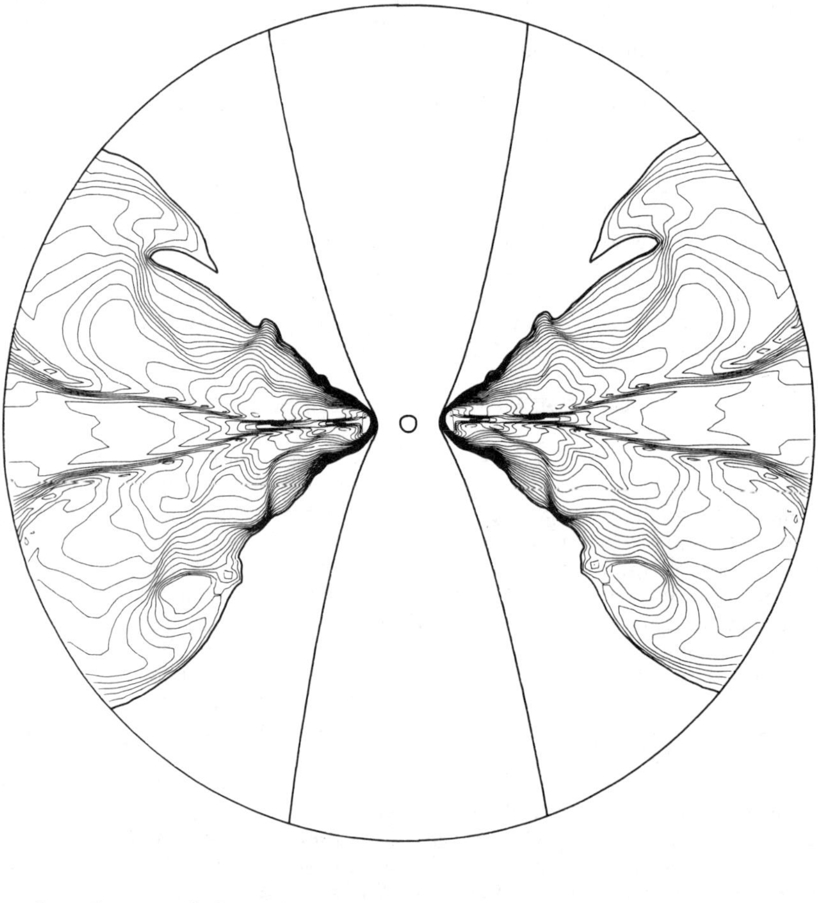

Fig. 9. The ℓ = 4.5 M thin inflow at t = 1650 M. The wedge of material in the equatorial plane is the inflowing gas, the wind above the disk has been shock heated at the accretion shock just outside the funnel wall. Note that there is no funnel wall standoff shock as there is in Figure 7. For this value of ℓ, no flow goes down the hole.

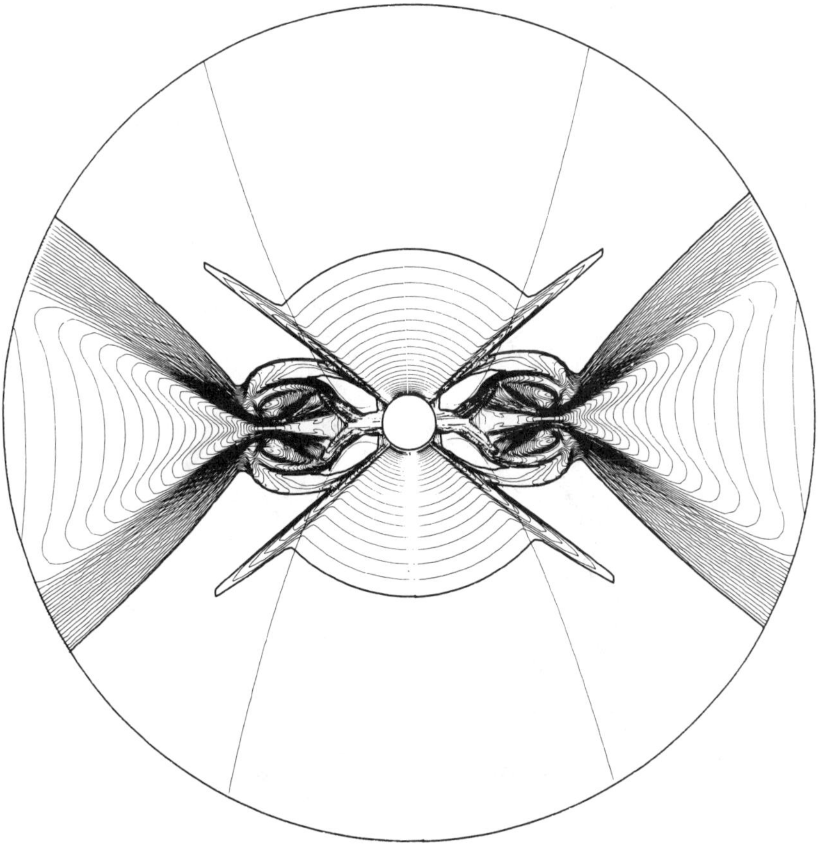

Fig. 10. The ℓ = 3.8 M thin inflow at t = 970 M in closeup (outer edge r = 30 M). Note that the funnel wall does not intersect the equator. Even so, the fluid shocks off the overhanging funnel wall. The flow cannot be reversed by the centrifugal acceleration and the entire flow is focused down the hole.

LIFE INSIDE 6M - NOZZLE ACCRETION INTO HOLES

We now consider what happens when we rerun the thick and thin inflows, but with a slightly lower specific angular momentum than the ℓ = 4.5 M used in the model described above. We will use a value of ℓ = 3.8 M which is below the marginally bound value of $\ell = \ell_{mb}$ = 4.0 M while staying above the marginally stable value of $\ell = \ell_{ms}$ = 3.46 M. As seen in Figure 3, the funnel wall is now open onto the hole and inflow into the hole becomes possible.

The first run with ℓ = 3.8 M has the same thin inflow outer boundary condition as that of the ℓ = 4.5 M run shown in Figure 9.

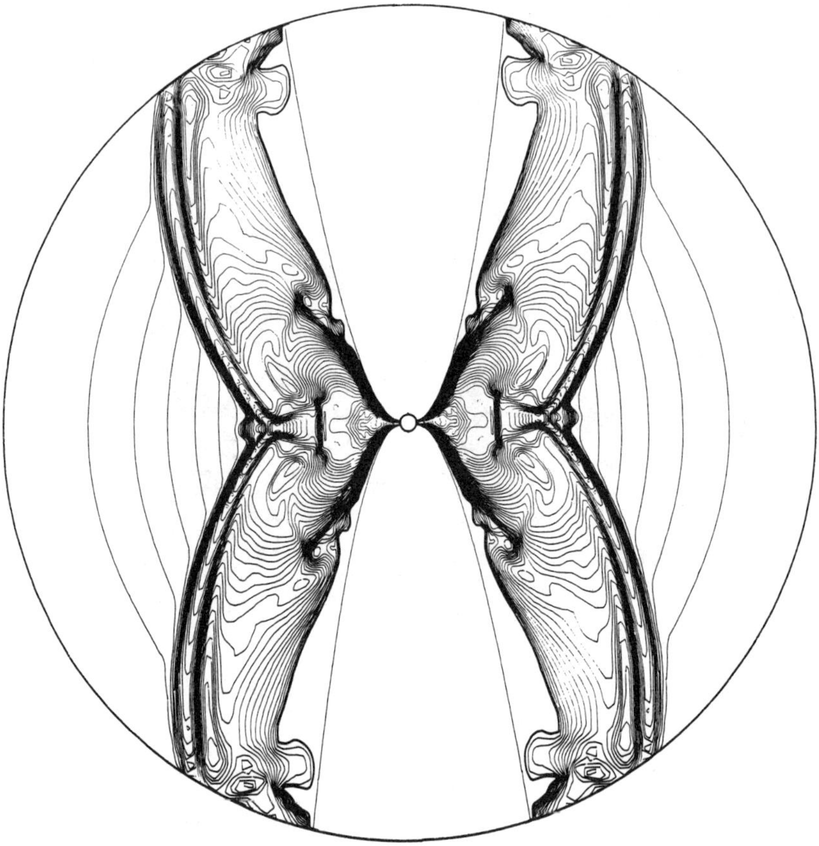

Fig. 11a. The ℓ = 3.8 M thick inflow at t = 1700 M. Compare to Figure 7 to see the great similarity.

However, as shown in Figure 10, the ℓ = 3.8 M flow is completely different. Such a fluid has a centrifugal barrier which turns over at r = 4.57 M (see Figure 2). To hit this barrier, the fluid's energy must drop below $-U_t$ = 0.959. As Figure 10 shows, the inflow proceeds smoothly from the outer boundary to a convergence point on the equator. Here an equatorial stream forms, similar to the ones noted in the ℓ = 4.5 M run above. A pair of criss-cross shocks form at r = 10 M where $-U_t$ is reduced to its minimum value of 0.987. Note that this is insufficient to permit the fluid to hit the centrifugal barrier. Interior to this point, the flow expands in a rarefaction valley. However, there is no terminal accretion shock to further reduce the energy $-U_t$ and therefore the flow continues, unhindered, into the hole. No thick disk, no hot wind, and no hollow jet is created.

In contrast, we now consider the effects of reducing ℓ below the marginally bound value when the flow is geometrically

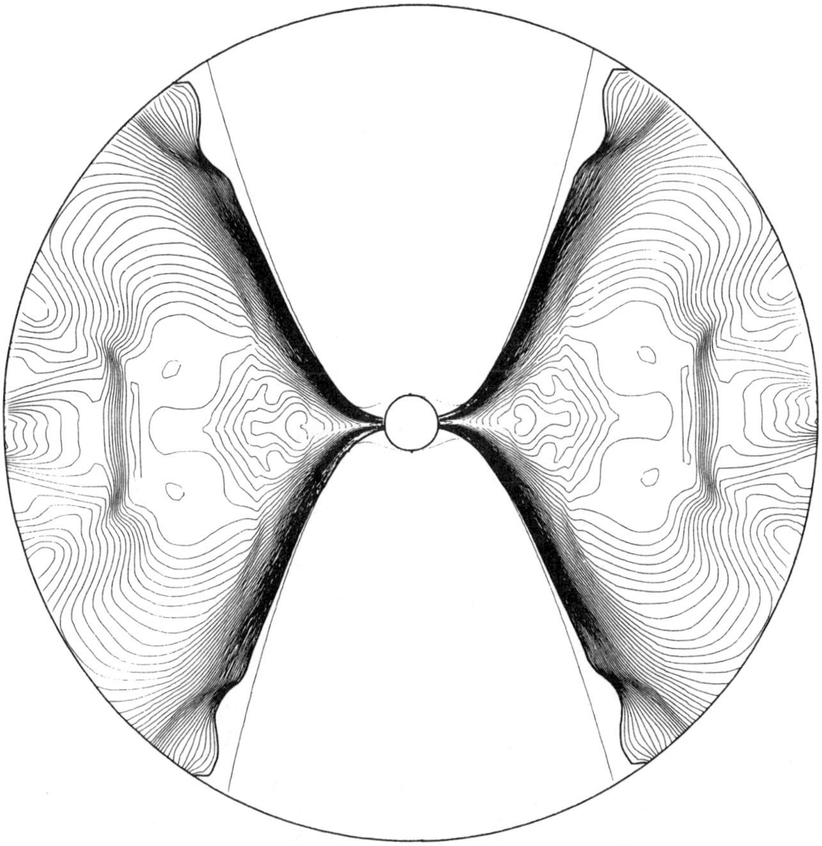

Fig. 11b. A closeup of Figure 11a with outer edge r = 30 M. Note that the funnel wall focus the flow into the hole through a "nozzle." About 4% of the inflow from r = 100 M goes down the hole.

thick. We will use the same thick inflow boundary conditions as in the ℓ = 4.5 M run shown in Figures 4-8, but with the lower ℓ = 3.8 M value for the specific angular momentum used in Figure 10. The formation of the funnel wall standoff shock and the rarefaction valley proceed as in the high angular momentum (ℓ = 4.5 M) thick inflow case described above. This is illustrated at time t = 1700 M in Figure 11a. Because of the greater convergence of fluid to the equator, the criss-cross shocks in the thick run are stronger than those in the thin inflow (Figure 10) by a factor of 10, as measured by pressure jump. In spite of this, it is still not strong enough to reduce $-U_t$ below the critical value. However, the enhanced pressure in the rarefaction valley pushes the fluid away from the equator and up against the overhanging funnel wall, increasing the centrifugal deceleration which in turn creates an accretion shock, allowing formation of a thick, hot disk. As in the ℓ = 4.5 M run (Figure 7), we see that the thickness of the inflow leads to drastically different accretion structures even though the angular momentum of the fluid is the same in Figures 10 and 11a.

Let us compare in more detail the two thick inflow runs of different angular momenta, represented by Figures 11 and 7. The single most important change is that inflow into the hole, proscribed in the ℓ = 4.5 M case, is now possible. A new feature, which we refer to as the "nozzle", forms at the inner edge of the disk. Here, an inward pointing pressure gradient drives inflow through the opening in the funnel wall and into the hole. A closeup of the fat disk and accretion nozzle is given in Figure 11b.

What are the characteristics of the flow through this nozzle? Inside of the accretion shock, the fluid velocities are subsonic. However, the inflow into the hole is supersonic, hence there is a sonic point within the nozzle. The nozzle flow begins at the inner edge of the disk at a radius of r = 5.5 M and the sonic point occurs at r = 4.4 M. It is interesting to note that for U_ϕ = 3.65 M, which characterizes the fluid in the nozzle, the unstable circular orbit (local effective potential maximum) is at r = 4.56 M. Thus, the flow goes transonic inside of the potential barrier peak.

The accretion rate through the nozzle and into the hole is constant after the nozzle flow has been established. In steady state the flow into the hole is 4% the amount of fluid sent onto the grid. Clearly, much more fluid is coming onto the grid than the nozzle can handle. This suggests that the nozzle is capable of supporting only a limited inflow rate. When the accretion rate exceeds this limit, fluid piles up and a thick disk forms.

The nozzle is quite similar to a deLaval nozzle[37-39]. The nozzle walls are provided by centrifugal confinement. Given a fixed entropy, energy hU_t and a transonic point in the nozzle, the mass flux is fixed. A larger accretion rate requires that the nozzle "walls" expand, resulting in centrifugal deceleration and the shock heating required to reduce U_t below the critical value.

In addition to its role in creating shocks and promoting disk formation, the nozzle plays an important role in the static fat disk model[19,20,36] by providing the mechanism for the final accretion into the hole. In this picture, viscous stress acts to lower the fluid's angular momentum sufficiently to permit flow past the maximum in the effective potential, i.e., through the funnel wall. The transonic nozzle observed in this model is exactly the hydrodynamics expected. Our calculation shows that such an inflow nozzle arises naturally in the accretion of low angular momentum fluid ($\ell < \ell_{mb}$), and demonstrates the viability of the idea that the final inflow from a thick disk into the hole is through just such a nozzle.

Of course, the formulation of a complete fat disk model involves viscosity and many questions remain unanswered. For example, can viscous stress increase the accretion rate above what can be handled by the nozzle? Presumably shocks would then occur driving outflow along the funnel wall in a manner analogous to that discovered in these calculations. Alternatively, viscosity along the inner edge of the disk might reduce the angular momentum sufficiently to open the nozzle up, thereby increasing the permitted inflow rate. These questions regarding what dynamics might be induced by viscosity can only be answered by a more complete model. Nevertheless, the hydrodynamic processes we have describe remain applicable in the flows which result.

CONCLUSIONS

Our aim in this paper has been to describe new hydrodynamic pheomena in black hole accretion flows and suggest alternative ways to look at the role of the funnel. First and foremost, these results are an extension of the Bondi spherical accretion problem to nonspherical inflows. But these calculations also suggest modifications to both the standard thin and fat disk paradigms.

In the Bondi solution, the angular momentum is zero; the inflow is purely radial. The fluid is adiabatically heated as it falls, but this heating is very inefficient. All the gas falls into the hole. The flow, initially subsonic at large r, passes through a transonic critical point at some radius and infalls supersonically with a velocity on the order of, but below, free fall. Since a black hole has no surface, the inflow does not end in an accretion shock; it merely passes into the hole without affecting the fluid behind it.

This scenario is qualitatively changed by even a small amount of angular momentum. Consider the canonically "insignificant" angular momentum, that less than the marginally stable value. While we did not specifically discuss this case here, we have performed a number of calculations which show that despite the lack of a stable circular orbit, even this low value of ℓ alters the inflow by creating an evacuated funnel along the orbital axis. Furthermore, standoff shocks are formed which focus the flow into the hole through the equatorial plane--a very nonspherical form of accretion. Whether there is any flow with angular momentum which resembles Bondi flow remains to be determined.

As ℓ increases to the point (ℓ_{ms}) where a stable circular orbit is possible, an orbital turning point is created (Figure 2). This centrifugal barrier begins to hinder inflow into the hole. If the fluid is marginally bound, and $\ell < \ell_{mb}$, enough kinetic energy may be retained to overcome this potential barrier. This occured in the thin inflow model (Figure 10). However, the funnel walls restrict the available space through which the fluid can flow into the hole. This creates an effective nozzle, and the fluid shocks when the inflow is greater than what the nozzle can carry. Such was the case in the thick inflow model with $\ell = 3.8$ (Figure 11). The nozzle continues to support flow into the hole at a fixed rate but, when the total inflow rate exceeds that level, fluid piles up and a hot disk forms. This disk drives outflow through the evacuated space between the static funnel wall and the inflowing fluid's surface.

When the angular momentum is greater than marginally bound, the nozzle closes (Figure 2) and none of the infalling fluid ends in the hole. The thin inflow model changes drastically from the $\ell < \ell_{mb}$ case, where all the flow goes down the hole, to a flow which blows away from the hole in a hot wind. The thick inflow model is very similar to the $l < l_{mb}$ case, except that now no flow goes into the hole through the nozzle.

We have emphasized only those features which arise from consideration of the fluid dynamics in the near hole region, yet the fluid dynamics alone are sufficient to produce interesting and important new phenomena. Historically, the move from the standard analytic thin disk theory to the comparative complexity of the analytic thick disk was achieved simply through the addition of pressure. In our models, we take this a step further by considering time-dependent flows. In each of these cases, important and unanticipated features emerge from the addition of new physics. We anticipate that this trend will continue as additional effects, including radiation transport, viscous torques, magnetic fields, and turbulence are added in the development of a realistic, complete theory. Such a theory would tell us which effects dominante and in what circumstances. However, even when additional physical processes are considered, the richness of the hydrodynamical pheomena we have reported on herein remains to be reckoned with.

ACKNOWLEDGEMENTS

We thank the Max-Planck-Institut für Physik and Astrophysik, Institut für Astrophysik, and its Director Kippenhahn for their hospitality, generous support, and the use of the CRAY-1 computer upon which these calculations were performed. The tireless and patient help of our coworkers Karl-Heinz Winkler and Michael Norman is gratefully acknowledged. In addition we thank Martin Rees, Mitch Begelman, Roger Blandford, James Wilson, Dean Sumi and William Press for useful discussions, and Richard Brandolino for assistance with software development for the data analysis, performed on the University of Illinois Vax and Image Processing System.

REFERENCES

1. Begelman, M. C., Blandford, R. D., Rees, M. J., Rev. Mod. Phys., **56**, 255 (1984).
2. Willis, B. et. al., Astrophys. J., **274**, 62 (1983).
3. Hawley, J. F., Smarr, L. L., Wilson, J. R., Astrophys. J., **277**, 296 (1984 Paper I).
4. Hawley, J. F., Smarr, L. L., Wilson, J. R., Astrophys. J. Suppl. Ser., **55**, 211 (1984 Paper II).
5. Wilson, J. R., Astrophys. J., **173**, 431 (1972).

6. Wilson, J. R., from Ann. N. Y. Acad. Sci., **262**, 123 (1975).
7. Wilson, J. R., in Proceedings of the Marcel Grossman Meeting ed. R. Ruffini (Amsterdam: North Holland, 1977), p. 393.
8. Wilson, J. R., in Proceedings of the Int. Sch. of Phys. Fermi Course LXV, eds. R. Giacconi and R. Ruffini (Amsterdam: North Holland, 1978), p. 644.
9. Hawley, J. F., Smarr, L. L., in Problems of Collapse and Numerical Relativity, eds. D. Bancel and M. Signore (Dordrecht: Reidel, 1984), p. 253.
10. Hawley, J. F., Ph.D. Thesis, University of Illinois, (1984).
11. Zabusky, N. J., Physics Today, July 1984, p. 36.
12. Ruffini, R. and Wilson, J. R., Phys. Rev. D, **12**, 2959 (1975).
13. Damour, T., Hanni, R. S., Ruffini, R., Wilson, J. R., Phys. Rev. D, **17**, 1518 (1978).
14. Sloan, J. H. and Smarr, L. L., in Numerical Astrophysics: A Festschrift in Honor of James R. Wilson eds. J. Centrella, R. Bower, and J. LeBlanc (Boston: Jones and Bartlett, 1984), p. 52.
15. Thorne, K. S., Macdonald, D., Mon. Not. R. Astron. Soc. **198**, 339 (1982).
16. Misner, C., Thorne, K., and Wheeler, J., Gravitation (San Francisco: Freeman, 1973).
17. Smarr, L. L., Taubes, C., Wilson, J. R., from Essays in General Relativity, ed. F. J. Tipler (New York: Academic Press, 1980), p. 157.
18. Shakura, N. I., Sunyaev, R. A., Astron. Astrophys. **24**, 337 (1973).
19. Kozłowski, M., Jaroszynski, M., Abramowicz, M. A., Astron. Astrophys. **63**, 209 (1978).
20. Abramowicz, M. A., Jaroszynski, M., and Sikora, M., Astron. Astrophys., **63**, 221 (1978).
21. Abramowicz, M. A., Acta Astron., **24**, 45 (1974).
22. Fishbone, L. G., Moncrief, V., Astrophys. J., **207**, 926 (1976).
23. Lynden-Bell, D., Phys. Scripta, **17**, 185 (1978).
24. Abramowicz, M. A., Piran, T., Astrophys. J. Lett., **241**, L7 (1980).
25. Sikora, M., Wilson, D. B., Mon. Not. R. Astron. Soc. **197**, 529 (1981).
26. Nobili, L., Turolla, R., Calvani, M., Lett. Nuovo Cimento, **35**, 335 (1982).
27. Piran, T., Astrophys. J. Lett., **257**, L23 (1982).
28. Rees, M. J., Begelman, M. C., Blandford, R. D., Phinney, E. S., Nature, **295**, 17 (1982).
29. Novikov, I., Thorne, K. S., in Black Holes, ed. B. DeWitt and C. DeWitt, (New York: Gordon and Breach, 1973), p. 343.
30. Norman, M. L., Smarr, L. L., Winkler, K.-H. A., Smith, M. D., Astro. Astrophys., **113**, 285 (1982).
31. Norman, M. L., Winkler, K.-H. A., Smarr, L. L., in Astrophysical Jets, ed. A. Ferrari, A.G. Pachokzyk (Reidel: Boston, 1983), p. 227.

32. Norman, M. L., Smarr, L. L., Winkler, K.-H. A., in Numerical Astrophysics: A Festschrift in Honor of James R. Wilson, ed. J. Centrella, R. Bowers, and J. LeBlanc (Boston: Jones and Bartlett, 1985), p. 88.
33. Smarr, L. L., Norman, M. L., Winkler, K.-H.A., Physica **12**, 83 (1984).
34. Smith, M. D., Norman, M. L., Smarr, L. L., Winkler, K.-H.A., Mon. Not Roy. Astron. Soc., submitted (1984).
35. Norman, M. L., Winkler, K.-H. A. in Astrophysical Radiation Hydrodynamics ed. K.-H.A. Winkler and M. L. Norman (Reidel: Boston, 1985), in press.
36. Abramowicz, M. A., Calvani, M., Nobili, L., Astrophys. J., **242**, 772 (1980).
37. Courant, R. and Friedrichs, K. O., Supersonic Flows and Shockwaves, (New York: Springer Verlag, 1948).
38. Norman, M. L., Smarr, L., Wilson, J. R., and Smith, M. D., Astrophys. J., **247**, 52 (1981).
39. Smith, M. D., Norman, M. L., Smarr, L. L., Wilson, J. R., Astrophys. J., **264**, 432 (1983).

THEORY OF AXISYMMETRIC MAGNETO-HYDRODYNAMIC FLOWS

R.V.E. Lovelace
Department of Astronomy and Department of Applied Physics,
Cornell University, Ithaca, N.Y. 14853

C. Mehanian, C. M. Mobarry, and M. E. Sulkanen
Department of Physics, Cornell University, Ithaca, N.Y. 14853

ABSTRACT

A derivation from the weak-gravity limit of general relativity is given of the basic equation for the flux function $\psi(r,z)$ for general, steady, axisymmetric, relativistic, ideal magneto-hydrodynamic flows around a black hole or a rotating magnetized star. One limit of this equation gives the Grad-Shafranov equation, which describes the equilibrium of axisymmetric fusion plasmas. Another limit gives, except for an additional term, the 'pulsar equation' of Scharlemann, Wagoner, and Michel for relativistic plasma flows around an aligned, rotating, magnetized neutron star.

Detailed applications of the theory are made to thin, magnetized disks around an aligned, rotating, magnetized star. The present work does not consider field-aligned currents or flows, so that the oppositely directed electromagnetic beams discussed in our earlier work are not included. Nevertheless, the magnetic forces may have significant dynamical effects on the disks. We have generated a family of self-consistent disk equilibria having, with increasing r, an X point and an O point in the equatorial plane. Inside the X point, which must lie within the speed-of-light radius of the star, the disk matter co-rotates with the star. Outside the X point, the angular rotation of the disk matter depends on the flux function $\psi(r,z)$.

I. INTRODUCTION AND SUMMARY

A disk-like distribution of gas arises naturally from the radiative cooling of a rotating gas cloud owing to the slow process of angular momentum removal. Of particular interest is the nature of the disks subsequent to the formation of a star or a black hole. The theory of the thin Keplerian accretion disks around stars or black holes has been extensively developed (Shapiro and Teukolsky 1983, and references therein). The key ingredient of the theory is the turbulent viscosity which allows the gradual outflow of angular momentum and the inflow of matter. The infalling matter loses its excess energy to radiation.

The influence of small-scale (<<r) ordered (Coroniti 1984, and references therein) and fine-scale turbulent magnetic fields (Eardley and Lightman 1975) has been studied in detail. Less attention has been given to the nature of disks under conditions where there is a highly ordered magnetic field. A wide range of astrophysical evidence points to the prevalence of well-ordered magnetic fields.

Ghosh and Lamb (1979a, 1979b) and Lamb (1984) have developed a detailed physical theory of accretion disks surrounding an aligned, rotating neutron star with an intrinsic magnetic dipole field. This theory provides a good fit to the main observational parameters for pulsating X-ray sources. However, the theory does not attempt to give a self-consistent derivation of the magnetic field structure.

The present work develops a general theory for the axisymmetric flows of an ideal relativistic magneto-hydrodynamic (MHD) fluid around a star or black hole. The theory leads to a basic, second-order, non-linear partial differential equation for the flux function $\psi(r,z)$ [in a cylindrical (r, ϕ, z) coordinate system]. This equation for ψ is derived from the basic equations of general relativity in the limit of weak gravitational fields. The function ψ is such that the poloidal (r,z) projection of a magnetic field line is described by $\psi(r,z) = \text{const}$. The equation for ψ involves five essentially arbitrary functions of ψ which arise from: (1) the conservation of mass (F); (2) the perfect electrical conductivity (G); (3) the conservation of angular momentum (H); (4) conservation of energy (J, which is Bernoulli's constant); and (5) the conservation of entropy (S). The identification of these functions follows from the work of Woltjer (1959), Mestel (1961, 1968), Heinemann and Olbert (1978), Zehrfeld and Green (1972), Chan and Henriksen (1980), and Aly (this volume).

In the absence of flow (F,G = 0), the equation for ψ becomes the Grad-Shafranov equation, which is the basic equation for the equilibrium of axisymmetric fusion plasmas such as Tokomaks (Grad and Rubin 1958, Shafranov 1966, Miyamoto 1976). For this reason we refer to our basic equation for ψ as the generalized Grad-Shafranov equation. In the absence of a magnetic field, the equation for ψ becomes the equation for the Stokes' stream function of a general axisymmetric flow (Scott and Lovelace 1982). In a further limit, the equation for ψ gives, except for an additional term, the 'pulsar equation' (Scharlemann and Wagoner 1972, and Michel 1973) for relativistic plasma flows around aligned, rotating, magnetized neutron stars.

Detailed applications of the theory are made in the calculation of thin, self-consistent magnetized disk equilibria around an aligned, rotating, magnetized star. For simplicity, these applications exclude the possibility of field aligned currents and flows (F = 0, H = 0). Thus, the oppositely directed electromagnetic beams proposed earlier as a model for extra-galactic double radio sources (Lovelace 1976, Blandford 1976) are not included. Nevertheless, the magnetic forces may have significant, non-perturbative dynamical effects.

In Section II we derive the generalized Grad-Shafranov equation for MHD equilibria with weak gravity and arbitrary relativistic flow from the basic equations of general relativity. Finally, in Section III, we apply the results of Section II to thin magnetized disks around an aligned, rotating magnetized star.

II. RELATIVISTIC MHD EQUILIBRIA

Here, a derivation is given of the basic equation governing axisymmetric magnetohydrodynamic (MHD) equilibria with relativistic flow and/or pressure and a weak gravitational field ($|\Phi_g|/c^2 \ll 1$). Our notation follows that of Landau and Lifshitz (1962). An application of the results of this section is given in Section III. The basic equations are:

(1) Particle conservation,

$$(nu^\alpha)_{;\alpha} = 0 \quad , \tag{1}$$

where n is the 'proper' number density, that is the density in the reference frame comoving with the fluid, u^α is the fluid 4-velocity with $g_{\alpha\beta}u^\alpha u^\beta = -1$ and $g_{\alpha\beta}$ the metric tensor, and where the semicolon denotes covariant differentiation.

(2) Energy-momentum conservation,

$$[T_f^{\alpha\beta} + T_{em}^{\alpha\beta}]_{;\alpha} = 0 \quad , \tag{2}$$

where T_f and T_{em} are the energy-momentum tensors for the fluid and the electromagnetic field, respectively. For the fluid,

$$T_f^{\alpha\beta} = nwu^\alpha u^\beta + pg^{\alpha\beta}, \tag{3}$$

where $w = [e + p]/n$ is the enthalpy per particle, e is the proper internal energy density, and p is the proper pressure.

(3) The Maxwell equations,

$$F^{\alpha\beta}_{;\alpha} = \frac{4\pi}{c} J_e^\beta \quad ,$$

$$F_{\beta\gamma,\alpha} + F_{\gamma\alpha,\beta} + F_{\alpha\beta,\gamma} = 0 \quad ,$$

and

$$T^{\alpha\beta}_{em\;;\alpha} = -F^\beta_\gamma J^\gamma_e \quad , \tag{4}$$

where $F^{\alpha\beta}$ is the electromagnetic field tensor, and $J_e^\alpha = (\rho_e c, J)$ is the four-vector electric current-density with $J^\alpha_{e;\alpha} = 0$. A comma denotes ordinary differentiation.

(4) The equation for perfect electric conductivity,

$$u_\alpha F^{\alpha\beta} = 0 \quad . \tag{5}$$

(5) The equation of state for an ideal gas,

$$S = k_B(\Gamma - 1)^{-1} \ln(pn^{-\Gamma}), \tag{6}$$

where S is the proper entropy per particle; $\Gamma = 5/3$ for non-relativistic motion in the comoving frame, or $\Gamma = 4/3$ for ultra-relativistic motion. Also, notice that we have $e = nmc^2 + (\Gamma - 1)^{-1}p$ and $w = mc^2 + \Gamma p[n(\Gamma - 1)]^{-1}$, where m is the particle rest mass.

By contracting u_β with equations (2), (4), and (5), and using the thermodynamic identity $dw = dp/n + TdS$, one obtains

$$(nS\, u^\alpha)_{;\alpha} = 0 \quad , \tag{7}$$

which is the equation for the adiabatic motion of the fluid. By contracting $g_{\delta\alpha} + u_\delta u_\alpha$ with equation (2) we obtain the Euler equation

$$nwu^\beta(u_\alpha)_{;\beta} = -p_{,\alpha} - u_\alpha u^\beta(p)_{,\beta} + F_{\alpha\beta}J_e^\beta/c \quad . \tag{8}$$

We now simplify equations (1), (7), and (8) to the limit of large distances from a spherically symmetric mass M, $r \gg 2GM/c^2$. Also recall that we are considering equilibria so that $\partial/\partial t = 0$. In a non-rotating (laboratory) reference frame the metric tensor can be expressed as $g_{\alpha\beta} = \eta_{\alpha\beta} + h_{\alpha\beta}$, with $\eta_{\alpha\beta} = (-1,1,1,1)$ along the main diagonal, $h_{\alpha\beta} = -2\Phi_g c^{-2}(1,1,1,1)$ with $|h_{\alpha\beta}| \ll 1$, and $\Phi_g = -GM/r$ (Misner, Thorne, and Wheeler 1970). We let $u^\alpha = \tilde{u}^\alpha + \delta u^\alpha$, where $\eta_{\alpha\beta}\tilde{u}^\alpha\tilde{u}^\beta = -1$ and $\tilde{u}_\alpha = \eta_{\alpha\beta}\tilde{u}^\beta$, so that to first order in h, $\delta u^\alpha = -(1/2)h^{\alpha\beta}\tilde{u}_\beta$. Also, we let

$$g \equiv 1 - \Phi_g/c^2 \quad , \tag{9}$$

where $g > 1$. Thus, $u^0 = \tilde{u}^0 g$ and $u^i = \tilde{u}^i/g$ for $i = 1, 2, 3$.

The weak field approximation to equation (1) is

$$(ng\tilde{u}^\alpha)_{,\alpha} = 0 \quad . \tag{10}$$

The approximation to equation (7) is:

$$(ng\tilde{u}^\alpha S)_{,\alpha} = 0 \quad . \tag{11}$$

The approximation to the Euler equation (8) is:

$$nw\tilde{u}^\beta(\tilde{u}_\alpha)_{,\beta} = -[p_{,\alpha} + \tilde{u}_\alpha\tilde{u}^\beta(p)_{,\beta}] + f_\alpha^g + f_\alpha^{em} \quad , \tag{12}$$

where f_α^g is the gravitational force,

$$f^g_\alpha = \tfrac{1}{2} nw\tilde{u}^\beta \tilde{u}^\gamma [h_{\beta\gamma,\alpha} - h_{\beta\alpha,\gamma}] \quad , \tag{13a}$$

with $\tilde{u}^\alpha f^g_\alpha = 0$, and where f^{em}_α is the electromagnetic force,

$$f^{em}_\alpha = \frac{a(\alpha)}{c} F_{\alpha\beta} J^\beta_e \quad , \tag{13b}$$

with $u^\alpha f^{em}_\alpha = 0$ and $a[\alpha] = [g^2, 1, 1, 1]$ for $\alpha = 0, 1, 2, 3$. Equation (12) includes all terms first order in Φ_g but neglects terms of higher order, such as $(\Phi_g)^2$ and $\Phi_g |\underset{\sim}{\nabla}\Phi_g|$.

In three-space notation, with the over-tildes dropped,

$$f^g_0 = nw\gamma^2 (\underset{\sim}{v} \cdot \underset{\sim}{\nabla})(\Phi_g/c^2) \quad ,$$

$$f^g_i = -nw\gamma^2 \underset{\sim}{\nabla}(\Phi_g/c^2)$$
$$-nw(\gamma^2 - 1)(\underset{\approx}{1} - \frac{\underset{\sim}{v}\underset{\sim}{v}}{v^2}) \cdot \underset{\sim}{\nabla}(\Phi_g/c^2) \quad , \tag{13c}$$

where $u^0 = \gamma$ and $u^i = \gamma \underset{\sim}{v}/c$, with $\underset{\sim}{v}$ the flow velocity, and $\gamma = (1 - v^2/c^2)^{-\frac{1}{2}}$ the Lorentz factor of the flow. Also, in three-space notation, and with the over-tildes dropped,

$$f^{em}_0 = \frac{a(0)}{c} F_{0\beta} J^\beta_e = -g^2 \underset{\sim}{E} \cdot \underset{\sim}{J}_e/c \quad ,$$
$$= +(\underset{\sim}{v} \times \underset{\sim}{B}) \cdot \underset{\sim}{J}_e/c^2 \quad ,$$
$$\underset{\sim}{f}^{em}_i = \frac{a(1)}{c} F_{i\beta} J^\beta_e = \rho_e \underset{\sim}{E} + (\underset{\sim}{J}_e \times \underset{\sim}{B})/c \quad ,$$
$$= -\rho_e (\underset{\sim}{v} \times \underset{\sim}{B})/(cg^2) + (\underset{\sim}{J}_e \times \underset{\sim}{B})/c \quad , \tag{13d}$$

Here, $\rho_e = J^0_e/c$ is the electric charge density and $\underset{\sim}{J}_e = J^i_e$ is the current density (in the laboratory reference frame); the perfect conductivity condition is

$$\underset{\sim}{E} + (\underset{\sim}{v} \times \underset{\sim}{B})/(cg^2) = 0 \quad ; \tag{13e}$$

the laboratory electromagnetic fields are

$$\underset{\sim}{E} = (E_x, E_y, E_z) = (F^{01}, F^{02}, F^{03})$$

and

$$\underset{\sim}{B} = (B_x, B_y, B_z) = g^4 (F^{23}, F^{31}, F^{12}) \quad ;$$

and the Maxwell equations are:

$$\nabla \cdot \mathbf{B} = 0 \quad ,$$

$$\nabla \times \mathbf{E} = 0 \quad ,$$

$$[\nabla \times (\mathbf{B}/g^2)]/g^2 = 4\pi \mathbf{J}_e/c \quad ,$$

$$[\nabla \cdot (g^2\mathbf{E})]/g^2 = 4\pi\rho_e \quad . \tag{14}$$

We have $E_\phi = 0$ owing to the axi-symmetry, and $\mathbf{v}_p = k(r,z)\mathbf{B}_p(r,z)$ owing to the perfect conductivity. Thus, the continuity equation (10) and $\nabla \cdot \mathbf{B} = 0$ imply

$$4\pi\rho k = F(\psi) \quad , \tag{15}$$

where

$$\rho \equiv nmg\gamma \quad ,$$

is an effective mass density in the laboratory reference frame, and m is the mean plasma particle rest mass, and

$$B_r = -\frac{1}{r}\frac{\partial\psi}{\partial z} \quad , \quad B_z = \frac{1}{r}\frac{\partial\psi}{\partial r} \quad . \tag{16}$$

Owing to the perfect conductivity we have

$$(v_\phi - kB_\phi)(rg^2)^{-1} = G(\psi) \quad . \tag{17}$$

Hence, the flow velocity can be expressed as

$$\mathbf{v} = \frac{F(\psi)}{4\pi\rho(r,z)}\,\mathbf{B}_p(r,z) + \left[\frac{F(\psi)}{4\pi\rho(r,z)}\,B_\phi(r,z) + rg^2G(\psi)\right]\hat{\boldsymbol{\phi}} \quad . \tag{18}$$

The toroidal component of the Euler equation (12), the thermodynamic identity $dw = dp/n + TdS$, and equations (10) and (11), imply angular momentum conservation on each flux surface,

$$rB_\phi/g^2 - FWrv\phi = H(\psi) \quad , \tag{19}$$

where

$$W \equiv wg\gamma(mc^2)^{-1}$$

is dimensionless and $W > 0$. In the non-relativistic limit $\gamma \to 1$, $w \to mc^2$, and $g \to 1$.

The 'zero' component of the Euler equation (12), and equations (10), (13), and (14) give Bernoulli's equation. First notice that

$$\nabla \cdot [nw\gamma^2\mathbf{v} + c(\mathbf{E} \times \mathbf{B})(4\pi g^2)^{-1}] = 0 \quad ,$$

which implies

$$[FW - GrB_\phi c^{-2}]g^{-2} = I(\psi) \quad , \qquad (20)$$

where $I(\psi)$ is a function only of ψ. This equation represents the constancy of the energy flux-density on each flux surface. The term FWg^{-2} represents the transport of energy by the matter, whereas the term $-GrB_\phi(cg)^{-2}$ represents the transport by the Poynting flux. As an example, in the dynamo model [Lovelace 1976, and Blandford 1976], the energy flux of the matter and that of the electromagnetic field are in the same direction ($F > 0$ and $GB_\phi > 0$) with the Poynting flux dominant near the central object.

Using equation (19) we rewrite equation (20) as

$$\frac{W}{g^2}\left[1 - \frac{g^2 Grv_\phi}{c^2}\right] = \frac{I}{F} + \frac{GH}{c^2 F} \quad ,$$
$$= 1 + J(\psi)/c^2 \quad , \qquad (21)$$

where $J(\psi) \equiv c^2 I/F + G\,H/F - c^2$. In the non-relativistic limit, $\gamma - 1 \simeq \underset{\sim}{v}^2/2c^2 \ll 1$, $W/g^2 \simeq (w/mc^2)(1 + \underset{\sim}{v}^2/2c^2 + \phi_g/c^2)$, and $w - mc^2 = m[\int(dp/\rho)]_S \ll mc^2$.

Also, we have $S = S(\psi)$ owing to equations (10) and (11). Thus, equation (16) implies

$$w = mc^2\left[1 + \frac{\Gamma(n/n_0)^{\Gamma-1}}{\Gamma - 1}\exp[S(\psi)(\Gamma-1)/k_B]\right] \quad , \qquad (22)$$

where n_0 is a reference density.

We have

$$rB_\phi = g^2[H + (rg)^2 GFW]\left[1 - \frac{g^2F^2W}{4\pi\rho}\right]^{-1} , \qquad (23a)$$

and

$$rv_\phi = \left[(rg)^2 G + \frac{g^2 FH}{4\pi\rho}\right]\left[1 - \frac{g^2F^2W}{4\pi\rho}\right]^{-1} . \qquad (23b)$$

The singularity of equations (23) at $g^2F^2W/(4\pi\rho) \equiv M_a^2 = 1$ is associated with the shear Alfven waves. In order to have a field-line or streamline which passes through the Alfven surface (where $M_a = 1$), we must have $v_\phi = g^2 rG$, $H = -(rg)^2 GFW$ and $B_\phi = 0$ at this surface. Also, we must have $d[(rg)^2W]/ds = 0$ at this surface if $d[g^2W/\rho]/ds \neq 0$, where s is the arc length along the field line.

We have from equation (21),

$$W = [1 + J/c^2]\left[1 - \frac{g^2 Grv_\phi}{c^2}\right]^{-1} \quad . \tag{24}$$

In general, $W > 0$. We define $x \equiv Grv_\phi(g/c)^2$. Thus, for situations where $1 + J/c^2 > 0$, we have $x < 1$. In general, we have $x^2 < (g^2rG/c)^2 < 1 + (B_\phi/B_p)^2$ from equation (17). Equations (23) and (24) evidently lead to a quadratic formula for rv_ϕ or for rB_ϕ. Also defining $A \equiv (rG/c)^2g^4 + FGHg^4(4\pi\rho c^2)^{-1}$ and $B \equiv (1 + J/c^2)F^2g^4(4\pi\rho)^{-1}$, we find

$$1 = \frac{A}{x} + \frac{B}{1 - x} \quad . \tag{25}$$

For, say, $A > 0$ and $B > 0$, and for a given point in space we may have: (i) no solution to (25) for x (an excluded region for the plasma); (ii) a single solution for x; or (iii) two solutions for x, $0 < x_1 < x_2 < 1$, and a corresponding pair of solutions for rB_ϕ. The fact that two solutions may exist is a consequence of special relativity.

The relativistic Grad-Shafranov equation is obtained from the poloidal component of the Euler equation (12) parallel to $\nabla\psi$. A fairly direct calculation from equations (9) and (12) gives

$$\begin{aligned}\frac{\rho c^2}{(\gamma g)^2}\nabla\psi \cdot \nabla W - (\rho W/g^2)\nabla\psi \cdot [\mathbf{v} \times (\nabla \times \mathbf{v})] = \\ + p\nabla\psi \cdot (\nabla S/k_B) \\ + \nabla\psi \cdot (\rho_e \mathbf{E} + \frac{1}{c}\mathbf{J}_e \times \mathbf{B}) \\ - 2\rho W(\nabla\psi \cdot \nabla\Phi_g) \quad .\end{aligned} \tag{26}$$

A further manipulation of equation (26) gives the <u>generalized Grad-Shafranov equation</u>:

$$\begin{aligned}\left[1 - \left(\frac{g^2rG}{c}\right)^2 - \frac{g^2WF^2}{4\pi\rho}\right]\Delta^*\psi - \frac{1}{2r^2}\nabla\left[\frac{(rg)^4G^2}{c^2}\right]\cdot\nabla\psi \\ - g^2F\,\nabla\left(\frac{FW}{4\pi\rho}\right)\cdot\nabla\psi + 2(\nabla\Phi_g/c^2)\cdot\nabla\psi \\ = 4\pi r^2g^4p(S'/k_B) \\ - g^4(H + Wrv_\phi F)[H' + (Wrv_\phi)F'] \\ -4\pi\rho r^2g^4[J' + (Wrv_\phi)G'] \quad .\end{aligned} \tag{27}$$

The functions dependent only on ψ, that is, F, G, H, J, and S, must be supplied. The primes denote, for example, $F' = dF/d\psi$, and $\Delta^* \equiv \nabla^2 - (2/r)(\partial/\partial r)$.

The mathematical nature of the operator on the left-hand-side of equation (27) can be extracted by straightforward but lengthy manipulations in which equations (15), (17), (19), (23b) and (24) are used to express $\nabla(W/\rho)$ in terms of $\nabla(\nabla\psi)^2$. Retaining only the highest (second) order derivatives of ψ, the left-hand-side of (27) is

$$P\left(A_{rr}\frac{\partial^2\psi}{\partial r^2} + A_{rz}\frac{\partial^2\psi}{\partial r\partial z} + A_{zz}\frac{\partial^2\psi}{\partial z^2}\right) ,$$

where

$$P \equiv 1 - \left(\frac{g^2 rG}{c}\right)^2 - \frac{g^2 WF^2}{4\pi\rho} ,$$

$$A_{rr} \equiv 1 - \frac{(v_p)^2(v_z)^2}{R} ,$$

$$A_{rz} \equiv \frac{2(v_p)^2 v_r v_z}{R} ,$$

$$A_{zz} \equiv 1 - \frac{(v_p)^2(v_r)^2}{R} ,$$

and

$$R \equiv v_a^2 C_s^2 f - v_p^2\left[C_s^2 + v_A^2 - v_a^2\left(\frac{g^2 rG}{c}\right)^2\right] + v_p^4 .$$

Here, $v_A \equiv [(B_p^2 + B_\phi^2)/(4\pi\rho g^2 W)]^{\frac{1}{2}}$ is introduced as the effective Alfven speed; $v_a = [B_p^2/(4\pi\rho g^2 W)]^{\frac{1}{2}}$ is the effective Alfven speed projected into the poloidal plane; $C_s \equiv (c_s/\gamma)[1 + (c_s/\gamma c)^2]^{-\frac{1}{2}}$ is the effective sound speed in the laboratory reference frame; $c_s = c[(\Gamma - 1)^{-1} + mc^2/(\partial p/\partial n|_S)]^{-\frac{1}{2}} < 3^{-\frac{1}{2}}c$ is the speed of sound in the comoving reference frame of the fluid; $f \equiv 1 - x + [(g^2 rG/c)^2 - x](\gamma)^{-2}$; $x \equiv Grv_\phi(g/c)^2$; and $v_p \equiv (v_r^2 + v_z^2)^{\frac{1}{2}}$. Notice that v_A and v_a depend implicitly on the flow speed in that $W \propto \gamma$. Also, from equation (17) we have $g^2 rG/c = (v_\phi/c) \mp (v_p/c)(B_\phi/B_p)$, where the minus (or plus) sign corresponds to $F > 0$ (or $F < 0$). In general, $(v_a)^2(g^2 rG/c)^2 < (v_A)^2$. In the non-relativistic limit, R coincides with the formula of Lovelace et al. (1984).

Equation (27) is elliptic (or hyperbolic) depending on whether the discriminant, $D \equiv (A_{rz})^2 - 4A_{rr}A_{zz}$, is negative (or positive). We readily find $D = -4[1 - (v_p^4/R)]$. For simplicity, consider the limit where $C_s << (v_p, v_a)$ so that

$$D = \frac{v_A^2 - v_a^2(g^2rG/c)}{v_p^2 + v_a^2(g^2rG/c)^2 - v_A^2} \quad .$$

as mentioned, the numerator of D is non-negative. For a given sign of F, the denominator of D vanishes for two values of v_p--one positive and the other negative--as given by

$$v_p^2 = v_a^2[1 + b^2 - [(v_\phi/c) \mp b(v_p/c)]^2] \quad ,$$

where $b \equiv B_\phi/B_p$. Only one of the v_p roots will be pertinent in a given situation corresponding to the fast magnetosonic wave of the plasma. For both roots, $|v_p| < v_A$. One root is larger and the other is smaller in magnitude than $v_a[1 + b^2 - (v_\phi/c)^2]^{\frac{1}{2}}[1 + b^2(v_a/c)^2]^{-\frac{1}{2}}$. For example, for $v_\phi > 0$ and $FB_\phi < 0$, the smaller magnitude root is pertinent.

A 'force-free' limit of equation (27) may be valid in the limited spatial regions of a global solution. In this limit, the inertial forces--proportional to the particle rest mass--are negligible compared with the electromagnetic forces. Noting that $F^2(4\pi\rho)^{-1} = (4\pi\rho)(\underset{\sim}{v}_p)^2(\underset{\sim}{B}_p)^{-2}$ is proportional to ρ, and neglecting the term $2\underset{\sim}{\nabla}\Phi_g \cdot \underset{\sim}{\nabla}\psi$, the force-free limit of (27) is

$$\left[1 - \left(\frac{g^2rG}{c}\right)^2\right]\Delta^*\psi - \frac{1}{2r^2}\left[\underset{\sim}{\nabla}\left(\frac{(rg)^4G^2}{c^2}\right)\right] \cdot \underset{\sim}{\nabla}\psi = -g^4HH' \quad . \qquad (28)$$

where $H(\psi) = rB_\phi/g^2$ and $(rg)^2G(\psi) = rv_\phi - g^2H\,\ell im[F(4\pi\rho)^{-1}]$. The limit is evidently indeterminate in so far as $H\,\ell im[F(4\pi\rho)^{-1}]$ is undefined. This indeterminance is removed, for example, if $\ell im(F/\rho) \neq 0$ and $H = 0$ (poloidal flow but no toroidal field), or if $F \equiv 0$ and $H \neq 0$ (a toroidal field but no poloidal flow). In the latter case, equation (28) with $g = 1$ is equivalent to the 'pulsar equation' (Michel 1973, and Scharlemann and Wagoner 1973) except that the 'pulsar equation' neglects without discussion the term $-[r^2/(2c^2)]\underset{\sim}{\nabla}(G^2) \cdot \underset{\sim}{\nabla}\psi$ on the left-hand-side of equation (28). This term arises from the space charge density at the boundary surface between the private magnetic flux of the pulsar and the external world, and from the volume distributed space charge density on the external field lines due to the non-constancy of G in this region. It cannot be neglected in a complete analysis. For F = 0 the flow velocity is $v = g^2 rG\phi$. Thus, the physical solutions of equation (28) must have $(g^2 rG/c)^2 < 1$.

A subcase of equation (27) of interest to us in the next section is that corresponding to <u>no</u> poloidal flow ($\underset{\sim}{v}_p = 0$ or $F(\psi) = 0$) and <u>no</u> toroidal magnetic field ($B_\phi = 0$ or $H(\psi) = 0$):

$$\left[1 - \left(\frac{g^2 rG}{c}\right)^2\right]\Delta^*\psi - \frac{1}{2r^2}\left[\underset{\sim}{\nabla}\left(\frac{(rg)^4 G^2}{c^2}\right)\right] \cdot \underset{\sim}{\nabla}\psi + 2(\underset{\sim}{\nabla}\Phi_g/c^2) \cdot \underset{\sim}{\nabla}\psi$$

$$= 4\pi\ r^2 g^4 p(S'/k_B)$$

$$- 4\pi\rho r^2 g^4[J' + W(rg)^2(G^2/2)'] \quad . \qquad (29)$$

In this case, only three functions of ψ, that is, G, J, and S, need be supplied. Notice that with F = 0, the flow velocity is $v = g^2 rG(\psi)\phi$, and that $(W/g^2)[1 - (g^2 rG/c)^2] = 1 + J(\psi)/c$. The magnitude of the flow velocity in a physical solution of (29) must be less than the speed of light; equivalently, $[g^2 rG(\psi)/c]^2 < 1$.

III. THIN RELATIVISTIC DISKS

Here, we apply the results of section II to magnetized disks, including the possibility of relativistic azimuthal motion, but excluding both poloidal flow and a toroidal field [$F(\psi) = 0$ and $H(\psi) = 0$ in equation (27)]. The central object is assumed to be an aligned, rotating magnetized star. The star is characterized by its mass, $M_\star$, magnetic moment, $m_\star > 0$, angular rotation rate, $\Omega_\star$, and radius $r_\star$. We consider the radial force balance in section IIIb, and specific disk equilibria in section IIIc.

The disk thickness Δz is assumed small compared with r; the conditions for this are then established.

A number of further assumptions are made:

(1) The disk has reflection symmetry about z = 0.

(2) The gravitational potential $\Phi_g(r,z)$ is due to the central object ($M_\star \gg 4\pi\sigma r^3/\Delta z$), where σ is the surface density of the disk.

(3) The pressure p in the comoving reference frame is small compared with mc^2.

(4) The equation of state of the plasma on a given flux surface is isothermal, with $p = nk_B T(\psi)$ so that the enthalpy per particle is

$$w = mc^2 + m[c_s(\psi)]^2 \ell n \left[\frac{\rho}{\gamma\rho_0}\right] \quad , \tag{30}$$

where $c_s = [k_B T(\psi)/m]^{\frac{1}{2}}$ is the Newtonian sound speed in the comoving reference frame, and ρ_0 is a constant reference density. Assumption (3) corresponds to $c_s^2 << c^2$.

(5) The variation of c_s with ψ is gradual in comparison with that of $J(\psi)$ and $G(\psi)$ so that the term $4\pi r^2 g^4 p(S'/k_B)$ in equation (29) is negligible.

(6) The gravitational field is weak, $g - 1 = -\Phi_g/c^2 << 1$. Consequently, the term $2(\nabla\Phi_g/c^2) \cdot \nabla\psi$ in equation (29) is negligible. In all of the other terms of (29), the g-factors are for the present retained.

a) Radial Force Balance of Disk

With a rearrangement of terms and the above condition (5), the Grad-Shafranov equation (29) is

$$\Delta^*\psi = -\frac{4\pi r}{c} J_\phi(r,z) = \frac{\gamma^2}{2r^2}\left[\nabla\left(\frac{(rg)^4 G^2}{c^2}\right)\right] \cdot \nabla\psi - \frac{4\pi r}{c} J_{\phi_d} \quad , \tag{31a}$$

where

$$J_{\phi_d}(r,z) \equiv \rho c r g^4 \gamma^2 [J' + W(rg)^4 (G^2/2)'] \quad , \tag{31b}$$

represents the part of the total azimuthal current-density, $J_\phi(r,z)$, which is proportional to the disk mass-density.

Following the steps of Lovelace et al. (1984), we integrate equation (31a) over the range of z encompassing the disk matter to obtain the radial force balance equation for the disk:

$$\sigma W g^2 \frac{v_\phi^2}{r} = \frac{g^4}{\gamma}\frac{dP}{dr} + \sigma W\left(2 - \frac{1}{\gamma^2}\right)\frac{\partial\Phi_g(r,0)}{\partial r}$$
$$- \frac{1}{c\gamma^2} J_{\phi_d}(r) B_z(r,0) \quad . \tag{32a}$$

In the approximation where $g = 1$, $v_\phi = rG(\psi)$, and $W = \gamma$, we have

$$\sigma\gamma \frac{v_\phi^2}{r} = \frac{1}{\gamma}\frac{dP}{dr} + \sigma\left(2\gamma - \frac{1}{\gamma}\right)\frac{\partial \Phi_g}{\partial r} - \frac{1}{c\gamma^2} J_{\phi_d}(r) B_z(r,0) \quad . \qquad (32b)$$

Here,

$$\sigma(r) \equiv \int_{-\infty}^{\infty} dz\, \rho(r,z) \quad ,$$

$$J_{\phi_d}(r) \equiv \int_{-\infty}^{\infty} dz\, J_{\phi_d}(r,z),$$

$$P \equiv [c_s(\psi_0)]^2 \sigma(r) \quad ,$$

with $\psi_0 \equiv \psi(r,0)$. In the order given the different terms in equation (32) represent the 'centrifugal force', the pressure force, the gravitational force, and the electromagnetic force. Equation (32) neglects the contribution from the first term on the right-hand-side of equation (31a) which arises from the volume and surface distributed space charge density. The prefactor of γ^{-2} of the electromagnetic force arises from the fact that this term combines the magnetic and electric forces, which nearly cancel.

b) Specific Disk Equilibria

Here, we apply the results of section IIIa to disk equilibria around an aligned, rotating magnetized star. For simplicity, we assume that the pressure force in the disk is negligible and that the gravity is weak. Thus, the basic equations are (31a) with $g = 1$ and (32b) with $P = 0$.

The determination of $\psi(r,z)$ from equations (31a) and (32b) is inherently complicated due to the first term on the right-hand-side of (31a) which arises from the space-charge-density. Here, we consider an iteration method for solving (31a) wherein we first solve for $\psi^{(0)}$, where

$$\Delta^* \psi^{(0)} = -\frac{4\pi r}{c} J_{\phi_d}^{(0)}(r)\delta(z) \quad , \qquad (33)$$

and where $J_{\phi_d}^{(0)}(r)$ satisfies equation (32b) with $B_z^{(0)} = r^{-1}\partial\psi^{(0)}/\partial r$. In subsequent iterations, $n = 1, 2, \ldots,$

$$\Delta^* \psi^{(n+1)} = \left[\frac{\gamma^2}{2r^2} \nabla\!\!\!\!\sim \left(\frac{r^4 G^2}{c^2}\right)\right]^{(n)} \cdot \nabla\!\!\!\!\sim \psi^{(n)} - \frac{4\pi r}{c} J_{\phi_d}^{(n)}(r)\delta(z) \quad , \qquad (34)$$

where $G^{(n)} = G[\psi^{(n)}]$ and $\gamma^{(n)} = \gamma[G^{(n)}]$. The dimensionless quantity which must be small compared with unity in order for this iteration to converge rapidly is the ratio of the magnetic field energy to the magnitude of the gravitational binding energy of the disk. In the present work, we assume that this ratio is small. Therefore, only equation (33) is used to obtain an approximation to the flux function. Hence, the zero superscripts are dropped. The Green's function solution of equation (33) for $\psi(r) = \psi(r,z = 0)$ is:

$$\psi(r) = \psi_*(r) + \psi_b(r) + \frac{2\pi}{c}\int_{r_*}^{\infty} dr' \; g(r,r')J_{\phi_d}(r') \quad . \tag{35}$$

Here, $\psi_*(r) = m_*/r$ is the flux function of the star (due to its internal currents), m_* its magnetic moment, and r_* its radius; and, $\psi_b(r) = m_b/r$ is introduced so as to account for the boundary condition that the magnetic flux due to the currents in the disk does not penetrate the star. The constant m_b is evidently given by

$$m_b = -\frac{2\pi r_*}{c}\int_{r_*}^{\infty} dr' g(r_*,r')J_{\phi_d}(r') \quad . \tag{36}$$

The radial force balance equation (32b) is rewritten as

$$\sigma(r) = \frac{(cr\gamma^3)^{-1}J_{\phi_d}(r)B_z(r,0)}{\left(2 - \frac{1}{\gamma^2}\right)\frac{GM_*}{r^3} - \Omega^2(\psi)} \quad , \tag{37}$$

Here, the gravitational potential of the star is $\Phi_g = -GM_*/r$; the more familiar notation $\Omega(\psi) = G(\psi)$ is adopted for the angular rotation rate of the disk matter; and $\gamma = [1 - (r\Omega/c)^2]^{-\frac{1}{2}}$.

The field lines which pass through the star define what we refer to as the 'private' flux-surfaces of the star. For these ψ-surfaces, $\Omega(\psi) = \Omega_*$, which is the angular rotation rate of the star. For the other, non-private flux-surfaces, $\Omega = \Omega(\psi)$.

There is evidently a large degree of arbitrariness in the solutions to equations (35)-(37). However, some basic physical constraints on the solutions follow by inspection of (35)-(37): Firstly notice that for $\Omega = \Omega_*$, the vanishing of the denominator of (37) corresponds to an exact balance of the centrifugal and gravitational forces acting on a mass element corotating with the star. The balance occurs at the 'centrifugal radius',

$$r_c \equiv \left(\frac{GM_*}{\Omega_*^2}\left[1 + \left(\frac{r_c\Omega_*}{c}\right)^2\right]\right)^{1/3} \quad . \tag{38}$$

At the centrifugal radius, we have either (i) $B_z(r_c,0) = 0$ (an O or an X point of the magnetic field at r_c) or (ii) $J_{\phi_d}(r_c) = 0$.

Secondly, notice that the private flux of the star cannot penetrate the disk beyond the '<u>light-cylinder radius</u>',

$$r_\ell \equiv \frac{c}{\Omega_*} \quad . \tag{39}$$

This is simply due to the fact that the disk matter on the private flux corotates with the star with azumuthal velocity $\Omega_* r$. Close to the star, the stellar magnetic field may be assumed dominant so that the private field lines of the star loop through the disk. With increasing r, the looping of the private field-lines through the disk must cease at some distance inside the light cylinder. Therefore, we conclude that there must be at least one X-point in the equatorial plane at a distance $r = r_X < r_\ell$.

For typical conditions we have

$$r_* << r_c << r_\ell \quad . \tag{40}$$

Thus, to a good approximation the gravitational force is $-GM_*/r$ (or it is negligible) and $r_c = [GM_*/\Omega_*]^{1/3}$ (Michel and Dessler 1981). We are interested in disk equilibria which extend from r_* to distances larger than r_ℓ. Therefore, we can neglect ψ_b in equation (35) under the weak limitation $B_{zd}(r_*,0) << m_*(r_*)^{-3}$, where B_{zd} is the magnetic field due to the disk current.

In order to facilitate the numerical determination of $\psi(r)$, we introduce dimensionless variables. The basic length scale is taken to be r_c, and the basic flux scale is taken to be $\psi_0 \equiv m_*/r_c$. The dimensionless variables are: $\hat{r} = r/r_c$; $\hat{\psi} = \psi/\psi_0$; $\hat{J}_{\phi_d} = (r_c)^2 J_{\phi_d}/(c\psi_0)$; $g = g/r_c$; $\hat{\Omega}^2 = (r_c)^3 \Omega^2/(GM_*)$; and $\hat{\sigma} = (r_c)^2 \sigma/M_d$, where M_d is the total disk mass.

In dimensionless variables, with over-carets implicit, equation (35) becomes

$$\psi(r) = r^{-1} + 2\pi \int_{r_*}^{\infty} dr' \, g(r,r') J_{\phi_d}(r') \quad . \tag{41}$$

Equation (37) becomes

$$\sigma(r) = \frac{\xi_0 (r\gamma^3)^{-1} J_{\phi_d}(r) B_z(r,0)}{[r^{-3} - \Omega^2(\psi)]} \quad , \qquad (42)$$

where

$$\xi_0 = \frac{v_0^2}{GM_*M_d} = \frac{m_*^2(\Omega_*)^{4/3}}{[GM_*]^{5/3}M_d} \qquad (43)$$

A radial virial equation can be obtained directly from equations (41) and (42). In dimensionless variables, it is:

$$2W_K + W_g + W_{dd} + W_{d*} = 0 \quad . \qquad (44)$$

Here,

$$W_K \equiv \pi \int_{r_*}^{\infty} r dr \sigma \gamma \, (r\Omega)^2 > 0 \quad ,$$

is the kinetic energy of the disk matter;

$$W_g \equiv -2\pi \int_{r_*}^{\infty} r dr \;\; r \frac{\partial \Phi_g}{\partial r} < 0 \quad ,$$

is the gravitational potential energy of the disk matter;

$$W_{dd} \equiv 2\pi \xi_0 \int_{r_*}^{\infty} r dr \gamma^{-2} \; r J_{\phi_d}(r) B_{zd}(r,0) > 0 \quad ,$$

is the self magnetic energy of the disk; and

$$W_{d*} \equiv -2\pi \xi_0 \int_{r_*}^{\infty} r dr \gamma^{-2} \; r J_{\phi_d}(r) r^{-3} \gtrless 0 \quad ,$$

is the magnetic interaction energy between the disk and the star. Also, $B_{zd} \equiv r^{-1}(\partial\psi_d/\partial r)$, where ψ_d is the flux-function arising from the disk current. For $z = 0$, ψ_d is given by the second term on the right-hand-side of equation (41).

The total magnetic energy of the disk is

$$W_m = W_{dd} + W_{d*} \quad . \qquad (45)$$

From equation (44), we conclude that

$$\xi \equiv \frac{W_m}{|W_g|} < 1 \quad , \tag{46}$$

Notice that the self-magnetic energy of the star, which exceeds W_m by a factor $\sim\xi_o(r_c/r_*)^3 >> 1$, does not enter in equation (44).

In order to numerically generate illustrative solutions to equations (41) and (42), we consider the simplest possible field configurat ion which is consistent with the above-mentioned physical constraints, and which extends from r_* to distances well beyond r_ℓ. Specifically, we consider the current density profile:

$$J_{\phi_d}(r) = K(r - 1)\exp(-r/\lambda) \quad , \tag{47}$$

where K and λ are dimensionless constants. The choice of $J_{\phi_d}(r=1) = 0$ means that an O or an X point need not be located at r = 1. Instead, we have, under the conditions discussed below, at X point at r_x ($\neq$1) and an O point at $r_{oh} > r_x$. Both points are in the equatorial plane. The field configuration can be denoted 'XO' following the convention of Lovelace et al. (1984).

Rather than directly using K and λ to characterize the disks, we use Φ_d and λ, where Φ_d is the (reduced) total flux of the disk,

$$\Phi_d \equiv \max_r[\psi_d(r,0)] \quad . \tag{48}$$

Thus, the total disk flux is $2\pi\Phi_d$. Owing to the simple functional form of $J_{\phi_d}(r)$ of (47), we have $\Phi_d \sim K\lambda^3$ for $\lambda >> 1$. Thus, equation (41) can be used to evaluate $\psi(r)$ for given values of Φ_d and λ.

For given (Φ_d, λ), a disk equilibrium can be readily calculated. Firstly, equation (41) is used to calculate $\psi(r)$, and thus to determine r_x and r_{oh}. Then, in order to obtain $\sigma(r)$ from equation (42), we need $\Omega(\psi)$. For $r < r_x$, the disk corotates with the star, so that $\Omega = 1$. For $r > r_x$, we consider on ψ as in section IIIe. That is,

$$\Omega^2 = \begin{cases} 1 & , \quad r < r_x, \\ \omega^2|\psi|^n & , \quad r > r_x, \end{cases} \tag{49}$$

where ω and n are dimensionless constants. Notice that in equation (49) both the numerator and the denominator vanish at r = 1 giving a finite $\sigma(1)$ value. At $r = r_{oh}$, the numerator of equation (42) vanishes, and, therefore, the denominator must also vanish as in section IIIe. The vanishing of the denominator implies

$$\omega^2 = |\psi(r_{oh})|^{-n}(r_{oh})^{-3} \quad . \tag{50}$$

For given Φ_d, λ, and n, we can now derive $\sigma(r)$.

The normalization of σ requires $2\pi\int r dr \sigma = 1$. In turn, this implies the value of ξ_0 in equation (43). For $\lambda >> 1$, it can be readily shown that $\xi_o \sim (\Phi_d)^{-2}$ and that

$$\frac{\Phi_d}{\min\,(\Phi_d)} \sim \frac{(GM_* M_d)^{\frac{1}{2}}}{|m_* \Omega_* / c|} \quad . \tag{51}$$

Using values pertinent to a neutron star, we find

$$\frac{\Phi_d}{\min\,(\Phi_d)} \sim 8 \times 10^5 \left(\frac{M_*}{M_\odot}\right)^{\frac{1}{2}} \left(\frac{M_d}{10^{-6} M_\odot}\right)^{\frac{1}{2}} \left(\frac{P_*}{1s}\right) \left[\frac{10^{30}\ \text{Gauss-cm}^3}{m_*}\right] , \tag{52}$$

where $P_* = 2\pi/\Omega_*$ is the period of the star.

For the numerical calculation of a disk equilibrium, we evidently need to specify (Φ_d, λ, r_ℓ, r_*, and n). So as to simplify the discussion, we fix $r_\ell = 10$, $r_* = 0.1$, and $n = 3$.

Figure 1 shows the radial, equatorial profiles of the basic parameters for a specific disk equilibrium. Notice that there is a narrow 'gap' in the surface mass density at $r = r_X$, and that the angular rotation rate of the disk jumps from its corotation value of unity for $r = r_X - 0$ to a much lower value for $r = r_X + 0$. (Disk equilibria without such a gap in $\sigma(r)$ can be generated; however, they require a singular behavior of the current density, $J_{\phi_d} \propto (r_X - r)^{-\frac{1}{2}}$, for r close to but less than r_X.]

Figure 2 shows the field lines in the poloidal plane for the same disk equilibrium as Figure 1. Notice that this field line configuration involves an inconsistency with the full Grad-Shafranov equation (31a) owing to the first term on the right-hand-side of the equation which is due to the space-charge-density. As mentioned below equation (29), the requirement that the plasma flow speed be less than the speed of light is $[g^2 rG/c]^2 < 1$. This space-charge contribution acts to prevent the private field-line (or flux surfaces) of the star from crossing the speed of light cylinder (Lovelace et al. 1984).

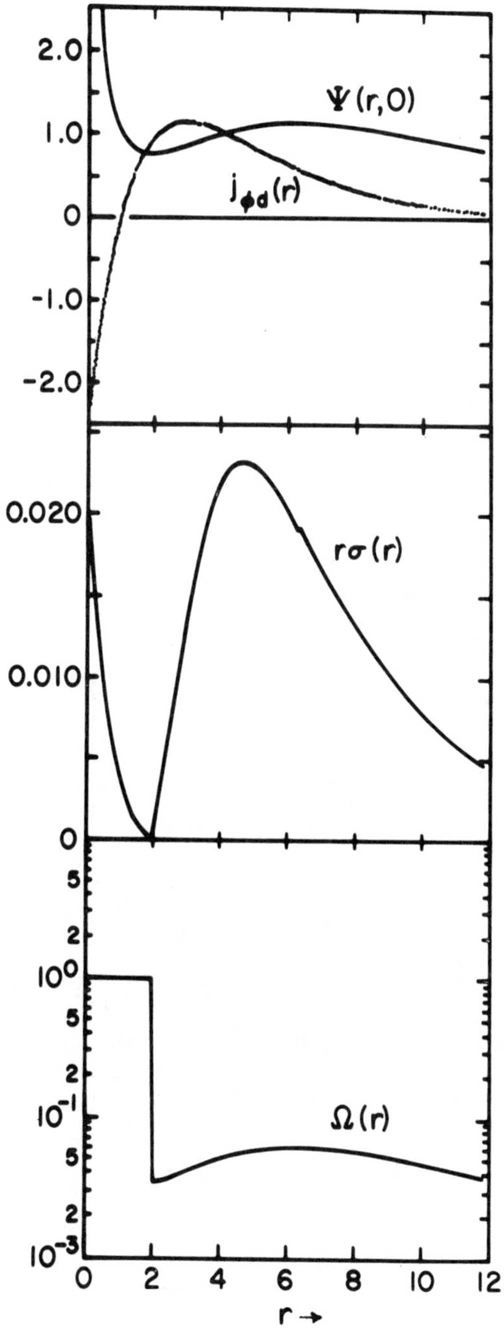

Figure 1. Radial, equatorial profiles of the total flux function (ψ), the disk current density (J_{ϕ_d}), the disk surface mass-density multiplied by r ($r\sigma$), and the angular rotation rate of the disk matter (Ω) for a disk equilibrium characterized by $\Phi_d = 1$ and $\lambda = 2$. All of the scales correspond to the dimensionless variables of IIb except for J_{ϕ_d}, which has been multiplied by 92.73. The disk matter corotates with the star out to the radius of the X point, $r_x = 1.97$. Just interior to the X point, the azumuthal velocity of the disk velocity is 6.83×10^{-3} c. The O point of the magnetic field is at $r_{oh} = 6.32$, where the azimuthal velocity has its Keplerian value, 3.98×10^{-2} c. The fraction of the total disk mass inside r_x is 6.59×10^{-2}.

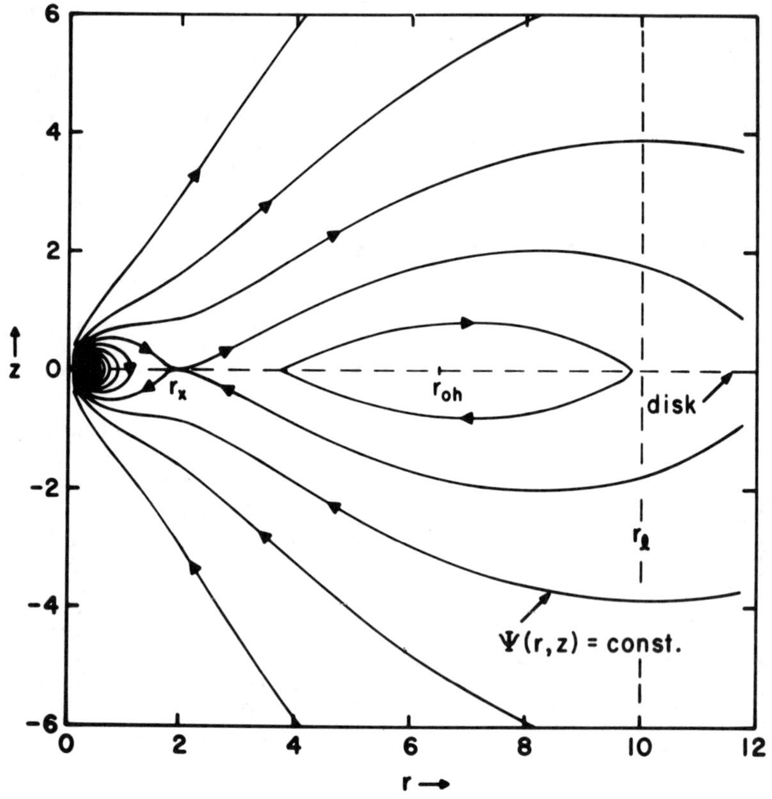

Figure 2. Flux surfaces for the same disk equilibrium as in Figure 1. The difference in the value of ψ between successive contours is $0.2\ \psi_0$, where $\psi_0 \equiv m_*/r_c$.

ACKNOWLEDGMENT

We thank F. W. Perkins for stimulating discussions and for the hospitality of the Princeton University Plasma Physics Laboratory where his work was begun. The work at Princeton was supported by the Department of Energy Contract No. DE-AC02-76-CH03073. We also thank J. J. Aly and F. K. Lamb for valuable discussions.

REFERENCES

R. D. Blandford, Mon. Not. Roy. Astron. Soc. 176, 465 (1976).

K. L. Chan and R. N. Henriksen, Ap. J. 241, 534 (1980).

F. V. Coroniti, 'Unstable Current Systems in Astrophysics', (M. R. Kundu, editor), Proc. of I.A.U. Symposium No. 111 (1984).

D. M. Eardley and A. P. Lightman, Ap. J. 200, 187 (1975).

P. Ghosh and F. K. Lamb, Ap. J. 232, 259 (1979a).

P. Ghosh and F. K. Lamb, Ap. J. 234, 296 (1979b).

H. Grad and H. Rubin, International Atomic Energy Agency Conference Proceedings 31 (Geneva), p. 190 (1958).

M. Heinemann, and S. Olbert, J. Geophys. Res. 83, 2457(1978).

F. K. Lamb, in 'Proceedings of the Los Alamos Workshop on Magnetospheric Phenomena in Astrophysics' (New York: American Institute of Physics), to appear.

L. D. Landau and E. M. Lifshitz, 'Classical Theory of Fields' (Oxford: Pergamon), Chapters 10, 11 (1962).

R.V.E. Lovelace, Nature 262, 649 (1976).

R.V.E. Lovelace, C. Mehanian, C. M. Mobarry and M. E. Sulkanen, 1984, Cornell University CRSR Report No. 818.

L. Mestel, Mon. Not. Roy. Astron. Soc. 122, 473 (1961).

L. Mestel, Mon. Not. Roy. Astron. Soc. 138, 359 (1968).

F. C. Michel, Ap. J. Letters 180, L133 (1973).

F. C. Michel and A. J. Dessler, Ap. J. 251, 654 (1981).

C. W. Misner, K. S. Thorne and J. A. Wheeler, 'Gravitation' (San Francisco: W. H. Freeman), Chapter 18 (1970).

K. Miyamoto, 'Plasma Physics for Nuclear Fusion' (Cambridge: MIT Press) (1976).

E. T. Scharlemann and R. V. Wagoner, Ap. J. 182, 951 (1973).

H. A. Scott and R.V.E. Lovelace, Ap. J. 252, 765 (1982).

V. D. Shafranov, Reviews of Plasma Physics 2 (New York: Consultants Bureau), p. 103 (1966).

S. L. Shapiro and S. A. Teukolsky, 'Black Holes, White Dwarfs, and Neutron Stars' (New York: Wiley), p. 428 (1983).

L. Woltjer, Ap. J. 130, 405 (1959).

H. P. Zehrfeld and B. J. Green, Nuclear Fusion 12, 569 (1972).

PARTICLE ACCELERATION BY ALFVEN WAVE TURBULENCE IN RADIO GALAXIES

Jean A. Eilek
Physics Department, New Mexico Tech, Socorro, NM 87801

ABSTRACT

Radio galaxies show evidence for acceleration of relativistic electrons locally within the diffuse radio luminous plasma. One likely candidate for the reacceleration mechanism is acceleration by magnetohydrodynamic turbulence which exists within the plasma. If Alfven waves are generated by a fluid turbulent cascade described by a power law energy-wavenumber spectrum, the particle spectrum in the presence of synchrotron losses will evolve towards an asymptotic power law which agrees with the particle spectra observed in these sources.

INTRODUCTION

This paper presents a short review of the particle acceleration problem in the particular case of extragalactic radio sources. These objects are thought to be due to directed outflows of plasma from active galactic nuclei (including quasars). The physical conditions pertaining to each source, such as the energy output and the state of the ambient extragalactic gas, determine the morphology and evolution of this flow, leading to the jets, lobes and tails which are observed. It must be noted that we observe only those parts of the plasma which are radio luminous. The luminous part is not necessarily a tracer of the entire flow. The radio luminosity is believed to arise from synchrotron radiation from a nonthermal distribution of relativistic electrons.

The ultimate energy source for the radio luminosity is probably the net (kinetic plus internal) convected energy of the outflowing plasma. The radiating electrons must be accelerated in situ in order to account for the surface brightness distribution, and in particular to overcome radiative and adiabatic energy losses. An important problem in radio source physics is to understand just how the outflowing energy is transferred to the relativistic electrons. Since the diffuse plasma is highly conducting, strong electric fields are unlikely to exist over large distances, and so one expects some type of stochastic process. Several mechanisms come to mind. Shocks almost certainly occur at some parts of the plasma flow in radio sources. Acceleration by high frequency plasma turbulence and also by magnetic reconnection is known to occur closer to home, namely in the solar atmosphere, solar wind and terrestrial magnetosphere; these mechanisms have not, to my knowledge, been considered in any detail in the radio galaxy environment. Finally, acceleration by low frequency, MHD wave modes will occur in the turbulent plasma.

In this paper I shall summarize the basic acceleration processes which have been suggested for radio galaxies, and shall focus on one particularly interesting case. The interaction of the electrons with Alfven waves is governed by the cyclotron resonance; when the Alfven waves are driven by Lighthill radiation from fully developed fluid turbulence, the feedback between the wave damping, particle acceleration and radiative losses leads to an asymptotic electron distribution which is just the "universal" power law spectrum observed in radio sources (a more detailed discussion of this effect is given in reference 1).

BACKGROUND

The relativistic electrons responsible for the radio luminosity must be accelerated locally. Several facts point to this conclusion. In many sources the particle lifetimes to synchrotron losses are a good bit less than the scale of the source divided by the outflow speed (indeed, less than the source size divided by lightspeed in a few cases). Classical double sources are most luminous at their outer ends, indicating some mechanism energizes the particles at the "working surface" where the directed flow runs into the external gas. In many expanding jets, the surface brightness does not obey the decline expected from adiabatic expansion in a constant velocity flow. In tails and wakes the particle spectrum does not steepen as rapidly as would be expected if radiative losses were the only important factor; in some cases the spectrum flattens at large distances from the nucleus. Many jets do not emerge fully luminous from the core, but rather "turn on" after some distance (the "gap"). All of these facts suggest that local acceleration of the radiating particles is needed. (Excellent reviews of radio galaxies can be found in references 2; of radio jets in particular, in reference 3.)

Thus, the global "acceleration problem" in radio galaxies involves understanding the nature of the large scale energy flow and its coupling to the turbulence, shocks or other "accelerators", as well as understanding the microphysics of the particle acceleration. An eventual goal of this work will be to understand the initial nature of the outflowing plasma. From this one would hope to gain a better understanding of the nature of the central energy source.

In particular, the overall energy flow in the source can be described as follows.

(a) The luminosity of the extended source is being supplied by some mechanism from the central engine; this is usually believed to be the convected energy of the outflowing plasma.

(b) This outflowing plasma must, by some means, make local accelerators -- shocks, turbulence or something else. The coupling between the energy source and these accelerators occurs when the (initially laminar?) outflow shocks and/or becomes turbulent. The efficiency of this initial coupling must be understood.

(c) These accelerators will transfer energy from the flow to the relativistic particles ("acceleration"), to the cool, inertial gas ("heating") and to the ambient extragalactic medium. The efficiency of this transfer is also important. Global energetics suggest that the total efficiency (of energy transfer from the directed outflow to the synchrotron radiation) is fairly high, perhaps on the order of one per cent.

(d) This chain of events will have a profuond effect on the structure and evolution of the source, and perhaps on its surroundings as well, through heating and disruption of the source, heating of the surroundings and through the slow energy depletion of the source.

The basic parameters of the radio luminous plasma can be can be estimated from the luminosity and polarization. The source sizes range from a few kpc (the width of some jets) to about a Mpc (setting these large radio galaxies among the the largest objects in the universe). The magnetic field $\approx 10^{-6}$ to 10^{-5} G, and the energy density in the relativistic particles $\approx 10^{-11}$ to 10^{-9} erg cm^{-3} (both estimated from minimizing the total energy density in these two components, subject to the observed luminosity and size). The relativistic electron distribution is a power law, $f(p) \propto p^{-s}$ with $4 \leq s \leq 5$, and the radiating electrons have typical Lorentz factors $\gamma \approx 10^2$ to 10^4 (or higher in the jets from which optical or X ray synchrotron emission has been detected). There also exists an inertially dominant but energetically unimportant cool gas component, with densities $\approx 10^{-4}$ to 10^{-2} cm^{-3}. The temperature and distribution function of this cool component is unknown, although some limits can be set on its temperature from the absence of radiation (either HII emission lines or bremsstrahlung). The directed flow speed of the jet is probably on the order of 10^3 to 10^4 km s^{-1}, although a few sources show evidence for relativistic speeds (with bulk Lorentz factors 2 to 5) in the core (the central few pc, observed with VLBI).

The plasma is believed to act as a fluid. The classical Coulomb scattering length is quite large compared to the size of the source, but the Larmor radii of the particles are much smaller than the source size. It is also possible that plasma microturbulence keeps the particle collision length small. Thus, the gas appears to be effectively collisional. A few sources (e.g., M87; reference 4) show features which are probably shocks. Further, the plasma is almost certainly turbulent. The indirect evidence for this includes the distribution and magnitude of linear polarization of the sources (which depends on the internal magnetic field direction through the local emissivity and through Faraday effects); laboratory analogs of fluid jets and current carrying beams; and the surface brightness of the sources, which give a strong visual impression of turbulence in many cases.

The nature of this turbulence is not easily observable, but several modes are probably represented. The large scale flow will develop fluid turbulence due to its interaction with the surrounding galactic and extragalactic gas. This turbulence usually develops as a cascade from large driving scales (such as the source scale, l_o, or the Taylor scale) down to small scales (the "dissipation length", $l_o/\mathrm{Re}^{3/4}$ where Re is the Reynolds number) with a power law spectrum in the inertial range between these two limits. The energy spectrum of the fluid turbulence often obeys either the Kolmogorov law (for the pure fluid case), $W_f(k) \propto k^{-5/3}$, or the Kraichnan spectrum (for the MHD case), $W_f(k) \propto k^{-3/2}$. However, "young" turbulence, driven at some low wavenumber, will initially show steeper spectra which flatten out to the Kolmogorov/Kraichnan form after about ten eddy times. Also, some recent simulations[4] have found k^{-3} spectra in MHD turbulence with asymptotic (v,B) correlations.

The longest scales of turbulence are observable with current interfometer resolution and sensitivity. Current limits are scales on the order of an arcsecond, corresponding to a few kpc, in the diffuse, low surface brightness plasma. Smaller scales are not observable at this time; they must be approached indirectly, such as by polarization modelling[5].

The plasma and magnetic field will support MHD wave modes, Alfven (A) and magnetosonic (MS) waves, in what is probably a stochastic situation. The MHD turbulence may be driven by edge instabilities and proceed via wave-wave interactions to form a cascade, as in the fluid case. It may also be driven internally by Lighthill radiation from the fluid turbulence[6]. Higher frequency (above the cool plasma cyclotron frequencies) plasma waves exist as well; except as a source of anomalous transport processes, such microturbulence shall not be considered further here.

PARTICLE ACCELERATION MECHANISMS IN RADIO GALAXIES

A model of the microphysics of particle aceleration in radio sources should account for the particle spectrum (the "universal" power law, $f(p) \propto p^{-s}$, with $4 \leq s \leq 5$), should determine the acceleration rate compared to the loss rates (synchrotron, inverse Compton on the background radiation, expansion) and should at least estimate the internal efficiency (the fraction of the energy input that goes into relativistic electrons rather than into heating the cooler, background gas).

A somewhat separate problem is the initial establishment of a "hot" component in the plasma; this is the so-called injection problem. The two component plasma, cool gas plus relativistic electrons, is not in thermal equilibrium. Either some internal process must go against thermal equilibrium in order to put most of the internal energy into a few energetic particles (for instance, boosting the tail of a Maxwellian distribution up to much higher energies), or the initial conditions of the jet flow must involve a very nonthermal distribution, which later gains a cooler component (such as by entrainment). A simpler problem than the injection problem is to understand how the already relativistic particles can maintain their high energy in the face of losses. Investigations of particle acceleration in radio sources have generally addressed this latter problem, and I shall limit myself to this "reacceleration" process.

a) Fermi acceleration

The foundation for stochastic acceleration models in astrophysics is the classic work of Fermi[7]. He considered test particle scattering from moving magnetic mirrors. If the mirror velocity is v_m and the time between collisions is t_{coll}, the rate of energy gain for a relativistic particle is

$$\frac{dp}{dt}=\alpha p \tag{1}$$

where the (energy independent) acceleration rate $\alpha \approx t_{coll} v_m/c$, for the case of approaching mirrors (first order process), or $\alpha \approx t_{coll} v_m^2/c^2$ for randomly moving mirrors (second order process). When this process is balanced against particle losses which are also energy independent (leakage, or expansion losses), an initially monoenergetic particle spectrum is broadened into a power law,

$$f(p) \propto p^{-s} \tag{2}$$

$$s=3+1/\alpha t_{loss}$$

Power laws initially steeper than this will be broadened to this spectrum; initially flatter power laws are unchanged. This process is attractive in that simple assumptions lead to power law spectra. However, the test particle model does not allow for feedback of the acceleration on the scattering clouds. In addition, the product αt_{loss} may well depend on local conditions and thus may have almost any value, rather than the range 0.5 to 1.0 required by the observed distributions. Recently some authors[8] have suggested that if the scattering clouds are magnetosonic waves, feedback process may restrict the product αt_{loss} to the desired range.

It should also be noted that the basic Fermi process in the presence of synchrotron losses will produce the above power law spectrum with a high energy (in fact exponential) cutoff. This cutoff will occur at the energy where the synchrotron losses,

$$\frac{dp}{dt}=\frac{4e^4B^2}{9m^4c^6}p^2 \tag{3}$$

exceed the acceleration rate.

b) Shock acceleration

An important application of the Fermi process is shock acceleration, as addressed in detail elsewhere in these proceedings[9]. Turbulence on either side of the shock will scatter streaming particles and act as converging mirrors, providing a first order Fermi process. For particle energies for which the scattering rate is energy independent, both the escape time from the shock region and the acceleration rate are related to this scattering rate. It is easy to show that the resulting power law depends only on the shock strength ($r = u_1/u_2$, where u_1 and u_2 are the upstream and downstream velocities),

$$f(p) \propto p^{-s} \tag{4}$$

$$s = 3r/(r-1)$$

The existence and energy dependence of the scattering centers (probably MHD or plasma waves), and the feedback of the acceleration on the shock structure, are areas of active investigation.

c) MHD turbulent acceleration

Finally, investigation of turbulent acceleration in radio galaxies has mainly addressed the two MHD modes, Alfven (A) and magnetosonic (MS) waves. The particle interaction with each mode is quite different.

The interaction with MS waves of wavenumber k occurs through the Landau resonance, $\Omega m = k_z v_z$ where "z" refers to the local magnetic field direction, v is the particle velocity and $\Omega = eB/mc$ is the nonrelativistic gyrofrequency. If the particle pitch angle distribution is fairly isotropic, this resonance allows all particles to "see" MS waves of essentially all wavelengths. The acceleration rate given by quasilinear theory for a particle of momentum p is[10]

$$\frac{dp}{dt} = a_1 \frac{v_m^2}{c} \frac{p}{B^2} \int_{k_{min}}^{k_{max}} k W_M(k) dk \tag{5}$$

where $\delta B^2 = \int W_m(k) dk$ is the total energy density of the MS waves and v_m is the wave speed; a_1 is an order unity constant. This is clearly an example of the Fermi process. The acceleration rate depends most importantly on the total wave intensity, and only secondarily on the spectrum. Consequences of this for radio galaxies have been addressed in reference 11.

RESONANT ACCELERATION BY ALFVEN WAVES

Particles interact with Alfven waves through the $(n=-1)$ cyclotron resonance,

$$pk_z = \Omega m/(\mu - v_A/c) \tag{6}$$

where μ is the cosine of the pitch angle. Thus, particles of a given energy see only a specific set of wavelengths, those at or shorter than their resonant wavelength, $\lambda_{res}(p) = 2\pi/k_{res}(p) \approx 2\pi p/\Omega m$. For electrons with $\gamma \approx 10^3$, this wavelength is on the order of 1 A.U.

Thus, the spectrum as well as the intensity of the Alfven waves is critical to the acceleration process. Modelling the wave spectrum requires knowledge of the driving and damping processes for the waves. I shall focus here on one particular case, that in which the Alfven waves are driven by Lighthill radiation from fully developed fluid turbulence. It is in this case that the wave-particle coupling in the presence of synchrotron losses for the particles leads asymptotically to the observed power law solutions.

Quasilinear theory gives the acceleration rate[10],

$$\frac{dp}{dt} = \frac{2\pi^2 e^2}{c^3 p} \int_{k_{res}(p)}^{k_{max}} W_A(k)\frac{1}{k}\left[1-\left(\frac{v_A}{c}+\frac{\Omega m}{pk}\right)^2\right]dk \tag{7}$$

if $W_A(k)$ is the energy density per wavenumber in Alfven waves; $k_{res}(p) \approx \Omega m/p$ from above; v_A is the Alfven speed. For the case of a power law wave spectrum, $W_A(k) \propto k^{-\nu}$ for $k > k_o$, the acceleration rate becomes

$$\frac{dp}{dt} = a_2 \frac{v_A^2}{c} \Omega m \frac{\delta B^2}{B^2}\left(\frac{k_o p}{\Omega m}\right)^{\nu-1} \tag{8}$$

where a_2 is another order-unity constant and δB^2 is now the total energy in the Alfven waves. This does not follow the usual Fermi criterion (equation 1), so that the particle spectrum resulting from this process must be investigated anew. In particular, comparing the synchrotron loss rate (equation 3) suggests that a wave spectrum $\propto k^{-3}$ could provide a balance against radiative losses at all energies for which resonant waves exist.

The wave spectrum depends on the balance of driving, damping and nonlinear processes at each wavenumber. For Alfven waves the most important damping is the acceleration of the relativistic particles. The damping rate at wavenumber k is[10]

$$\gamma_{cyc}(k)=\frac{4\pi^2 e^2 v_A^2}{c^3 k}\int_{p_{res}(k)}^{p_{max}} p^2\left[1-\left(\frac{v_A}{c}+\frac{\Omega m}{pk}\right)^2\right]\frac{\partial f(p)}{\partial p}\,dp \tag{9}$$

Other possible wave damping mechanisms are cyclotron resonant heating of the thermal gas, or nonlinear Landau damping. The former is insignificant except at the highest wavenumbers, while the latter is small for low wave intensities.

The wave driving is harder to estimate. A likely choice in the presence of fully developed fluid turbulence is driving of the MHD waves by the eddies in the fluid turbulence. In a process akin to Lighthill radiation of sound waves, each turbulent eddy will drive MHD waves of period comparable to the eddy period. In a radio source this will generate Alfven waves internally rather than from the edge, thus avoiding limb brightening problems.

One particular example of this was investigated analytically in reference 1. There, the fluid cascade was assumed to have the form

$$W_f(k)\ \alpha\ k^{-m} \tag{10}$$

where m may range from 3/2 (Kraichnan model) through 5/3 (Kolmogorov) up to, say, 2 ("young" turbulence). The wave spectrum at low intensity is then determined by the balance of driving and damping at each wavenumber,

$$\frac{dW(k)}{dt}=I(k)-\gamma_{cyc}(k)\,W(k) \tag{11}$$

where the driving function is assumed to be $I(k)\ \alpha\ k^{-s_t}$; $s_t = 3(m-1)/(3-m)$ ranges from 1 (Kraichnan) to 1.8 ("young"). This describes direct driving of waves by MHD turbulence, in a process akin to Lighthill radiation of sound waves by fluid turbulence[6]. If the particle spectrum is also a power law, the equilibrium wave spectrum is

$$W(k)\ \alpha\ k^{3-s-s_t} \tag{12}$$

Note, the response time for the waves is short compared to other relevant times in the problem, so that the waves will quickly reach a steady state (which may depend on the particle spectrum).

The evolution of the particle spectrum can be modelled using quasilinear formalism as[10]

$$\frac{\partial f(p)}{\partial t}=\frac{1}{p^2}\frac{\partial}{\partial p}\left[p^2 D(p)\frac{\partial f}{\partial p}+p^4 S f(p)\right] \tag{13}$$

where synchrotron losses are $S = 4B^2e^4/m^4c^6$; the stochastic acceleration coefficient is[10]

$$D(p) = \frac{2\pi^2 e^2 v_A^2}{c^3} \int_{k_{res}(p)}^{k_{max}} W_A(k) \frac{1}{k} \left[1 - \left(\frac{v_A}{c} + \frac{\Omega m}{pk}\right)^2\right] dk \qquad (14)$$

The coupling between the waves and the particles is evident. One would expect feedback from this coupling to drive arbitrary initial wave and particle distributions asymptotically to a final state, in which the wave and particle distributions are non-evolving. In particular, if we look for self similar solutions, we find that solutions exist only if $\nu = 3$. Only this particular wave spectrum allows the radiative loss time, $t_{sy}(p)$, and the acceleration time, $t_a(p)$, to have the same functional form with p. Further, the self similar solutions at large (pt) approach a power law with $s = 6 - s_t$.

Thus, this example suggests that this particular acceleration mechanism (including fluid turbulent driving and in the presence of synchrotron losses) will drive the particles toward a power law spectrum with exponent s in the range 4 to 5, which is just the observed range. Numerical experiments (1) confirm this. Further, the ratio of acceleration time to loss time is

$$\frac{t_a(p)}{t_{sy}(p)} = \frac{1}{s}\frac{S}{G} \qquad (15)$$

independently of p for this asymptotic case. The term G is proportional to the wave driving amplitude divided by the particle number density, and is defined by $D(p) = Gp^{\nu}$. Thus, the wave particle coupling will also drive this ratio to unity, through the inverse dependence on the particle intensity.

DISCUSSION

The theory of particle reacceleration in radio galaxies seems to be in decent shape at the microphysical level. Given the caveat that only low level turbulence (including that needed for shock acceleration) can be modelled with current quasilinear theories, mechanisms which can either create the observed power law electron distributions (say, in shocks or in bursts of strong turbulence) or maintain these distributions (as with Alfven turbulence, which along with synchrotron losses will drive the distribution to the same form) have been proposed. Intelligent estimates can now be made of the energetics and efficiency of these mechanisms, of the particular conditions under which each will occur and the particle energy ranges that will be affected.

The next important step for radio galaxy physics may be to re- investigate the large scale dynamics and evolution of the sources. The initial model of extended radio-luminous plasma being supplied energy by a directed outflow from the galactic (or quasar) nucleus seems essentially correct and has provided a valuable basic picture. The high quality of recent data, as well as the relation of this problem to fundamental problems such as the nature of the quasar energy source and the nature of galaxy evolution, now justifies more detailed and specific models. In particular, a fruitful area for investigation may be the several morphological or dynamical properties that appear to correlate with source power[12]. For instance:

a) High power sources tend to be edge brightened[13], with sharp boundaries and well defined hot spots at their leading edges. Lower power sources tend to be edge darkened, with rapidly spreading jets and with surface brightnesss declining steadily away from the nucleus as the flow becomes increasingly flocculent and (apparently) disrupted.

b) There seems to be a trend for higher power radio galaxies to have jets with lower relative luminosities compared to the lobes (although this trend may not continue for quasar jets[14]).

c) The "sidedness" of the jets, the ratio of the surface brightness on the two sides, is clearly related to the source power, in that the more powerful sources tend to be one sided.

d) Electromagnetic effects may be relatively more important in higher power sources. The lower power sources appear to be confined by the external cluster gas pressure; the higher power sources seem to be the ones which show overpressure and may require magnetic self-pinch confinement.

Morphological and dynamical trends such as these must be related to the global model of mass/energy transport in the jet flow, and especially to the creation and side effects of the turbulence and/or shocks which couple the flow energy to the radio luminosity. For instance, the strength and rate of development of large scale turbulence may depend on the source power (modulated by the flow Mach number, for instance), thus relating the "leakiness" of the low power jets and their edge-darkened morphology to their internal physics, in particular to the onset of fluid turbulence which will both brighten and disrupt the flow. On the other hand, electrodynamic effects may be relatively more important in higher power (higher net current?) sources, so that the details of source confinement, internal microphysical processes and also the effect of the radio source on the ambient extragalactic plasma may be quite different. Such wild speculations as these must be supported or disproved by calculations and self-consistent models; this author believes the observational situation is now good enough to warrant such efforts.

REFERENCES

1. J. A. Eilek and R. N. Henriksen, Astrophys. J., 227, 820 (1984).
2. D. S. De Young, Ann. Rev. Astron. Aastophys., 14, 447 (1976); G. Miley, Ann. Rev. Astron. Astrophys., 18, 165 (1980); M. C. Begelman, R. D. Blandford, and M. J. Rees, Rev. Mod. Phys., 56, 255 (1984).
3. A. H. Bridle and R. A. Perley, Ann. Rev. Ast. Ap., 22, 319 (1984).
4. R. Grappin, A. Pouquet and J. Leorat, Astron. Astrophys., 126, 51 (1984).
5. S. R. Spangler, Astrophys. J., 261, 310 (1982); J. A. Eilek, Bull. Amer. Astron. Soc., 16, 954 (1985).
6. S. Kato, Pul. Astron. Soc. Japan, 20, 59 (1972); R. F. Stein, Astrophys. J., 203, 313 (1981).
7. E. Fermi, Phys. Rev., 75, 1169 (1949).
8. B. J. Burn, Astron. Astrophys., 45, 435 (1975); A. Achterberg, Astron. Astrophys., 76, 276 (1979).
9. R. D. Blandford, these proceedings; A. R. Bell, Mon. Not. R. Astron. Soc., 182, 147 (1978); R. D. Blandford and J. P. Ostriker, Astrophys. J. Lett., 221, L20 (1978).
10. R. M. Kulsrud and A. Ferrari, Astrophys. Space Sci., 12, 302 (1971); C. Lacombe, Astron. Astrophys. 54, 1 (1977); J. A. Eilek, Astrophys. J., 230, 373 (1979).
11. G. V. Bicknell and D. B. Melrose, Astrophys. J., 262, 511 (1982).
12. A. H. Bridle, in The Physics of Energy Transport in Extragalactic Radio Sources, ed. A. H. Bridle and J. A. Eilek (NRAO, Green Bank, 1984), p. 1.
13. B. A. Fanaroff andl J. M. Riley, Mon. Not. R. Astron. Soc., 167, 31p (1974).
14. F. N. Owen and J. J. Puschell, Astron. J., in press (1985).

MAGNETIC HELICITY IN ASTROPHYSICS

George Field
Center for Astrophysics, Cambridge, Massachusetts 02138

ABSTRACT

Magnetic helicity, which is conserved in ideal MHD, measures the linkage of magnetic lines of force. After a brief review, I consider the helicity of a stellar magnetic field and show that a modified definition of helicity, called relative helicity, enables one to compute the relative amount of linkage of two field configurations above the photosphere even though the subphotospheric field is unobservable. I also show how the relative helicity changes as the result of motions of photospheric material. Then I consider the dissipation of helicity and show that in certain situations it is dissipated much more slowly than magnetic energy. Finally, I propose a variational principle for static problems, and show that in the presence of finite pressure, helicity conservation yields the usual equation of magnetostatic equilibrium. Woltjer's[1] conclusion that the result is a force-free field is a result of his neglect of pressure.

Section 1. A Brief Review

Many plasma physicists get along nicely without ever thinking about magnetic helicity, a concept first mentioned by Elsasser[2], further developed by Woltjer[1], and explained in physical terms by Moffatt[3]. Physically, helicity measures the linkage of lines of force within a specified volume; in ideal MHD it is conserved for each constituent tube of force because lines of force cannot be broken. Mathematically, it can be calculated from

$$H = \int_V \mathbf{A} \cdot \mathbf{B} d^3x, \tag{1}$$

where $\mathbf{A} = \text{curl}^{-1}\mathbf{B}$ is the vector potential and V is any volume bounded by a magnetic surface (one on which $\mathbf{B} \cdot \mathbf{n} = 0$). The reason that it is not vital to understand magnetic helicity, even though it is conserved in ideal MHD, is that conservation of flux implies conservation of helicity, but not vice-versa.

A small band of plasma physicists is hooked on helicity. One reason for this is Taylor's[4] conjecture that a Tokomak configuration undergoes turbulent relaxation to a state in which only the helicity of the whole plasma column (not that of any tube of force, as in ideal MHD) is conserved. This conjecture leads, through a theorem in Woltjer's[1] paper, to the conclusion that the final state is a constant-α force-free field, about the simplest configuration known to man. Since Tokomaks are observed to relax to approximately such a state[5], people are interested in helicity.

My own interest in helicity stems from the fact that helicity conservation is a concise way of stating a topological invariant, and I have the feeling that topology is basic in understanding things. Certainly topology of magnetic fields is behind many topics of interest in astrophysics, such as magnetic reconnection and particle acceleration.

Getting back to the definition of H, why must V be bounded by magnetic surfaces? Recall that the potential **A** representing a field **B** is not completely defined; adding the gradient of any function χ will do just as well. Doing so is called a gauge transformation:

$$\Delta \mathbf{A} = \mathbf{A}' - \mathbf{A} = \nabla\chi. \tag{2}$$

The physical quantity **B** is invariant under gauge transformations like (2), so we require H to be as well. This means that the gauge variation of H,

$$\begin{aligned}\Delta H = \Delta \int_V \mathbf{A}\cdot\mathbf{B}d^3x = \int_V (\Delta\mathbf{A})\cdot\mathbf{B}d^3x = \int_V (\nabla\chi)\cdot\mathbf{B}d^3x \\ = \int_V \nabla\cdot(\chi\mathbf{B})d^3x = \int_{\partial V} \chi\mathbf{B}\cdot d\mathbf{S},\end{aligned} \tag{3}$$

must vanish for all χ. This will be true in general only if $\mathbf{B}\cdot\mathbf{n} = 0$ on the surface ∂V.

The simplest application is a configuration made up of individual closed tubes of force, like a perfect dipole, or a Tokomak for which the toroidal winding number is a rational number times the poloidal winding number. Consider any two such tubes of force, (1) and (2). Their contribution to H is

$$H_{12} = \int_{V_1} \mathbf{A}\cdot\mathbf{B}d^3x + \int_{V_2} \mathbf{A}\cdot\mathbf{B}d^3x, \tag{4}$$

where V_1 and V_2 are the respective volumes. Since the tubes are thin, we write

$$d^3x = \Delta\sigma ds, \tag{5}$$

where $\Delta\sigma$ is the transverse area and ds is arc length, and

$$\mathbf{B} = B\mathbf{u}. \tag{6}$$

Hence

$$\mathbf{A}\cdot\mathbf{B}d^3x = (B\Delta\sigma)(\mathbf{A}\cdot\mathbf{u}ds) = (B\Delta\sigma)(\mathbf{A}\cdot ds). \tag{7}$$

In V_1, $B\Delta\sigma$ is the flux $\Delta\Phi_1$ of the tube, which is constant along the tube, so we take this quantity outside the integral, to obtain

$$\int_{V_1} \mathbf{A}\cdot\mathbf{B}d^3x = B\Delta\sigma \int_{V_1} \mathbf{A}\cdot ds = \Delta\Phi_1 \int_{A_1} \mathbf{B}\cdot d\mathbf{S}, \tag{8}$$

where A_1 is a surface of which the tube forms an edge. The last integral in (8) is $\Delta\Phi_2$, the flux of tube (2) which links tube (1). Since the second term in equation (4) contributes a similar term,

$$H_{12} = 2\Delta\Phi_1\Delta\Phi_2, \tag{9}$$

and from the construction, it is apparent that H is the sum of (9) over all pairs of tubes.

The next simplest case is a configuration composed of nested toroidal surfaces. Identify one of these surfaces by the toroidal flux ψ_T within it, and a neighboring surface by $\psi_T + d\psi_T$. Let ψ_P be the poloidal flux threading the hole created by the surface. The helicity attributable to the annulus $\psi_T, \psi_T + d\psi_T$ is that due to the poloidal flux outside the annulus (ψ_P) linking the toroidal flux $d\psi_T$ within it ($\psi_P d\psi_T$) plus that due to the poloidal flux within the annulus ($-d\psi_P$, negative because ψ_P increases inward) linking the toroidal flux $\psi_T(-\psi_T d\psi_P)$, so that

$$dH = \psi_P d\psi_T - \psi_T d\psi_P. \tag{10}$$

If we define the twist T by

$$T = -\frac{d\psi_P}{d\psi_T}, \tag{11}$$

it follows that

$$H = 2\int_{\circ}^{\Phi} T\psi_T d\psi_T, \tag{12}$$

where Φ is the total toroidal flux. If T is independent of ψ_T, corresponding to a unformly twisted configuration, then

$$H = T\Phi^2. \tag{13}$$

In general, magnetic field lines wander ergodically throughout a volume. It is possible to generalize the concept of linkage to this case (Arnol'd[6]), but we won't pursue that here.

We have stated that H is conserved in ideal MHD. To prove this, write Maxwell's equations in terms of the field tensor

$$F_{\alpha\beta} = \partial_\alpha A_\beta - \partial_\beta A_\alpha, \tag{14}$$

where

$$A^\alpha = (A^\circ, \mathbf{A}) \tag{15}$$

is the four-potential. The homogeneous Maxwell equations in tensor form are

$$\partial_\alpha {}^*F^{\alpha\beta} = 0, \tag{16}$$

where

$${}^*F^{\alpha\beta} = \frac{1}{2}\epsilon^{\alpha\beta\mu\nu}F_{\mu\nu} \tag{17}$$

is the dual of $F^{\alpha\beta}$. If we define a four-vector H^α by

$$H^\alpha = A_\beta {}^*F^{\alpha\beta}, \tag{18}$$

we find that its divergence is

$$\partial_\alpha H^\alpha = (\partial_\alpha A_\beta)^* F^{\alpha\beta} + A_\beta \partial_\alpha {}^*F^{\alpha\beta}. \tag{19}$$

The last term vanishes by (16), and the first gives

$$\partial_\alpha H^\alpha = \frac{1}{c}\frac{\partial}{\partial t}(\mathbf{A}\cdot\mathbf{B}) + \nabla\cdot(\mathbf{E}\times\mathbf{A} + A^\circ\mathbf{B}) = \frac{1}{2}F_{\alpha\beta}\,{}^*F^{\alpha\beta} = -2\mathbf{E}\cdot\mathbf{B}. \quad (20)$$

This is one of the two quadratic Lorentz invariants of the field. Because

$$\mathbf{E} = -\frac{\mathbf{v}}{c}\times\mathbf{B} \quad (21)$$

in ideal MHD, $\mathbf{E}\cdot\mathbf{B}$ vanishes, and (20) is a conservation law. If we substitute (21) into (20), we have

$$\frac{\partial}{\partial t}(\mathbf{A}\cdot\mathbf{B}) + \nabla\cdot[(cA^\circ - \mathbf{A}\cdot\mathbf{v})\mathbf{B} + (\mathbf{A}\cdot\mathbf{B})\mathbf{v}] = 0. \quad (22)$$

Suppose ∂V is a magnetic surface ($\mathbf{B}\cdot\mathbf{n} = 0$) across which there is no flow ($\mathbf{v}\cdot\mathbf{n} = 0$). Then if we integrate (22) over V, the term in brackets vanishes by Gauss's theorem, and the conservation of helicity takes the form

$$\frac{\partial H}{\partial t} = 0. \quad (23)$$

If we consider a case in which the surface ∂V moves with the fluid (so that if it is initially a magnetic surface it remains one, because the field is frozen in), we can manipulate the last term in (22) to yield

$$\nabla\cdot[(\mathbf{A}\cdot\mathbf{B})\mathbf{v}] = \mathbf{v}\cdot\nabla(\mathbf{A}\cdot\mathbf{B}) + (\mathbf{A}\cdot\mathbf{B})\nabla\cdot\mathbf{v} = \mathbf{v}\cdot\nabla(\mathbf{A}\cdot\mathbf{B}) - \frac{1}{\rho}(\mathbf{A}\cdot\mathbf{B})\frac{d\rho}{dt}. \quad (24)$$

Then (22) can be written

$$\rho\frac{d}{dt}\frac{\mathbf{A}\cdot\mathbf{B}}{\rho} + \nabla\cdot[(cA^\circ - \mathbf{A}\cdot\mathbf{v})\mathbf{B}] = 0. \quad (25)$$

If we integrate over a volume bounded by a magnetic surface moving with the fluid, the second term vanishes and so

$$\frac{d}{dt}\int_V \mathbf{A}\cdot\mathbf{B}d^3x = \frac{d}{dt}\int_V \left(\frac{\mathbf{A}\cdot\mathbf{B}}{\rho}\right)\rho d^3x = \int_V \rho\frac{d}{dt}\left(\frac{\mathbf{A}\cdot\mathbf{B}}{\rho}\right)d^3x = 0 \quad (26)$$

because

$$\frac{d}{dt}(\rho d^3x) = 0. \quad (27)$$

Thus, the helicity of the field which is frozen into the fluid is constant, and the conservation of helicity is equivalent to a constant of the motion.

In summary, helicity measures field-line linkage. As such, it would be expected to be conserved in ideal MHD, and it is. Is this fact useful in astrophysical situations? What follows is my work along with that of a recent Harvard Ph.D., Mitchell Berger[7], directed toward answering that question.

Section 2. Helicity of a Stellar Magnetic Field

One phenomenon of current interest is the dissipation of solar and stellar magnetic fields, leading to coronal heating and particle acceleration. In the sun, at least, coronal fields are characterized by

$$\beta = \frac{8\pi p}{B^2} << 1. \tag{28}$$

As a result, the fields are nearly force-free, so that

$$\mathbf{J} \times \mathbf{B} = \frac{c}{4\pi}(\nabla \times \mathbf{B}) \times \mathbf{B} \simeq 0. \tag{29}$$

Hence

$$\nabla \times \mathbf{B} = \alpha \mathbf{B}, \tag{30}$$

where α is a function which need not be constant. The divergence of (30) shows that

$$\mathbf{B} \cdot \nabla \alpha = 0, \tag{31}$$

so that α is a constant along each field line in the force-free region.

Equation (29) tells us that only field-aligned currents can be present if $\beta << 1$ and gravitational forces are unimportant. It is well-known that of all the fields having the same $\mathbf{B} \cdot \mathbf{n}$ on a simply-connected surface (here, the photosphere), the potential field (with $\mathbf{J} = 0$) has the minimum energy. It follows that magnetic energy over and above that of the potential field, which is thought to be the source of coronal heating and particle acceleration, must be associated with currents, and, because of the above argument, these currents must be field-aligned.

Let us go back to (20). In ideal MHD, $\mathbf{E} \cdot \mathbf{B} = 0$, but if we include dissipation, which is required for coronal heating,

$$\mathbf{E} = -\frac{\mathbf{v}}{c} \times \mathbf{B} + \eta \mathbf{J} \tag{32}$$

where η is the local resistivity (which must be much greater than the classical value if one is to account for the required dissipation). Hence the source term for magnetic helicity in equation (20) is no longer zero, but

$$-2c\eta \mathbf{J} \cdot \mathbf{B} \tag{33}$$

Thus, dissipation of field-aligned currents, which may be present in a force-free region, also change the helicity.

In order to pursue the connection between helicity and coronal heating, we must first define helicity more carefully for a star. Earlier, we stated that helicity is defined only for volumes bounded by magnetic surfaces. While a star together with its surrounding region is such a volume, the envelope of the star alone (defined as the optically thin layers) is definitely not such a region, because magnetic flux crosses the surface of the star. Berger and Field have addressed this problem by defining a relative helicity which is gauge invariant and whose rate of change can be calculated.

The idea is this. Let the exterior of the star be V_a and its interior be V_b; the fields and potentials in those regions are labelled (a) and (b), respectively. In principle we can observe a field $\mathbf{B}_1$ in V_a, but it is hopeless to observe $\mathbf{B}_1$ in V_b, so the helicity of the entire field configuration (1),

$$H_1 = \int_{V_a} \mathbf{A}_{1a} \cdot \mathbf{B}_{1a} d^3x + \int_{V_b} \mathbf{A}_{1b} \cdot \mathbf{B}_{1b} d^3x, \tag{34}$$

which is gauge invariant and which measures the linkage of the entire configuration (1), is impossible to determine.

To overcome this problem, we consider another field configuration (2), which is the same as $\mathbf{B}_1$ in V_b (so that $\mathbf{B}_{2b} = \mathbf{B}_{1b}$) but not necessarily the same in V_a. The difference in helicities is

$$\begin{aligned} \Delta H = H_1 - H_2 &= \int_V (\mathbf{A}_1 \cdot \mathbf{B}_1 - \mathbf{A}_2 \cdot \mathbf{B}_2) d^3x \\ &= \int_V (\mathbf{A}_1 - \mathbf{A}_2) \cdot (\mathbf{B}_1 + \mathbf{B}_2) d^3x + \int_V (\mathbf{A}_2 \cdot \mathbf{B}_1 - \mathbf{A}_1 \cdot \mathbf{B}_2) d^3x. \end{aligned} \tag{35}$$

By integrating by parts, we can show that the second term on the right vanishes. Because $\mathbf{A}_1$ and $\mathbf{A}_2$ represent the same field in V_b, we have

$$\mathbf{A}_1 - \mathbf{A}_2 = \nabla \xi \tag{36}$$

in V_b, where ξ is some function. The contribution of V_b to (35) is thus

$$\int_{V_b} (\mathbf{B}_1 + \mathbf{B}_2) \cdot \nabla \xi d^3x = \int_{V_b} \nabla \cdot [\xi(\mathbf{B}_{1b} + \mathbf{B}_{2b})] \, d^3x = \int_{\partial V_b} \xi(\mathbf{B}_{1b} + \mathbf{B}_{2b}) \cdot d\mathbf{S}. \tag{37}$$

The fact that $\nabla \cdot \mathbf{B} = 0$ implies that

$$\mathbf{B}_{1b} \cdot \mathbf{n} = \mathbf{B}_{1a} \cdot \mathbf{n} \tag{38}$$

and

$$\mathbf{B}_{2b} \cdot \mathbf{n} = \mathbf{B}_{2a} \cdot \mathbf{n} \tag{39}$$

on ∂V_b, whence (37) becomes

$$- \int_{\partial V_a} \xi(\mathbf{B}_{1a} + \mathbf{B}_{2a}) \cdot d\mathbf{S}. \tag{40}$$

Now H_1 and H_2 are gauge invariant, so ΔH is also and we are free to choose any gauge. If we adopt the Coulomb gauge ($\nabla \cdot \mathbf{A} = 0$), the "uncurling" of $\mathbf{A}$ yields

$$\mathbf{A}(\mathbf{x}) = -\frac{1}{4\pi} \int_V d^3x' \frac{\mathbf{r}}{r^3} \times \mathbf{B}(\mathbf{x}'). \tag{41}$$

Because $\mathbf{B}_{1b} = \mathbf{B}_{2b}, \mathbf{A}_1 - \mathbf{A}_2$ has a contribution only from V_a, so the integral over V_a in (35) depends only on the fields in V_a. By the same argument, (36) shows that ξ also depends only on the fields in V_a, and therefore the contribution of V_b to (35), given by (40), does also. We conclude that ΔH defined in (35) is independent of the (unobservable) fields in V_b; it is this property that makes ΔH useful.

If, further, we let

$$\mathbf{B}_{2a} = \mathbf{P}_a, \tag{42}$$

the potential field in V_a (which is completely determined by $\mathbf{B}_a \cdot \mathbf{n}$ on ∂V_a) we can define a "relative helicity"

$$H_R = H(\mathbf{B}_a, \mathbf{B}_b) - H(\mathbf{P}_a, \mathbf{B}_b) \tag{43}$$

which is gauge invariant, completely determined by the observable field $\mathbf{B}_a$ and its normal on the surface, and independent of the unobservable field $\mathbf{B}_b$. Particularly simple formulas are obtained if

$$\mathbf{B}_b = \mathbf{P}_b, \tag{44}$$

the potential field in the region V_b. If ∂V is a plane or a sphere, it was shown by Berger and Field[8] that

$$H(\mathbf{P}_a, \mathbf{P}_b) = 0 \tag{45}$$

and hence, that

$$H_R = H(\mathbf{B}_a, \mathbf{P}_b). \tag{46}$$

Because $\mathbf{P}_b$ can be calculated from $\mathbf{B} \cdot \mathbf{n}$ on ∂V, it is clear at once from (46) that H_R depends only on the observed field, $\mathbf{B}_a$.

Applying (20) to H_R, Berger[9] shows that

$$\frac{\partial H_R}{\partial t} = -2c \int_{V_a} \eta \mathbf{J} \cdot \mathbf{B} d^3x + 2c \int_{\partial V_a} (\mathbf{A}_p \times \mathbf{E}) \cdot d\mathbf{S}, \tag{47}$$

where $\mathbf{A}_p$ is the vector potential such that

$$\mathbf{P}_b = \nabla \times \mathbf{A}_p. \tag{48}$$

If the photospheric fields obey ideal MHD, the second term can be expressed in terms of the photospheric velocity field $\mathbf{v}$ by

$$2 \int_{\partial V_a} (\mathbf{A} \cdot \mathbf{B}) v_r dS - 2 \int_{\partial V_a} (\mathbf{A} \cdot \mathbf{v}) B_r dS. \tag{49}$$

The first term in (49) represents the upwelling of field which is already twisted; we will not discuss it further here. The second represents the twisting of fields in V_a due to fluid motions parallel to the surface of the star. Berger shows that if the footpoints of two flux tubes with fluxes Φ_1 and Φ_2 orbit around each other with a relative angular velocity $\dot{\theta}_{12}$, while each tube spins on its axis at angular frequency ω, the second term in (49) becomes

$$-\frac{1}{2\pi}(\omega_1\Phi_1^2 + \omega_2\Phi_2^2 + 2\dot{\theta}_{12}\Phi_1\Phi_2), \tag{50}$$

so that there are both spin and orbit contributions to the flow of helicity through the photosphere. Observational data on solar photospheric motions (Weart[10]) suggest that the helicity transmitted into the corona in this way is substantial. Since helicity measures twist, which in the force-free conditions of corona implies field-aligned currents, the energy of the coronal field can be thereby increased.

The magnetic energy stored in the corona can dissipate by resistivity. According to (47), the same effect will dissipate helicity. It has been conjectured by Taylor[4] that the net helicity in such a situation will change very little during a time when the energy changes significantly. In the next section we show how this conjecture applies to a stellar corona.

Section 3. Dissipation of Helicity

Let us define the magnetic viscosity by

$$\nu_M = \frac{c^2\eta}{4\pi}. \tag{51}$$

In terms of it, the volume rate of relative helicity dissipation is, by (47),

$$\frac{\partial H_R}{\partial t} = -\frac{8\pi}{c}\int\limits_{V_a} \nu_M J_{\prime\prime} B d^3x, \tag{52}$$

where

$$J_{\prime\prime} = \frac{c}{4\pi}(\nabla \times \mathbf{B}) \cdot \frac{\mathbf{B}}{B}. \tag{53}$$

is the field-aligned current. Recall the Cauchy-Schwartz inequality: if f and g are square-integrable over V,

$$\left[\int\limits_V fgd^3x\right]^2 \le \left[\int\limits_V f^2d^3x\right]\left[\int\limits_V g^2d^3x\right]. \tag{54}$$

If we let

$$f = \nu_M^{\frac{1}{2}} B \tag{55}$$

and

$$g = \nu_M^{\frac{1}{2}} J_{\prime\prime}, \tag{56}$$

then (54) implies that

$$\left[\int\limits_V \nu_M J_{\prime\prime} B d^3x\right]^2 \le \left[\int\limits_V \nu_M B^2 d^3x\right]\left[\int\limits_V \nu_M J_{\prime\prime}^2 d^3x\right]. \tag{57}$$

If we define a weighted mean magnetic viscosity by

$$\overline{\nu}_M = \frac{\int\limits_V \nu_M B^2 d^3x}{\int\limits_V B^2 d^3x}, \tag{58}$$

and introduce the total magnetic energy,

$$E_M = \frac{1}{8\pi}\int\limits_V B^2 d^3x, \tag{59}$$

and the rate of Ohmic dissipation due to the field-aligned current,

$$\dot{E}_{M_{\parallel}} = -\frac{4\pi}{c^2}\int\limits_V \nu_M J_{\parallel}^2 d^3x, \tag{60}$$

we find from (57) that

$$\dot{H}_R^2 \leq -128\pi^2 \overline{\nu}_M E_M \dot{E}_{M\parallel}. \tag{61}$$

Certainly $E_M \leq E$, the total energy, including internal, kinetic, and magnetic energies. If we ignore any positive contributions to E (such as adiabatic compression of coronal gas)

$$-\dot{E}_{M_{\parallel}} \leq -\dot{E}, \tag{62}$$

Hence (61) implies that

$$\dot{H}_R^2 \leq -128\pi^2 \overline{\nu}_M E\dot{E}, \tag{63}$$

which can be integrated to give the change in helicity in time Δt:

$$|\Delta H_R| \leq 8\pi\sqrt{2}\overline{\nu}_M^{\frac{1}{2}} \int\limits_0^{\Delta t} dt(-E\dot{E})^{\frac{1}{2}}. \tag{64}$$

Consider a problem in which E(t) varies from $E(t=0) = E_i$ to $E(t=\Delta t) = E_f$, so the values of E at the end points of the integral can be be regarded as fixed. The value of the integral in (64) will depend upon the detailed variation of E with time, but we may use the calculus of variations to find a maximum value for it. The integral is of the form

$$I = \int\limits_0^{\Delta t} dt F(E, \dot{E}), \tag{65}$$

where

$$F(E,\dot{E}) = (-E\dot{E})^{\frac{1}{2}}. \tag{66}$$

The Euler-Lagrange equation for the variational problem is

$$\frac{d}{dt}\left(\frac{\partial F}{\partial \dot{E}}\right) = \frac{\partial F}{\partial E}. \tag{67}$$

For this case, (67) becomes

$$\frac{d}{dt}\left[-E(-E\dot{E})^{-\frac{1}{2}}\right] = -\dot{E}(-E\dot{E})^{-\frac{1}{2}}, \tag{68}$$

or

$$\frac{d}{dt}(-E\dot{E})^{-\frac{1}{2}} = 0. \tag{69}$$

Hence

$$E\dot{E} = \text{ const.} \tag{70}$$

The solution to (70) which satisfies the initial and final conditions is

$$E^2 = E_i^2 + (E_f^2 - E_i^2)\frac{t}{\Delta t}. \tag{71}$$

When this is inserted into the integral in (65) we find that

$$I \leq \frac{1}{\sqrt{2}}(E_i^2 - E_f^2)^{\frac{1}{2}}(\Delta t)^{\frac{1}{2}}. \tag{72}$$

When this is inserted into (64) we find that

$$|\Delta H_R| \leq 8\pi\left[\overline{\nu}_M(E_i^2 - E_f^2)\Delta t\right]^{\frac{1}{2}}. \tag{73}$$

We interpret (73) as giving the minimum time to change the helicity by ΔH_R:

$$\Delta t \geq \frac{D^2}{\overline{\nu}_M}. \tag{74}$$

where from (73) we define D as

$$D = \frac{|\Delta H|}{8\pi(E_i^2 - E_f^2)^{\frac{1}{2}}}. \tag{75}$$

To interpret (75), we set

$$|\Delta H| \simeq |H| \tag{76}$$

and

$$E_f << E_i, \tag{77}$$

so that

$$D \simeq \frac{|H|}{8\pi E_i} = \frac{|\int_V \mathbf{A}\cdot\mathbf{B}d^3x|}{\int_V B^2 d^3x}. \tag{78}$$

Because $\mathbf{A} = \text{ curl}^{-1}\mathbf{B}$, D is determined by the macroscopic length scales involved in the magnetic field configuration. Thus, for a coronal loop of twisted magnetic field whose radius is R and length is ℓ, Berger[8] shows that if there is about one twist between the footpoints,

$$D \sim \frac{1}{2}R. \tag{79}$$

Hence even if the energy dissipation is occuring on very short length scales within the tube (due, perhaps, to tearing-mode instabilities) the length scale which determines helicity dissipation is still the macroscopic dimension of the configuration. Correspondingly, the minimum time scale calculated from (75) is very large; in the specific case of a twisted coronal flux tube,

$$\Delta t \geq 10^6 (\overline{\nu}_M / \nu_{M,cl})^{-1} R_9^2 T_6^{\frac{3}{2}} \text{ years,} \tag{80}$$

where

$$\nu_{M,cl} = 7000 T_6^{-\frac{3}{2}} \tag{81}$$

is the classical magnetic viscosity, $R_6 = R/10^9$cm, and $T_6 = T/10^6$K. In fact, the expression (80) may underestimate Δt for the following reason. While it is true that $\overline{\nu}_M$ may exceed $\nu_{M,cl}$ by a substantial factor, this would be expected to be true only within relatively small diffusion regions, where plasma turbulence driven by high current densities leads to anomalous resistivity. The proper way to correct for this in (81) is to replace $\overline{\nu}_M$ by

$$\overline{\nu}_M = \frac{V_D}{V} \nu_{M,D} \tag{82}$$

where $\nu_{M,D}$ is the (enhanced) magnetic viscosity in the diffusion region. Hence (81) becomes

$$\Delta t \geq 10^6 (\nu_{M,D} / \nu_{M,cl})^{-1} (V_D/V)^{-1} R_9^2 T_6^{\frac{3}{2}} \text{ years.} \tag{83}$$

Thus, although $\nu_{M,D}/\nu_{M,cl}$ may be very large ($\sim 10^6$), bringing the value of Δt down, V_D/V is probably very small, raising it again.

Estimates of the rate of helicity flow into the corona based upon (50) applied to the observed motions of the footpoints of magnetic loops indicate that the helicity of the corona can change on far shorter time scales than it can be dissipated according to (83); if the radius of a flux tube is 10^9 cm, motions at 1 km sec^{-1} can double the relative helicity of the tube in only a few hours. Berger[9] concludes that the helicity content of coronal fields is essentially completely determined by the flow of helicity in through the photosphere and its flow out via the solar wind.

Section 4. Lagrangians Based Upon Helicity Conservation

Since the helicity of each elementary flux tube is conserved in ideal MHD, one might suppose that helicity would be a useful constraint in variational principles for magnetostatics and magnetohydrodynamics. Indeed it is; although Woltjer[1] initiated discussion of this constraint, one can go further as follows.

It is desired to find a Lagrangian for a static system which is stationary to first-order changes in the positions of fluid elements and in the vector potential, when the static system obeys

$$\nabla p = \frac{1}{c}\mathbf{J} \times \mathbf{B}, \tag{84}$$

the equation of magnetostatic equilibrium. Such a Lagrangian will be of the form

$$L = \int_V \mathcal{L} d^3x, \tag{85}$$

where $\mathcal{L}$ is the Lagrangian density, and where the displacements $\delta\mathbf{x}$ and variations in vector potential $\delta\mathbf{A}$ vanish on the boundary ∂V, taken to be at ∞.

There are three contributions to $\mathcal{L}$: the matter Lagrangian is

$$\mathcal{L}_M = U, \tag{86}$$

the internal energy per unit volume. If the fluid element is displaced by $\delta\mathbf{x}$, it is a standard result that $\delta\mathcal{L}_M$ is equivalent to $(\doteq)$*

$$\delta\mathcal{L}_M \doteq -\nabla p \cdot \delta\mathbf{x} \tag{87}$$

if the displacements conserve mass and entropy. The electromagnetic Lagrangian with $\mathbf{E} = 0$ is

$$\mathcal{L}_{EM} = \frac{1}{8\pi}B^2 = \frac{1}{8\pi}(\nabla \times \mathbf{A})^2, \tag{88}$$

whose variation is equivalent to

$$\delta\mathcal{L}_{EM} \doteq \frac{1}{4\pi}\delta\mathbf{A} \cdot \nabla \times \mathbf{B} \tag{89}$$

So far, the discussion is completely conventional.

Now we seek to impose ideal MHD conditions by requiring that the helicity of each tube of force is conserved. If the lines of force are closed, the helicity of the ith tube is

$$\Delta H_i = \int_{V_i} \mathbf{A} \cdot \mathbf{B} d^3x. \tag{90}$$

Since this quantity is conserved, it should be stationary under variations satisfying the conditions of ideal MHD. As usual, such a constraint can be imposed by using Lagrange multipliers. Hence, we add to the Lagrangian the constraint term

$$L_c = -\Sigma_i\lambda_i\Delta H_i = -\Sigma_i\lambda_i \int_{V_i} \mathbf{A} \cdot \mathbf{B} d^3x, \tag{91}$$

* Two variations are equivalent if they differ by a divergence, which by Gauss' theorem does not contribute to L because variations vanish on the boundary.

where the λ_i is the Lagrange multiplier for the ith tube of force. If we let the volume of each tube shrink and increase the number of tubes without limit, (91) becomes the integral

$$L_c = -\int_V \lambda \mathbf{A} \cdot \mathbf{B} d^3x, \tag{92}$$

where, according to the method of construction, λ is constant on each tube of force. That is,

$$\mathbf{B} \cdot \nabla \lambda = 0. \tag{93}$$

Interestingly, condition (93) assures that L_c is gauge invariant, as follows. The gauge variation of L_c is

$$\begin{aligned} \Delta L_c &= -\int_V \lambda \mathbf{B} \cdot \nabla \chi d^3x = -\int_V \lambda \nabla \cdot (\chi \mathbf{B}) d^3x \\ &= -\int_V \nabla \cdot (\lambda \chi \mathbf{B}) d^3x + \int_V \chi \mathbf{B} \cdot \nabla \lambda d^3x \\ &= -\int_{\partial V} \lambda \chi \mathbf{B} \cdot d\mathbf{S} + \int_V \chi \mathbf{B} \cdot \nabla \lambda d^3x. \end{aligned} \tag{94}$$

The first term vanishes because ∂V is at ∞, and the second term vanishes because of (93).

The variation of $\mathcal{L}_c = \lambda \mathbf{A} \cdot \mathbf{B}$ is equivalent to

$$\delta \mathcal{L}_c \doteq -(\nabla \lambda \cdot \delta \mathbf{x})(\mathbf{A} \cdot \mathbf{B}) - 2\lambda \mathbf{B} \cdot \delta \mathbf{A} - (\nabla \lambda \times \mathbf{A}) \cdot \delta \mathbf{A}. \tag{95}$$

Collecting terms in $\delta \mathcal{L}$, we find that

$$\delta \mathcal{L} \doteq -\delta \mathbf{x} \cdot [\nabla p + (\mathbf{A} \cdot \mathbf{B}) \nabla \lambda] - \delta \mathbf{A} \cdot \left[2\lambda \mathbf{B} + \nabla \lambda \times \mathbf{A} - \frac{1}{4\pi} \nabla \times \mathbf{B} \right]. \tag{96}$$

Because $\delta \mathbf{x}$ and $\delta \mathbf{A}$ are arbitrary and independent in this formulation, each bracket must individually vanish:

$$\nabla p + (\mathbf{A} \cdot \mathbf{B}) \nabla \lambda = 0; \tag{97}$$

$$2\lambda \mathbf{B} + \nabla \lambda \times \mathbf{A} - \frac{1}{4\pi} \nabla \times \mathbf{B} = 0. \tag{98}$$

To eliminate λ, we cross-multiply (98) with $\mathbf{B}$ and add to (97) to obtain

$$\begin{aligned} &\nabla p + (\mathbf{A} \cdot \mathbf{B}) \nabla \lambda - \frac{1}{4\pi} (\nabla \times \mathbf{B}) \times \mathbf{B} + (\nabla \lambda \times \mathbf{A}) \times \mathbf{B} \\ &\quad = \nabla p - \frac{1}{4\pi} (\nabla \times \mathbf{B}) \times \mathbf{B} + (\mathbf{A} \cdot \mathbf{B}) \nabla \lambda - (\mathbf{A} \cdot \mathbf{B}) \nabla \lambda + (\mathbf{B} \cdot \nabla \lambda) \mathbf{A} = 0. \end{aligned} \tag{99}$$

By virtue of (93), (99) is equivalent to the general equation of magnetohydrostatic equilibrium (84).

It is interesting to compare the present derivation with Woltjer's.[1] He neglected the contribution of the internal energy U to $\mathcal{L}$, and therefore obtained $p = 0$ in (96) and (97). As a consequence, (97) then implies that $\nabla\lambda = 0$, or $\lambda =$ constant. Equation (98) then implies that

$$\nabla \times \mathbf{B} = 8\pi\lambda\mathbf{B} \tag{100}$$

- a constant-α force-free field, as he concluded, with $\alpha = 8\pi\lambda$. The fact that the field is force-free should be no surprise, as the neglect of internal energy eliminates any force capable of opposing the Lorentz force. The fact that α is constant is more subtle. From the present derivation we learn that if p is finite, no matter how small, one obtains (84), not (100). As p approaches zero, (84) implies that

$$\nabla \times \mathbf{B} \simeq 8\pi\lambda\mathbf{B}, \tag{101}$$

but there is no implication that λ is constant. I believe that imposing the condition that λ is constant at the outset is unwarranted.

Section 5. Helicity and Energy of Force-Free Fields

Equation (30) implies that

$$\nabla \times \mathbf{B} = \alpha\nabla \times \mathbf{A}, \tag{102}$$

and therefore, if $\alpha =$ constant,

$$\mathbf{B} = \alpha\mathbf{A} + \nabla\phi, \tag{103}$$

so that

$$B^2 = \alpha\mathbf{A}\cdot\mathbf{B} + \nabla\cdot(\phi\mathbf{B}). \tag{104}$$

Hence, the magnetic energy in a volume V in which the field is a constant-α force-free field is a constant times the magnetic helicity in that volume, plus a surface term. Berger discusses this relationship in his thesis[6]. I have shown (unpublished notes) that the surface term is related to the surface currents which occur at the boundary between force-free and non-force-free regions. I have also extended the discussion to variable-α fields.

Here I want to discuss the relationship between energy and magnetic helicity from a completely different point of view. I start by showing that magnetic energy itself can be interpreted as a type of helicity.

By a vector identity,

$$\nabla\cdot(\mathbf{A}\times\mathbf{B}) = B^2 - \mathbf{A}\cdot\nabla\times\mathbf{B} = B^2 - \frac{4\pi}{c}\mathbf{J}\cdot\mathbf{A}, \tag{105}$$

so that the magnetic energy in V is given by

$$E_M = \frac{1}{8\pi}\int_V B^2 d^3x = \frac{1}{2c}\int_V \mathbf{J}\cdot\mathbf{A}d^3x + \frac{1}{8\pi}\int_{\partial V}(\mathbf{A}\times\mathbf{B})\cdot d\mathbf{S}. \tag{106}$$

It is probably well-known by many of you, but I only recently discovered for myself an interpretation of (106). As in the case of magnetic helicity density $\mathbf{A} \cdot \mathbf{B}$, the vector multiplying $\mathbf{A}$ in the first term on the right-hand side of (106) is divergence-free: the lines of J ("current lines"), like those of $\mathbf{B}$, cannot end in a case like ideal MHD where accumulations of free charge are not present. As in the case of magnetic helicity, consider a case in which the current lines are closed. Consider one current line (i), and a small tube of variable cross section $\Delta\sigma$ centered on it, such that

$$\Delta I_i = J_i \Delta\sigma = \text{ constant.} \tag{107}$$

Such a tube may be called a current tube. The contribution of the ith current tube to (106) is

$$\frac{1}{2c} \int\limits_{\Delta V_i} \mathbf{J} \cdot \mathbf{A} d^3x = \frac{1}{2c} \int\limits_{\Delta V_i} (J_i \Delta\sigma)(\mathbf{A} \cdot \mathbf{u}) ds = \frac{1}{2c} \Delta I_i \int\limits_{\Delta V_i} \mathbf{A} \cdot ds = \frac{1}{2c} \Delta I_i \Phi_i, \tag{108}$$

where Φ_i is the magnetic flux linking the ith current tube. Hence

$$\frac{1}{2c} \int\limits_V \mathbf{J} \cdot \mathbf{A} d^3x = \frac{1}{2c} \sum_i \Phi_i \Delta I_i, \tag{109}$$

showing that this term, like magnetic helicity, is a sum of linkages, and is therefore itself a helicity. Note, however, that although Φ_i is conserved in ideal MHD, I_i is not, so that we would not expect magnetic energy to be conserved in ideal MHD (and, of course, it is not). Thus, it is not clear what applications this interpretation has in general.

If the field is force-free, this interpretation does have application, for then

$$\mathbf{J} = \frac{c}{4\pi} \nabla \times \mathbf{B} = \frac{c}{4\pi} \alpha \mathbf{B} \tag{110}$$

is related to another divergence-free field, $\mathbf{B}$, and

$$\frac{1}{2c} \int\limits_V \mathbf{J} \cdot \mathbf{A} d^3x = \frac{1}{8\pi} \int\limits_V \alpha \mathbf{A} \cdot \mathbf{B} d^3x \tag{111}$$

is related to the magnetic helicity, as we had hoped originally [see eq. (104)].

If the current lines are closed, so are the lines of force in this case, and since α is constant and equal to α_i on the ith line of force because of (31),

$$\frac{1}{2c} \int\limits_V \mathbf{J} \cdot \mathbf{A} d^3x = \frac{1}{8\pi} \sum_i \alpha_i \int\limits_{V_i} \mathbf{A} \cdot \mathbf{B} d^3x = \frac{1}{8\pi} \sum_i \alpha_i H_i, \tag{112}$$

and the relationship to magnetic helicity is manifest. Note that each H_i is gauge invariant because it relates to a closed tube of force, which is bounded by a magnetic surface.

Now consider the surface term in (106). Naturally, the results depend upon what surface is chosen. In an application to a toroidal system, we choose a toroidal magnetic surface outside of all plasma currents, so that on the surface

$$\mathbf{B} = \nabla\varphi. \tag{113}$$

As φ is not single valued, we introduce a boundary of $\partial V, \partial(\partial V)$, which is composed of a closed contour C on ∂V which winds around once in each direction the short way ($\pm C_s$), together with a closed contour which winds around once in each direction the long way ($\pm C_\ell$). If we integrate (113) around C_s we get

$$\int_{C_s} \mathbf{B}\cdot d\mathbf{s} = \frac{4\pi}{c} I_T = \int_{C_s} \nabla\varphi\cdot d\mathbf{s} = \Delta_s\varphi, \tag{114}$$

where I_T is the toroidal current threading C_s and $\Delta_s\varphi$ is the jump in φ at the point where C_ℓ intersects C_s. Similarly,

$$\int_{C_\ell} \mathbf{B}\cdot d\mathbf{s} = \frac{4\pi}{c} I_P = \int_{C_\ell} \nabla\varphi\cdot d\mathbf{s} = \Delta_\ell\varphi, \tag{115}$$

where I_P is the poloidal current threading C_ℓ and $\Delta_\ell\varphi$ is the jump at the point where C_ℓ intersects C_s.

Now

$$\int_{\partial V} (\mathbf{A}\times\mathbf{B})\cdot d\mathbf{S} = \int_{\partial V} (\mathbf{A}\times\nabla\varphi)\cdot d\mathbf{S} = -\int_{\partial V} \nabla\times(\varphi\mathbf{A})\cdot d\mathbf{S} + \int_{\partial V} \varphi\mathbf{B}\cdot d\mathbf{S}. \tag{116}$$

The second term vanishes because ∂V is a magnetic surface. We apply Stokes' theorem to the first term, using the fact that $\partial(\partial V) = C$, getting

$$
\begin{aligned}
\int_{\partial V} (\mathbf{A} \times \mathbf{B}) \cdot d\mathbf{S} &= -\int_{C} \varphi \mathbf{A} \cdot d\mathbf{s} \\
&= -\int_{C_s} [\varphi(C_s) - \varphi(-C_s)]\, \mathbf{A} \cdot d\mathbf{s} \\
&\quad - \int_{C_\ell} [\varphi(C_\ell) - \varphi(-C_\ell)]\, \mathbf{A} \cdot d\mathbf{s} \\
&= \Delta_\ell \varphi \int_{C_s} \mathbf{A} \cdot d\mathbf{s} - \Delta_s \varphi \int_{C_\ell} \mathbf{A} \cdot d\mathbf{s}.
\end{aligned}
\tag{117}
$$

Here we have used the fact that $\varphi(C_s)$ differs from $\varphi(-C_s)$ everywhere along C_s by a constant, $-\Delta_\ell \varphi$. Since

$$
\int_{C_s} \mathbf{A} \cdot d\mathbf{s} = \Phi_T, \tag{118}
$$

the toroidal magnetic flux inside ∂V, and

$$
\int_{C_\ell} \mathbf{A} \cdot d\mathbf{s} = \Phi_P, \tag{119}
$$

the poloidal magnetic flux threading the hole in ∂V, with the help of (114) and (115) we can write (117) in the form

$$
\int_{\partial V} (\mathbf{A} \times \mathbf{B}) \cdot d\mathbf{S} = \frac{4\pi}{c}(I_P \Phi_T - I_T \Phi_P). \tag{120}
$$

Substituting this and (112) into (106) yields

$$
E_M = \frac{1}{8\pi} \sum_i \alpha_i H_i + \frac{1}{2c}(I_P \Phi_T - I_T \Phi_P). \tag{121}
$$

A relationship similar to this for the constant-α case was derived by a different method by Rieman.[11] I have also shown that it holds if ∂V lies within the plasma-current region (and is therefore a surface $\alpha =$ constant), but then I and Φ must include currents and fields in the plasma as well as those outside the plasma.

Equation (121) provides an interesting connection between energy and helicity which might be useful in some applications.

ACKNOWLEDGEMENTS

It is a pleasure to acknowledge many rewarding conversations with Mitch Berger, and the hospitality of John Bahcall and Harry Woolf at The Institute for Advanced Study, where some of this work was done.

REFERENCES

1. L. Woltjer, Proc. Nat. Acad. Sci. 44, 489 (1958).
2. W.M. Elsasser, Rev. Mod. Phys. 23, 135 (1956).
3. H.K. Moffatt, J. Fluid Mech. 35, 117 (1969).
4. J.B. Taylor, Phys. Rev. Lett. 33, 1139 (1974).
5. J.B. Taylor, in Pulsed High Beta Plasmas, ed. D. Evans, (Pergamon Press, 1975), p. 59.
6. V.I. Arnol'd, "Asymptotic Hopf Invariant and Its Applications", in Proc. Summer School in Different Equations (Armenian SSR Academy of Sciences, Erevan, 1974).
7. M.A. Berger, Ph.D. Thesis (Harvard University, Cambridge, Massachusetts, 1984).
8. M.A. Berger and G.B. Field, J. Fluid Mech. 147, 133 (1984).
9. M.A. Berger, Geophys. Astrophys. Fluid Dynamics 30, 79 (1984).
10. S. Weart, Astrophys. J. 177, 271 (1972).
11. A. Rieman, "Effect of Induced Wall Currents on Taylor Relaxation", in Proc. Reversed-Field Pinch Workshop (Los Alamos National Laboratory, 1980), p. 276.

SIMULATIONS OF COLLISIONLESS SHOCKS

Kevin B. Quest
Los Alamos National Laboratory, Los Alamos, NM. 87544

ABSTRACT

A problem of critical importance to space and astrophysics is the existence and properties of high-Mach-number (HMN) shocks. In this letter we present the results of simulations of perpendicular shocks with Alfvén Mach number 22. We show that the shock structure is a sensitive function of resistivity, becoming turbulent when the resistivity is too low. We discuss the problem of electron heating, and the extension of our results to higher Mach numbers.

INTRODUCTION

A shock is a nonlinear supersonic compressive wave across which ordered flow energy is converted into disordered thermal energy and magnetic energy. In a collisionless shock, the mean free path of ion-electron and electron-electron collisions is very long, so dissipation is provided by anamolous wave-particle scattering and ion reflection. The best known example of a collisionless shock is the earth's bow shock, formed when the supersonic, super-Alfénic solar wind runs into the earth's magnetosphere.[1] Other astrophysical examples are the low Mach number shocks produced when an early-type star first turns on,[2] and the high Mach number shocks driven by supernova remnants.[3]

In recent years, the ISEE satellite program has stimulated an intense theoretical and observational program in collisionless shock physics. By combining theory, simulations, and observations, a good understanding of the Earth's bow shock has emerged but there remain, of course, many unanswered questions. One of the most important of these is the nature of high-Mach-number (HMN) shocks. At present, the fastest shock observed within the solar system was at Jupiter, and had a magnetosonic (fast mode) Mach number of 12.[4] The lack of higher Mach numbers have led to the suggestions that (1) there is a "2nd critical" Mach number, above which the shock physics changes radically, or (2) that the solar wind, which generates the planetary bow shocks, rarely exceeds magnetosonic Mach numbers above 10.

The Jovian bow shock observed by Russell et al.[4] is displayed in Figure 1. The magnetic field (in units of $\gamma = 10^{-5}$ Gauss) is plotted as a function of time (hours and minutes). Upstream of the shock (right), the magnetic field strength is less than one γ, the plasma β (ratio of plasma to magnetic pressure) is 2.9, and the shock-normal angle θ_{Bn} is 78°. The magnetic field at the shock is observed to rise (overshoot) well above its mean downstream value.

Particle observations,[5] theory,[6] and simulations[7] have shown that this overshoot is a consequence of a reflected ion beam which is

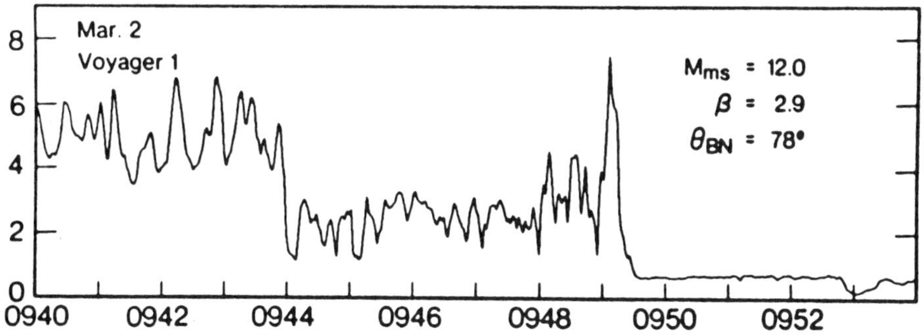

Fig 1. Magnetic field (10^{-5} G) vs. time (hours and minutes) for an outbound Jovian bow shock crossing (from Ref. 4).

the primary source of dissipation for collisionless shocks at these Mach numbers.

The structure of HMN shocks is also a topic of interest to astrophysics. Observations of the x-ray emissions from young super nova remnants show the existance of a well defined spherical shell of hot electrons.[8] This shell is believed to be the downstream wake of an outward propagating shock, with Alfvén Mach number as high as 1000.[9] If true, then there is no apparent upper limit to the speed of an astrophysical shock, and additionally, such shocks strongly heat the electrons. Predicting how such heating occurs, and what the relative ion to electron temperature ratio is downstream of the shock, is an unresolved issue of plasma astrophysics.

In this paper we present numerical simulations of HMN shocks. We review the results of Quest,[10] who showed that in the absence of resistivity HMN shocks are unsteady, and compare these results with shock simulations including electron dissipation.

SIMULATIONS

The numerical model we will use is a one dimensional (in x) electromagnetic hybrid code which follows the individual particle orbits of the ions and treats the electrons as a resistive fluid. Because the electrons are massless, plasma oscillations are suppressed, and the neglect of the displacement current in Ampere's Law eliminates light waves. As a consequence, large spatial and temporal steps are possible, which allows following the evolution of the shock for several ion gyroperiods. This model has been described in detail previously.[11] Plasma is continuously injected from the left-hand boundary (x=0) and moves in the positive x-direction (see Fig. 2a). When the plasma hits the right-hand boundary (x=L) it is reflected and a shock is launched (see Fig.

2b). As the shock continues to propagate through the box, its separation distance from the wall becomes greater than the downstream ion gyroradius, effectively separating piston and shock heated plasma (Fig. 2c). The simulation run is continued until the

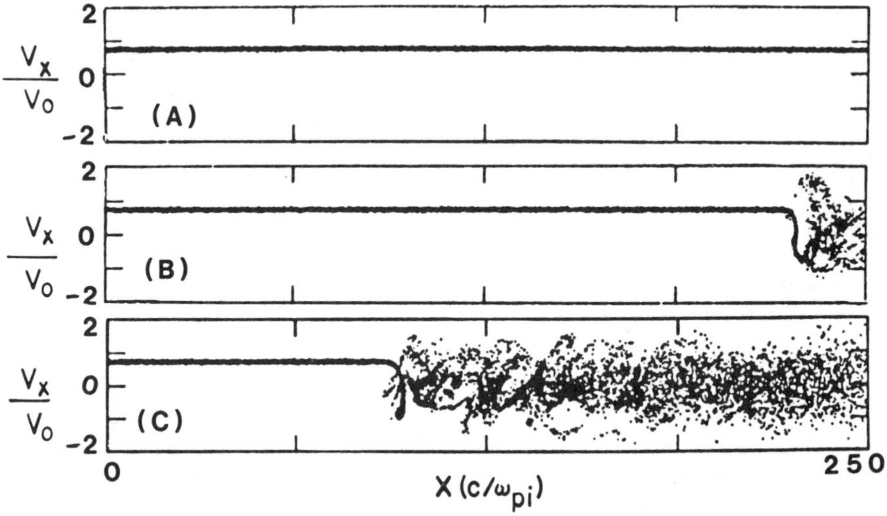

Fig 2. V_x - x phase space initially (A), and at later times (B-C) in the simulation run.

shock generated magnetic turbulence decays before reaching the right boundary, insuring that piston effects do not govern the shock.

The shock simulation we will examine[10] is perpendicular ($\underline{B} = B_z\hat{z}$), propagates at an Alfvén Mach number of 22, and has an upstream electron and ion β of 0.5 respectively, where β is the ratio of thermal pressure to magnetic pressure. The upstream ratio of ion plasma frequency (ω_{pi}/ω_{ci}) is 2 x 10^4 and the resistivity η is between 1.5 - 12 x $10^{-4}\omega_{pi}^{-1}$. A check of the average downstream values show that $n_d/n_u \cong 3$, $B_d/B_u \cong 3$ and $T_{id}/(1/2 M_i V_o^2) \cong 0.5$ where u denotes upstream, d downstream, and V_o the shock speed. These results are consistent with the 2-dimensional Rankine-Hugoniot relations, with most of the shock energy being deposited in the ions.

Because the shock speed is well above the critical Alfvén Mach number, (approximately 3 for these upstream conditions) dissipation by electron heating alone is insufficient to stop shock steepening, and ion reflection results. Simulations of resistive perpendicular shocks with $3 < M_A < 10$ and β = 1 have shown that, in this Mach range, the shock structure is reasonable steady.[7] A fraction of the incoming ions is reflected by a potential barrier and magnetic ramp at the shock front. These ions gyrate in front of the shock, gaining energy from the $\underline{E} \times \underline{B}$ electric field, and are carried downstream. After thermalizing with the directly transmitted ions, a heated downstream population results. As the Mach number is increased, the reflection process continues to be the dominant source of dissipation, but can be highly oscillatory, depending on the magniture of the resistivity.

If the resistive diffusion length (proportional to the resistivity and inversely proportional to the upstream flow speed) is set much smaller than a spatial cell size, then it is not possible to stop shock ramp steepening by resistive dissipation. Under these conditions we find that the shock exhibits a periodicity (1/3 of an upstream gyroperiod) in which the shock steepens, breaks and overturns, and then steepens again. During this cycle, roughly all of the ions are transmitted through the shock, followed by a brief period of total reflection.

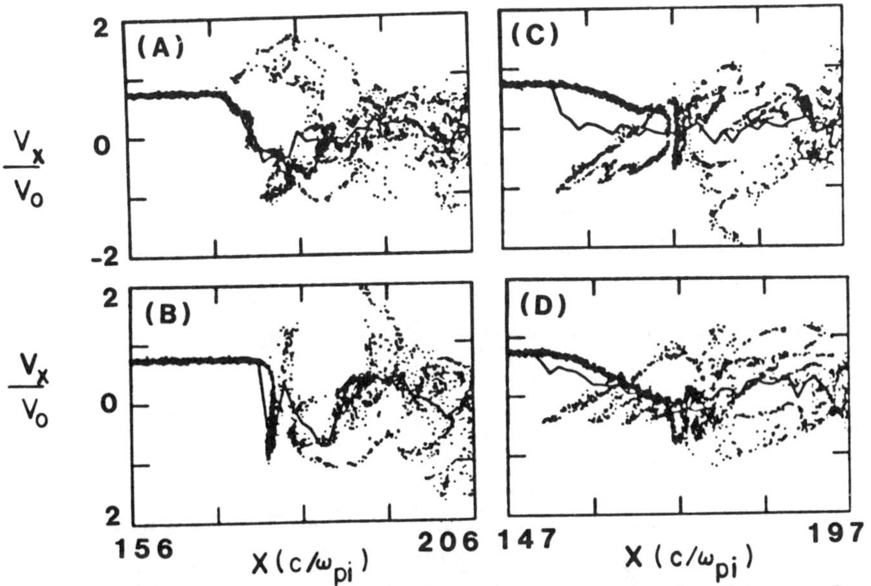

Fig 3. V_x - x phase space at 4 times during a wave breaking cycle. Resistivity η is set to 0 for this run.

In Fig. 3a we show a close-up of the V_x - x phase space after the shock has advanced roughly 1/3 of the way into the simulation box. The solid line is the average value of V_x. At this time the shock transition consists of a smooth ramp, with an energetic ion population downstream. The energetic ions are the result of the previous reflection cycle. In Fig. 3b we see the ramp has steepened to its minimum thickness and is starting to reflect the incoming ions. In Fig. 3c the ions have been reflected, travel upstream some distance, then turn around and head downstream. The ramp thickness is now very broad and in completing the cycle will steepen because of an E field which accelerates particles in the negative x direction. This returns us to 3a.

The behavior of the shock is very turbulent, and reminiscent of earlier shock studies in which a periodic formation and destruction of the shock was observed.[12] An important difference in our results, however, is that the minimum shock thickness (just before breaking) is numerically determined by the cell size. There is no resistive ion scattering or electron inertia in the code, so wave steepening cannot be balanced by dispersion or ion diffusion. Thus, our results demonstrate the process by which the shock will heat ions downstream in the absence of anomalous scattering (by wave breaking), but we are unable to predict details of the structure such as the magnitude of the turbulent magnetic overshoot. Such specifics will require 2-dimensional particle codes, which include both cross-field instabilities and electron inertia.

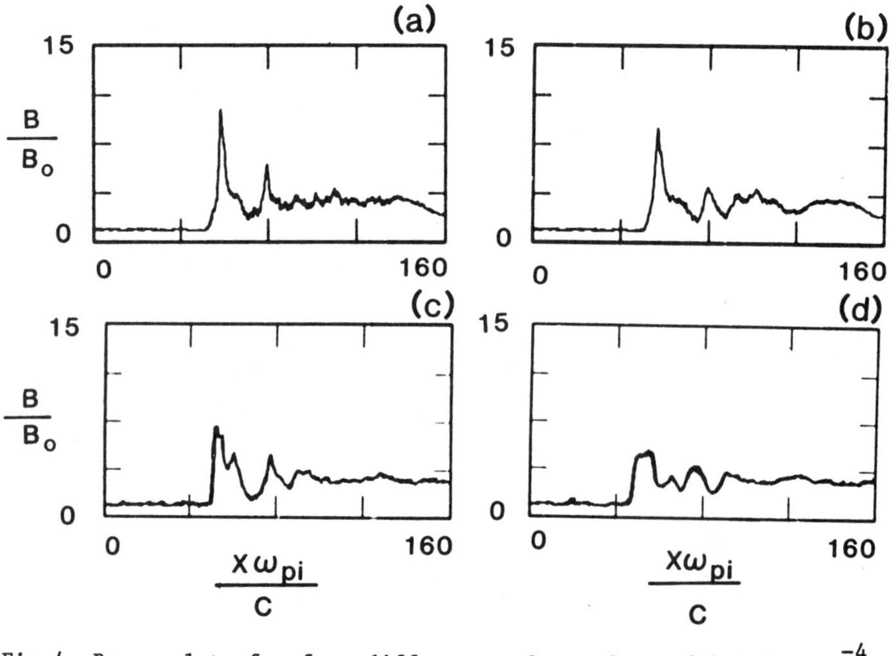

Fig 4. B - x plots for four different values of η = (A) 1.5×10^{-4}, (B) 3×10^{-4}, (C) 6×10^{-4}, and (D) 1.2×10^{-3}.

As the resistive diffusion length is increased to a magnitude greater than a cell size, the temporal behavior of the shock becomes much quieter. In figure 4a and 4b we show the magnetic field profile for runs with the resistivity set at 1.5 and 3×10^{-4} ω_{pi}^{-1}. While the magnetic overshoots are large (~10 times the upstream value in 4a), they are quite stationary, varying by less than 6% during the course of the run. These results indicate that there exists a range of finite resistivity over which stationary solutions, similar to those examined by Leroy et al., at lower Mach numbers. In fact, a simple set of jump conditions, with the fraction of reflected ions set as a free parameter, yield predictions quite similar to the above two runs. As the resistivity is increased further, the average magnetic overshoot and the fraction of reflected ions continue to decrease (see Fig. 4c, 4d), but the magnitude of RMS deviations increase strongly (13% for the magnetic overshoot). The problem is that increasing the resistivity decreases the fraction of the reflected ions. In order to maintain a steady state the additional dissipation must come from the heated electrons. There is a limit, however, to the amount of total electron heating (see for example, arguments in Leroy et al.[7]). When the mean number of reflected ions becomes too small, the shock structure oscillates.

CONCLUSIONS

Given the various classes of solutions for high-Mach number shocks, which will actually apply? The answer to that question will depend on the efficiency of wave-particle instabilities heating the shocked plasmas. Observationally, most shocks with plasma $\beta \simeq 1$ appear quite stationary, even at the higher Mach numbers. By contrast, shocks with $\beta >> 1$ are very unsteady.[13] It is tempting to speculate that in the former case ($\beta \simeq 1$), the lower-hybrid drift instability acts to smooth the shock structure, while in the latter ($\beta >> 1$), the mode is stabilized, resulting in cyclic shock steepening and wave breaking. An observational study is currently in progress to resolve these issues.[14] As the Mach number continues to increase, so does the amount of resistivity required to maintain a specified shock ramp thickness. It seems likely that for sufficiently fast shocks the resistivity will not keep up, and wave breaking will result. Another important point is that because of the one-dimensionality of the simulation and the suppression of short-wavelength oscillations the thermalization of the downstream shocked plasma is likely to be quite different from what has been presented here. Since the gyrational energy is perpendicular to the B field, anisotropy driven ion instabilities will generate large fluctuating fields. Current driven modes and beam modes could be destabilized, peaking at short wavelengths and driving the electrons resistive. For extremely high Mach numbers, the large fluctuating fields generated by the ion anisotropy could be absorbed by the resistive electrons, resulting in strong electron heating. Clearly, a great deal of work is required to clarify these and other issues raised by these simulations.

This work was supported by NASA Solar Terrestrial Theory Grant 10-23727 and the U.S. Department of Energy. The author would like to acknowledge useful comments by D. Winske, C. Goodrich, and D. Forslund on an earlier version of this manuscript.

REFERENCES

1. N. F. Ness, C. S. Scearce, and J. B. Seek, J. Geophys. Res., 69, 3531 (1964)

2. L. Spitzer, in Physical Processes in the Interstellar Medium, (Wiley, New York, 1978).

3 R. McCray, and T. P. Snow, Ann., Rev. Astron. Astrophys. 17, 213 (1979).

4. C. T. Russell, M. N. Hoppe, and W. A. Livsey, Nature 296, 45 (1982).

5. N. Sckopke, G. Paschmann, S. J. Bame, J. T. Gosling, and C. T. Russell, J. Geophys. Res. 88, 6121 (1983).

6. M. M. Leroy, Phys. Fluids 26, 2742 (1983).

7. M. M. Leroy, D. Winske, C. C. Goodrich, C. S. Wu, and K. Papadopoulos, J. Geophys. Res. 87, 5081 (1982); D. W. Forslund, K. B. Quest, J. U. Brackbill, and K. Lee, J. Geophys. Res. 89, 2142 (1984).

8. A. R. Bell, MRAS 179, 573 (1977).

9. C. F. McKee and D. J. Hollenbach, Ann. Rev. Astron. Astrophys. 18, 219 (1980).

10. K. B. Quest, Phys. Rev. Lett. 54, 1872 (1985).

11. D. Winske and M. M. Leroy, in Computer Simulations of Space Plasmas, edited by H. Matsumoto and T. Sato (Reidel, Boston, 1985).

12. D. Biskamp and H. Welter, Nucl. Fusion 12, 663 (1972).

13. V. Formisano, C. T. Russell, J. D. Means, E. W. Greenstadt, F. L. Scarf, and M. Neugebauer, J. Geophys. Res. 80, 2013 (1975).

14. D. Winterhalter (private communication).

AIP Conference Proceedings

		L.C. Number	ISBN
No. 1	Feedback and Dynamic Control of Plasmas – 1970	70-141596	0-88318-100-2
No. 2	Particles and Fields – 1971 (Rochester)	71-184662	0-88318-101-0
No. 3	Thermal Expansion – 1971 (Corning)	72-76970	0-88318-102-9
No. 4	Superconductivity in *d*- and *f*-Band Metals (Rochester, 1971)	74-18879	0-88318-103-7
No. 5	Magnetism and Magnetic Materials – 1971 (2 parts) (Chicago)	59-2468	0-88318-104-5
No. 6	Particle Physics (Irvine, 1971)	72-81239	0-88318-105-3
No. 7	Exploring the History of Nuclear Physics – 1972	72-81883	0-88318-106-1
No. 8	Experimental Meson Spectroscopy –1972	72-88226	0-88318-107-X
No. 9	Cyclotrons – 1972 (Vancouver)	72-92798	0-88318-108-8
No. 10	Magnetism and Magnetic Materials – 1972	72-623469	0-88318-109-6
No. 11	Transport Phenomena – 1973 (Brown University Conference)	73-80682	0-88318-110-X
No. 12	Experiments on High Energy Particle Collisions – 1973 (Vanderbilt Conference)	73-81705	0-88318-111–8
No. 13	π-π Scattering – 1973 (Tallahassee Conference)	73-81704	0-88318-112-6
No. 14	Particles and Fields – 1973 (APS/DPF Berkeley)	73-91923	0-88318-113-4
No. 15	High Energy Collisions – 1973 (Stony Brook)	73-92324	0-88318-114-2
No. 16	Causality and Physical Theories (Wayne State University, 1973)	73-93420	0-88318-115-0
No. 17	Thermal Expansion – 1973 (Lake of the Ozarks)	73-94415	0-88318-116-9
No. 18	Magnetism and Magnetic Materials – 1973 (2 parts) (Boston)	59-2468	0-88318-117-7
No. 19	Physics and the Energy Problem – 1974 (APS Chicago)	73-94416	0-88318-118-5
No. 20	Tetrahedrally Bonded Amorphous Semiconductors (Yorktown Heights, 1974)	74-80145	0-88318-119-3
No. 21	Experimental Meson Spectroscopy – 1974 (Boston)	74-82628	0-88318-120-7
No. 22	Neutrinos – 1974 (Philadelphia)	74-82413	0-88318-121-5
No. 23	Particles and Fields – 1974 (APS/DPF Williamsburg)	74-27575	0-88318-122-3
No. 24	Magnetism and Magnetic Materials – 1974 (20th Annual Conference, San Francisco)	75-2647	0-88318-123-1

No. 25	Efficient Use of Energy (The APS Studies on the Technical Aspects of the More Efficient Use of Energy)	75-18227	0-88318-124-X
No. 26	High-Energy Physics and Nuclear Structure – 1975 (Santa Fe and Los Alamos)	75-26411	0-88318-125-8
No. 27	Topics in Statistical Mechanics and Biophysics: A Memorial to Julius L. Jackson (Wayne State University, 1975)	75-36309	0-88318-126-6
No. 28	Physics and Our World: A Symposium in Honor of Victor F. Weisskopf (M.I.T., 1974)	76-7207	0-88318-127-4
No. 29	Magnetism and Magnetic Materials – 1975 (21st Annual Conference, Philadelphia)	76-10931	0-88318-128-2
No. 30	Particle Searches and Discoveries – 1976 (Vanderbilt Conference)	76-19949	0-88318-129-0
No. 31	Structure and Excitations of Amorphous Solids (Williamsburg, VA, 1976)	76-22279	0-88318-130-4
No. 32	Materials Technology – 1976 (APS New York Meeting)	76-27967	0-88318-131-2
No. 33	Meson-Nuclear Physics – 1976 (Carnegie-Mellon Conference)	76-26811	0-88318-132-0
No. 34	Magnetism and Magnetic Materials – 1976 (Joint MMM-Intermag Conference, Pittsburgh)	76-47106	0-88318-133-9
No. 35	High Energy Physics with Polarized Beams and Targets (Argonne, 1976)	76-50181	0-88318-134-7
No. 36	Momentum Wave Functions – 1976 (Indiana University)	77-82145	0-88318-135-5
No. 37	Weak Interaction Physics – 1977 (Indiana University)	77-83344	0-88318-136-3
No. 38	Workshop on New Directions in Mossbauer Spectroscopy (Argonne, 1977)	77-90635	0-88318-137-1
No. 39	Physics Careers, Employment and Education (Penn State, 1977)	77-94053	0-88318-138-X
No. 40	Electrical Transport and Optical Properties of Inhomogeneous Media (Ohio State University, 1977)	78-54319	0-88318-139-8
No. 41	Nucleon-Nucleon Interactions – 1977 (Vancouver)	78-54249	0-88318-140-1
No. 42	Higher Energy Polarized Proton Beams (Ann Arbor, 1977)	78-55682	0-88318-141-X
No. 43	Particles and Fields – 1977 (APS/DPF, Argonne)	78-55683	0-88318-142-8
No. 44	Future Trends in Superconductive Electronics (Charlottesville, 1978)	77-9240	0-88318-143-6
No. 45	New Results in High Energy Physics – 1978 (Vanderbilt Conference)	78-67196	0-88318-144-4
No. 46	Topics in Nonlinear Dynamics (La Jolla Institute)	78-57870	0-88318-145-2

No. 46	Topics in Nonlinear Dynamics (La Jolla Institute)	78-57870	0-88318-145-2
No. 47	Clustering Aspects of Nuclear Structure and Nuclear Reactions (Winnepeg, 1978)	78-64942	0-88318-146-0
No. 48	Current Trends in the Theory of Fields (Tallahassee, 1978)	78-72948	0-88318-147-9
No. 49	Cosmic Rays and Particle Physics – 1978 (Bartol Conference)	79-50489	0-88318-148-7
No. 50	Laser-Solid Interactions and Laser Processing – 1978 (Boston)	79-51564	0-88318-149-5
No. 51	High Energy Physics with Polarized Beams and Polarized Targets (Argonne, 1978)	79-64565	0-88318-150-9
No. 52	Long-Distance Neutrino Detection – 1978 (C.L. Cowan Memorial Symposium)	79-52078	0-88318-151-7
No. 53	Modulated Structures – 1979 (Kailua Kona, Hawaii)	79-53846	0-88318-152-5
No. 54	Meson-Nuclear Physics – 1979 (Houston)	79-53978	0-88318-153-3
No. 55	Quantum Chromodynamics (La Jolla, 1978)	79-54969	0-88318-154-1
No. 56	Particle Acceleration Mechanisms in Astrophysics (La Jolla, 1979)	79-55844	0-88318-155-X
No. 57	Nonlinear Dynamics and the Beam-Beam Interaction (Brookhaven, 1979)	79-57341	0-88318-156-8
No. 58	Inhomogeneous Superconductors – 1979 (Berkeley Springs, W.V.)	79-57620	0-88318-157-6
No. 59	Particles and Fields – 1979 (APS/DPF Montreal)	80-66631	0-88318-158-4
No. 60	History of the ZGS (Argonne, 1979)	80-67694	0-88318-159-2
No. 61	Aspects of the Kinetics and Dynamics of Surface Reactions (La Jolla Institute, 1979)	80-68004	0-88318-160-6
No. 62	High Energy e^+e^- Interactions (Vanderbilt, 1980)	80-53377	0-88318-161-4
No. 63	Supernovae Spectra (La Jolla, 1980)	80-70019	0-88318-162-2
No. 64	Laboratory EXAFS Facilities – 1980 (Univ. of Washington)	80-70579	0-88318-163-0
No. 65	Optics in Four Dimensions – 1980 (ICO, Ensenada)	80-70771	0-88318-164-9
No. 66	Physics in the Automotive Industry – 1980 (APS/AAPT Topical Conference)	80-70987	0-88318-165-7
No. 67	Experimental Meson Spectroscopy – 1980 (Sixth International Conference, Brookhaven)	80-71123	0-88318-166-5
No. 68	High Energy Physics – 1980 (XX International Conference, Madison)	81-65032	0-88318-167-3
No. 69	Polarization Phenomena in Nuclear Physics – 1980 (Fifth International Symposium, Santa Fe)	81-65107	0-88318-168-1

No. 70	Chemistry and Physics of Coal Utilization – 1980 (APS, Morgantown)	81-65106	0-88318-169-X
No. 71	Group Theory and its Applications in Physics – 1980 (Latin American School of Physics, Mexico City)	81-66132	0-88318-170-3
No. 72	Weak Interactions as a Probe of Unification (Virginia Polytechnic Institute – 1980)	81-67184	0-88318-171-1
No. 73	Tetrahedrally Bonded Amorphous Semiconductors (Carefree, Arizona, 1981)	81-67419	0-88318-172-X
No. 74	Perturbative Quantum Chromodynamics (Tallahassee, 1981)	81-70372	0-88318-173-8
No. 75	Low Energy X-Ray Diagnostics – 1981 (Monterey)	81-69841	0-88318-174-6
No. 76	Nonlinear Properties of Internal Waves (La Jolla Institute, 1981)	81-71062	0-88318-175-4
No. 77	Gamma Ray Transients and Related Astrophysical Phenomena (La Jolla Institute, 1981)	81-71543	0-88318-176-2
No. 78	Shock Waves in Condensed Mater – 1981 (Menlo Park)	82-70014	0-88318-177-0
No. 79	Pion Production and Absorption in Nuclei – 1981 (Indiana University Cyclotron Facility)	82-70678	0-88318-178-9
No. 80	Polarized Proton Ion Sources (Ann Arbor, 1981)	82-71025	0-88318-179-7
No. 81	Particles and Fields –1981: Testing the Standard Model (APS/DPF, Santa Cruz)	82-71156	0-88318-180-0
No. 82	Interpretation of Climate and Photochemical Models, Ozone and Temperature Measurements (La Jolla Institute, 1981)	82-71345	0-88318-181-9
No. 83	The Galactic Center (Cal. Inst. of Tech., 1982)	82-71635	0-88318-182-7
No. 84	Physics in the Steel Industry (APS/AISI, Lehigh University, 1981)	82-72033	0-88318-183-5
No. 85	Proton-Antiproton Collider Physics –1981 (Madison, Wisconsin)	82-72141	0-88318-184-3
No. 86	Momentum Wave Functions – 1982 (Adelaide, Australia)	82-72375	0-88318-185-1
No. 87	Physics of High Energy Particle Accelerators (Fermilab Summer School, 1981)	82-72421	0-88318-186-X
No. 88	Mathematical Methods in Hydrodynamics and Integrability in Dynamical Systems (La Jolla Institute, 1981)	82-72462	0-88318-187-8
No. 89	Neutron Scattering – 1981 (Argonne National Laboratory)	82-73094	0-88318-188-6
No. 90	Laser Techniques for Extreme Ultraviolt Spectroscopy (Boulder, 1982)	82-73205	0-88318-189-4

No. 91	Laser Acceleration of Particles (Los Alamos, 1982)	82-73361	0-88318-190-8
No. 92	The State of Particle Accelerators and High Energy Physics (Fermilab, 1981)	82-73861	0-88318-191-6
No. 93	Novel Results in Particle Physics (Vanderbilt, 1982)	82-73954	0-88318-192-4
No. 94	X-Ray and Atomic Inner-Shell Physics – 1982 (International Conference, U. of Oregon)	82-74075	0-88318-193-2
No. 95	High Energy Spin Physics – 1982 (Brookhaven National Laboratory)	83-70154	0-88318-194-0
No. 96	Science Underground (Los Alamos, 1982)	83-70377	0-88318-195-9
No. 97	The Interaction Between Medium Energy Nucleons in Nuclei – 1982 (Indiana University)	83-70649	0-88318-196-7
No. 98	Particles and Fields – 1982 (APS/DPF University of Maryland)	83-70807	0-88318-197-5
No. 99	Neutrino Mass and Gauge Structure of Weak Interactions (Telemark, 1982)	83-71072	0-88318-198-3
No. 100	Excimer Lasers – 1983 (OSA, Lake Tahoe, Nevada)	83-71437	0-88318-199-1
No. 101	Positron-Electron Pairs in Astrophysics (Goddard Space Flight Center, 1983)	83-71926	0-88318-200-9
No. 102	Intense Medium Energy Sources of Strangeness (UC-Sant Cruz, 1983)	83-72261	0-88318-201-7
No. 103	Quantum Fluids and Solids – 1983 (Sanibel Island, Florida)	83-72440	0-88318-202-5
No. 104	Physics, Technology and the Nuclear Arms Race (APS Baltimore –1983)	83-72533	0-88318-203-3
No. 105	Physics of High Energy Particle Accelerators (SLAC Summer School, 1982)	83-72986	0-88318-304-8
No. 106	Predictability of Fluid Motions (La Jolla Institute, 1983)	83-73641	0-88318-305-6
No. 107	Physics and Chemistry of Porous Media (Schlumberger-Doll Research, 1983)	83-73640	0-88318-306-4
No. 108	The Time Projection Chamber (TRIUMF, Vancouver, 1983)	83-83445	0-88318-307-2
No. 109	Random Walks and Their Applications in the Physical and Biological Sciences (NBS/La Jolla Institute, 1982)	84-70208	0-88318-308-0
No. 110	Hadron Substructure in Nuclear Physics (Indiana University, 1983)	84-70165	0-88318-309-9
No. 111	Production and Neutralization of Negative Ions and Beams (3rd Int'l Symposium, Brookhaven, 1983)	84-70379	0-88318-310-2

No. 112	Particles and Fields – 1983 (APS/DPF, Blacksburg, VA)	84-70378	0-88318-311-0
No. 113	Experimental Meson Spectroscopy – 1983 (Seventh International Conference, Brookhaven)	84-70910	0-88318-312-9
No. 114	Low Energy Tests of Conservation Laws in Particle Physics (Blacksburg, VA, 1983)	84-71157	0-88318-313-7
No. 115	High Energy Transients in Astrophysics (Santa Cruz, CA, 1983)	84-71205	0-88318-314-5
No. 116	Problems in Unification and Supergravity (La Jolla Institute, 1983)	84-71246	0-88318-315-3
No. 117	Polarized Proton Ion Sources (TRIUMF, Vancouver, 1983)	84-71235	0-88318-316-1
No. 118	Free Electron Generation of Extreme Ultraviolet Coherent Radiation (Brookhaven/OSA, 1983)	84-71539	0-88318-317-X
No. 119	Laser Techniques in the Extreme Ultraviolet (OSA, Boulder, Colorado, 1984)	84-72128	0-88318-318-8
No. 120	Optical Effects in Amorphous Semiconductors (Snowbird, Utah, 1984)	84-72419	0-88318-319-6
No. 121	High Energy e^+e^- Interactions (Vanderbilt, 1984)	84-72632	0-88318-320-X
No. 122	The Physics of VLSI (Xerox, Palo Alto, 1984)	84-72729	0-88318-321-8
No. 123	Intersections Between Particle and Nuclear Physics (Steamboat Springs, 1984)	84-72790	0-88318-322-6
No. 124	Neutron-Nucleus Collisions – A Probe of Nuclear Structure (Burr Oak State Park - 1984)	84-73216	0-88318-323-4
No. 125	Capture Gamma-Ray Spectroscopy and Related Topics – 1984 (Internat. Symposium, Knoxville)	84-73303	0-88318-324-2
No. 126	Solar Neutrinos and Neutrino Astronomy (Homestake, 1984)	84-63143	0-88318-325-0
No. 127	Physics of High Energy Particle Accelerators (BNL/SUNY Summer School, 1983)	85-70057	0-88318-326-9
No. 128	Nuclear Physics with Stored, Cooled Beams (McCormick's Creek State Park, Indiana, 1984)	85-71167	0-88318-327-7
No. 129	Radiofrequency Plasma Heating (Sixth Topical Conference, Callaway Gardens, GA, 1985)	85-48027	0-88318-328-5
No. 130	Laser Acceleration of Particles (Malibu, California, 1985)	85-48028	0-88318-329-3
No. 131	Workshop on Polarized ^{3}He Beams and Targets (Princeton, New Jersey, 1984)	85-48026	0-88318-330-7
No. 132	Hadron Spectroscopy–1985 (International Conference, Univ. of Maryland)	85-72537	0-88318-331-5

No. 133	Hadronic Probes and Nuclear Interactions (Arizona State University, 1985)	85-72638	0-88318-332-3
No. 134	The State of High Energy Physics (BNL/SUNY Summer School, 1983)	85-73170	0-88318-333-1
No. 135	Energy Sources: Conservation and Renewables (APS, Washington, DC, 1985)	85-73019	0-88318-334-X
No. 136	Atomic Theory Workshop on Relativistic and QED Effects in Heavy Atoms	85-73790	0-88318-335-8
No. 137	Polymer-Flow Interaction (La Jolla Institute, 1985)	85-73915	0-88318-336-6
No. 138	Frontiers in Electronic Materials and Processing (Houston, TX, 1985)	86-70108	0-88318-337-4
No. 139	High-Current, High-Brightness, and High-Duty Factor Ion Injectors (La Jolla Institute, 1985)	86-70245	0-88318-338-2
No. 140	Boron-Rich Solids (Albuquerque, NM, 1985)	86-70246	0-88318-339-0
No. 141	Gamma-Ray Bursts (Stanford, CA, 1984)	86-70761	0-88318-340-4
No. 142	Nuclear Structure at High Spin, Excitation, and Momentum Transfer (Indiana University, 1985)	86-70837	0-88318-341-2
No. 143	Mexican School of Particles and Fields (Oaxtepec, México, 1984)	86-81187	0-88318-342-0